Maize

Maize is widely cultivated throughout the world due to its high-yield potential. The economic and nutritional value of maize grains is associated with its starch content, protein, fibre, bioactive compounds, and minerals. Maize is used worldwide for the preparation of health-benefiting, antioxidant-rich, fortified products and dietary supplements.

Maize: Nutritional Composition, Processing, and Industrial Uses explores the status of maize in terms of its production, nutritional composition, biofortification, processing methods, health benefits, maize-based products, and storage. This book also emphasizes the key features of maize grains which make it an ideal crop for industrial use. It covers all aspects of recent research about maize and provides updated information.

Key features:

- Discusses information related to the chemistry of maize components
- Highlights comprehensive information on the physical and milling properties of maize
- Explains the structural, functional, and antioxidant properties of maize flour
- Provides the latest scientific development in the modification of maize starch
- Explores various maize-based food products and their storage
- Examines maize protein, scenarios, and quality improvement through bio-fortification

In-depth information is provided regarding the various health-benefiting nutrient components of maize flour, offering meaningful information for product formulation. This book unfolds the potential of maize grains for industrial use.

Cereals: Science and Processing Technology

Series Editors:
Sneh Punia Bangar and Manoj Kumar

Maize: Nutritional Composition, Processing, and Industrial Uses
Edited by Sukhvinder Singh Purewal, Pinderpal Kaur, Sneh Punia Bangar, Kawaljit Singh Sandhu, Surender Kumar Singh, and Maninder Kaur

Maize

Nutritional Composition, Processing, and Industrial Uses

Edited by
Sukhvinder Singh Purewal
Pinderpal Kaur
Sneh Punia Bangar
Kawaljit Singh Sandhu
Surender Kumar Singh
Maninder Kaur

CRC Press
Taylor & Francis Group
Boca Raton London New York

CRC Press is an imprint of the
Taylor & Francis Group, an **informa** business

First edition published 2023
by CRC Press
6000 Broken Sound Parkway NW, Suite 300, Boca Raton, FL 33487-2742

and by CRC Press
4 Park Square, Milton Park, Abingdon, Oxon, OX14 4RN

CRC Press is an imprint of Taylor & Francis Group, LLC

Library of Congress Cataloging-in-Publication Data

Names: Purewal, Sukhvinder Singh, editor. | Kaur, Pinderpal, editor. | Punia, Sneh, editor. | Sandhu, Kawaljit Singh, editor. | Singh, Surender Kumar, editor. | Kaur, Maninder, editor.
Title: Maize : nutritional composition, processing, and industrial uses / edited by Sukhvinder Singh Purewal, Pinderpal Kaur, Sneh Punia, Kawaljit Singh Sandhu, Surender Kumar Singh, Maninder Kaur.
Description: Boca Raton : CRC Press, 2023. | Series: Cereals science and processing technology | Includes bibliographical references and index.
Identifiers: LCCN 2022015467 (print) | LCCN 2022015468 (ebook) | ISBN 9781032127927 (hardback) | ISBN 9781032156675 (paperback) | ISBN 9781003245230 (ebook)
Subjects: LCSH: Corn. | Corn--Utilization. | Corn--Nutrition.
Classification: LCC SB191.M2 M1224 2023 (print) | LCC SB191.M2 (ebook) | DDC 633.1/5--dc23/eng/20220330
LC record available at https://lccn.loc.gov/2022015467
LC ebook record available at https://lccn.loc.gov/2022015468

ISBN: 9781032127927 (hbk)
ISBN: 9781032156675 (pbk)
ISBN: 9781003245230 (ebk)

DOI: 10.1201/9781003245230

Typeset in Century Old Style
by Deanta Global Publishing Services, Chennai, India

Contents

CONTENTS

Series preface

The *Cereals: Science and Processing Technology Series* is a collection of works about all the major cereals (wheat, rice, maize, oat, barley, sorghum) and millets (pearl, finger, foxtail, barnyard, kodo, little, proso millet) grown around the world. It focuses on chemistry, nutritional/compositional properties, anti-nutritional components, processing aspects, storability, beneficial health effects, disease prevention activities, and areas for future research. The Series aims to cover the most relevant production and distribution channels of functional food ingredients to understand the positioning of the industry to meet the challenge of functional food properties. Understanding the potentiality of cereals and millets provides a basis for better utilization of crops and the matters for more research in the field of agriculture sciences. The Series aims at academics teaching undergraduate and graduate students in the disciplines of cereal-based foods to present and discuss the details and characteristics of the many factors involved, such as technological advances and demand factors, and the trend of consumption of healthy foods.

Preface

Maize is the third most important member of the *Poaceae* family after wheat and rice. It has its own importance in terms of human nutrition and health benefits. It is known as the queen of cereals. Worldwide, maize is commonly known as Makai (Punjab, India), Makkai (Gujrat, India), Bhutta (Bengal, India), Toumorokoshi (Japan), Mielie (Africa), and Gaudume makka (Persia). Maize is considered the common man's food because of its affordable price. Detailed descriptions in this book related to maize, its components, products, and commercial uses explain its potential to assist in human welfare. Comprehensive knowledge of maize will definitely fill in the remaining gaps between the previous and latest scientific findings.

Chapter 1 describes the physical and functional attributes of maize along with milling and byproducts. **Chapter 2** discusses the various macro- and micronutrients of maize along with their detailed biochemistry. Further, different types of processing applied to maize are also discussed. **Chapter 3** discusses worldwide production details, micro- and macronutrients, and the effects of different processing methods on maize components. This chapter will be helpful from an industrial point of view, as it explains the processing effect on the nutritional profile of maize, therefore helpful during the development of different maize-based food products. Starch is the major component of the grains; therefore, discussion on this component will show its versatility in food and non-food industries. **Chapter 4** describes starch extraction methods, morphological features, physico-chemical, pasting, and rheological properties, along with food and non-food uses of maize starch. Despite being an important macromolecule, native starch has some limitations, and to overcome these restrictions and to achieve desirable features in starch, different modification processes being used are discussed in **Chapter 5**. Protein is another macromolecule present in cereal grains and has its own importance in human nutrition. **Chapter 6** describes maize protein in detail, their classification, extraction, purification, quality improvement, and applications. Further, bioactive metabolites are also gaining interest from consumers, as they help to combat oxidative stress and reduce its effects. Information about different bioactive compounds present in maize grains along with extraction methods, effects of processing methods on phenolic compounds, and health benefits are discussed in **Chapter 7**. In addition to specific nutrients, maize also has a unique blend of pigments. **Chapter 8** describes pigment accumulation, its biosynthesis, extraction, different products, and health benefits. Further, to overcome the nutritional deficiency in grains, bio-fortification processes are being used effectively. The detailed information related to quality protein maize, bio-fortification approaches, plant breeding strategies, genetic engineering, advantages, and limitations are discussed in **Chapter 9**. Detailed information about the nutritional profile, quality improvement, and effect of processing methods, applications, and health benefits is necessary to promote any crop on a commercial scale. Besides these parameters, storage is another important aspect which must be explored to make the crop available even in the off-season. Therefore, the importance of storage, post-harvest management, factors affecting storage, and storage structures have been described in **Chapter 10**.

About the editors

Sukhvinder Singh Purewal, PhD, is presently working as a Young Scientist (DST SYST Scheme) in the Department of Food Science & Technology, Maharaja Ranjit Singh Punjab Technical University, Bathinda, Punjab, India. His areas of interests include solid state fermentation, bioactive compounds from natural resources, antioxidants, and fruit processing. He has presented his research at various national and international conferences and has published more than 40 research papers and book chapters in national and international journals and books. Dr. Purewal has published one authored book with CRC Press/Taylor & Francis Group. He is an active member of the Association of Microbiologists of India (AMI), Mycological Society of India (MSI), Association of Food Scientists and Technologists (AFSTI), Mysore, India, and Indian Science Congress Association (ISCA). Dr. Purewal received his M.Sc., M.Phil., and doctorate degree in biotechnology from Chaudhary Devi Lal University, Sirsa, Haryana, India. He also worked on a major UGC, Delhi-sponsored research project from 2012–2015. He was awarded Best Paper Presentation in national as well as international conferences.

Pinderpal Kaur is currently a PhD Research Scholar in the Department of Food Science and Technology, Maharaja Ranjit Singh Punjab Technical University, Bathinda, Punjab, India. Her areas of interest include antioxidants from natural resources, starch modifications, and cereal technology. She has published more than ten research papers in journals of international repute. She has actively participated in national and international conferences. She is an active life member of the Association of Food Scientists and Technologists (AFSTI), Mysore.

Sneh Punia Bangar, PhD, is presently working as a researcher in the Department of Food, Nutrition and Packaging Sciences, Clemson University, Clemson, SC, USA. Dr. Punia is involved in mandated research activities of the institution and has expertise in extraction and functional characterization of antioxidants, starch, their modifications, and functional products. She has presented her research at various national and international conferences and has published more than 100 research papers and book chapters in national and international journals and books. Dr. Punia has published two edited books and one authored book with CRC Press/Taylor & Francis Group. She also serves as the reviewer for various international journals. She earned her MSc and PhD degrees in food science and technology from Chaudhary Devi Lal University, Sirsa, Haryana, India.

Kawaljit Singh Sandhu, PhD, is presently working as an associate professor in the Department of Food Science and Technology, Maharaja Ranjit Singh Punjab Technical University, Bathinda, Punjab, India. He was awarded a postdoctoral fellowship from Korea University, Seoul, South Korea. With more than 15 years of research and teaching experience, his research interests are focused on starch, starch modification, antioxidants, nano-technology, and drug delivery. Dr. Sandhu received several research projects from UGC, DST, and MOFPI, New Delhi. He was given the Young Scientist Award from the Association of Food Scientists and Technologists (AFSTI), Mysore, India. Dr. Sandhu is an active member AFSTI, India, the Association of Microbiologists of India, and the Korean Society of Food Science and Technology, South Korea. He has published more than 60 research papers in various national and international journals.

Surender Kumar Singh, PhD, is Principal Scientist at the National Centre for Integrated Pest Management, NCIPM, New Delhi, India. He has three international and seven national patents to his credit. During 27 years of research, nine pieces of technology have been transferred to the industry. Dr. Singh received several research projects from the Government of India. He received many awards during national and international conferences. He is an active member of the *Indian Journal of Entomology*, *Indian Journal of Plant Protection*, *Annals of Plant Protection*, *Journal of Biological Control*, *Journal of Oilseed Research*, and *Journal of Insect Science*. He has published eight books.

Maninder Kaur, PhD, is associate professor in the Department of Food Science and Technology, Guru Nanak Dev University, Amritsar, India. She has published more than 60 research articles in various international and national peer-reviewed journals. Dr. Kaur is a member of various scientific societies and associations, including being an active member of the Association of Food Scientists and Technologists (AFSTI), Mysore, India, and the Punjab Academy of Sciences. She also serves as the reviewer for various international journals. She was awarded a postdoctoral fellowship from Korea University, Seoul, South Korea. Dr. Kaur's specialization is characterization of biomacromolecules from different botanical sources, their modifications, and applications. Dr. Kaur has supervised three PhD students.

List of contributors

Dr. Adeleke Omodunbi Ashogbon
Department of Chemical Sciences
Adekunle Ajasin University
Akungba-Akoko, Ondo State,
    Nigeria

Dr. Goran Bekavac
Institute of Field and Vegetable
    Crops
National Institute of the Republic of
    Serbia
Novi Sad, Serbia

Dr. Preeti Birwal
Department of Processing and
    Food Engineering
Punjab Agricultural University
Ludhiana, India

Dr. Ivica Đalović
Institute of Field and Vegetable
    Crops
National Institute of the Republic of
    Serbia
Novi Sad, Serbia

Dr. E. Dilipan
Department of Biotechnology
Selvam College of Technology
    (Affiliated with Anna University)
Ponnusamy Nagar, Namakkal,
    Tamil Nadu

Dr. Anuradha Dutta
Department of Food and Nutrition
Govind Ballabh Pant University of
    Agriculture and Technology
Pantnagar, Udham Singh Nagar,
    Uttarakhand, India

Dr. Bojana Filipčev
University of Novi Sad
Institute of Food Technology
Novi Sad, Serbia

Manas K. Jha
Hexagon Geosystems
Gurugram, Haryana

Dr. Gurveer Kaur
Department of Processing and
 Food Engineering
Punjab Agricultural University
Ludhiana, India

Dr. Maninder Kaur
School of Agricultural
 Biotechnology
Punjab Agricultural University
Ludhiana, India

Dr. Maninder Kaur
Department of Food Science and
 Technology
Guru Nanak Dev University
Amritsar, India

Dr. Pinderpal Kaur
Department of Food Science and
 Technology
Maharaja Ranjit Singh Punjab
 Technical University
Bathinda, Punjab, India

Dr. Ramandeep Kaur
Department of Chemistry
Punjab Agricultural University
Ludhiana, India

Dr. Ramandeep Kaur
KVK Budh Singh
Punjab Agricultural University
Ludhiana, India

Dr. Pooja Manchanda
School of Agricultural
 Biotechnology
Punjab Agricultural University
Ludhiana, India

Dr. Dilip Kumar Markandey
Central Pollution Control Board
Delhi

Dr. Zvonko Nježić
University of Novi Sad
Institute of Food Technology
Novi Sad, Serbia

Dr. Ricardo Tadeu Paraginski
Instituto Federal de Educação,
 Ciência e Tecnologia Farroupilha
 (IFFar)
Campus Santo Augusto
RS, Brazil

Dr. Vania Zanella Pinto
Food Engineering and Graduate
 Program in Food Science and
 Technology
Universidade Federal da Fronteira
 Sul (UFFS)
Campus Laranjeiras do Sul
PR, Brazil

Dr. Milica Pojić
University of Novi Sad
Institute of Food Technology
Novi Sad, Serbia

Dr. Sneh Punia Bangar
Department of Food, Nutrition and
 Packaging Sciences
Clemson University
Clemson, South Carolina, USA

Dr. Sukhvinder Singh Purewal
Department of Food Science and
  Technology
Maharaja Ranjit Singh Punjab
  Technical University
Bathinda, Punjab, India

Dr. Raj Kumar Salar
Department of Biotechnology
Chaudhary Devi Lal University
Sirsa, India

Dr. Kawaljit Singh Sandhu
Department of Food Science and
  Technology
Maharaja Ranjit Singh Punjab
  Technical University
Bathinda, Punjab, India

Dr. Sandhya
Department of Processing and
  Food Engineering
Punjab Agricultural University
Ludhiana, India

Dr. Jaya Sharma
Central Pollution Control Board
Delhi, India

Dr. Olivera Šimurina
University of Novi Sad
Institute of Food Technology
Novi Sad, Serbia

Dr. Arashdeep Singh
Department of Food Science and
  Technology
Punjab Agricultural University
Ludhiana, India

Dr. Sukriti Singh
Department of Food Technology
Uttaranchal University
Dehradun, Uttarakhand, India

Dr. Surender Kumar Singh
National Centre for Integrated Pest
  Management
Pusa Campus
New Delhi, India

Dr. Anil Kumar Siroha
Department of Food Science and
  Technology
Chaudhary Devi Lal University
Sirsa, India

Dr. Shweta Suri
Department of Food and Nutrition
Govind Ballabh Pant University of
  Agriculture and Technology
Pantnagar, Udham Singh Nagar,
  Uttarakhand, India

Dr. Urvashi
Department of Chemistry
Punjab Agricultural University
Ludhiana, India

Arihant Yuvraaj
Institute of Nuclear Medicine and
  Allied Sciences
Timarpur, New Delhi
New Delhi, India

Chapter 1

Physical and milling properties of maize

Bojana Filipčev, Ivica Đalović, Zvonko Nježić,
Olivera Šimurina, Goran Bekavac, and Milica Pojić

DOI: 10.1201/9781003245230-1

1.1 Introduction

Maize (*Zea mays* L.) is an annual plant of an average height of 2.5 m, with yellow or white grainy fruit. It belongs to the genus *Zea*, the family Poaceae (grasses), and the order Cyperales. The genus *Zea* consists of four species of which *Z. mays* L. is economically important and has a number of hybrids that differ from one another with respect to chemical composition and grain structure (Shah et al., 2016). The other *Zea* species (teosintes) are wild grasses native to Central America and Mexico. Maize is widely cultivated all over the world, and each year its production increases more than that of any other grain product (IGCMR, 2013). Today, maize is mostly grown in the United States of America (about 40%) and China (about 20%); other top producers are Brazil, Argentina, Indonesia, Ukraine, India, Mexico, Indonesia, and France (FAOSTAT, 2020). Maize is mainly used as animal feed, as a raw material in industry, and, to a lesser extent, as human food (especially in developing countries). Because of a growing world population and an increased need for food, it is predicted that the production of maize will have surpassed the production of wheat and rice by 2050 (Rosengrant et al., 2008; Ortiz et al., 2010). The volume of maize production in the world is mostly attributed to the development of technology and the seed industry, increased agro-efficiency, innovative maize food, and technical maize products, and, principally, innovation and increased production of bioethanol and biodiesel (Bekrić & Radosavljević, 2008).

Maize kernel is composed of four main fractions: the kernel root (tip cap, 1–2%, mainly cellulose); pericarp (hull, 5.5–6%, mainly cellulose); germ (embryo, 10–14%, containing mostly oil, proteins, and carbohydrates); and endosperm (82%, containing mostly starch, proteins, and fats) (Gulati et al., 1996). Because maize kernels are rich in starch (60–75%), industrial maize production is oriented toward obtaining this starch, whereas the germ is treated as a by-product. About 80–84% of the total kernel oil is present in the germ, followed by 12% in the aleurone and 5% in the endosperm (Rajendran et al., 2012). The oil content in maize grains can be genetically controlled. After a long selection process, the kernel oil content can be increased by up to 20% (Dudley, 2007). Maize with an oil content level above 6% is designated "high oil maize". The maize germ is the most important part of the kernel for oil production. Maize germ contains 35–56% oil and linoleic acid being the most common fatty acid (49–61.9%) (Ni et al., 2016). In addition, maize germ contains about 1–3% phosphatides, 1% sterols, and 1.5% FFAs. Nowadays, numerous maize grain products are used in the food, pharmaceutical, chemical, and textile industries; thus, after, processing, there is practically no loss. Maize germ oil is especially important because of its use in human foods and biodiesel production.

1.2 Physical properties

The physical properties of maize grains are a group of important quality attributes, in addition to the chemical composition and cellular structure of the grains. They are exceptionally important from the aspect of maize grain processing and influence the handling of grains from harvesting to processing. Determination of physical properties is important for proper design of equipment used in grading, handling, processing, storage, packaging, and transport. Physical properties of grains include many parameters: hardness, density, kernel weight, size, shape, surface area, coefficient of friction, colour, moisture content, and dielectric and thermal properties (García-Lara et al., 2019; Barnwal et al., 2012). There are different indicators that are used to measure some of these properties (Table 1.1). Other physical properties are also important in maize quality evaluation such as pericarp damage and presence of stress cracks as indicators of hidden

Table 1.1 Major physical properties of maize grains and measurement indicators

Physical property	Indicator
Hardness	Visually observed areas of vitreous/floury endosperm Milling power/grinding resistance Breakage susceptibility Yield of various milling fractions
Density	Bulk density (test weight) True density Porosity
Kernel weight/size	1000-kernel weight
Moisture	Water content Moisture shrink

damage that may make grains more susceptible to breakage (Paulsen et al., 2019). Pericarp damage related to superficial injuries that affect outer layers of pericarp are not problematic. However, deeper lesions into the aleurone layer that expose the endosperm and germ making them susceptible to fungal deterioration are a matter of concern (Paulsen et al., 2019).

1.2.1 Hardness

From the aspect of industrial processing, the most important physical property is grain hardness because it influences many other traits such as: bulk density, some nutritive properties, susceptibility to breakage and dust formation, milling properties, yields of prime products, feasibility for production of specialty products, etc. (Paulsen et al., 2019; Radosavljević et al., 2001). The main determinant of grain hardness is the ratio of floury to vitreous endosperm. According to Radosavljević et al. (2001) there are no unambiguous definitions of hardness as a term and two explanations exist. According to the first one, hardness is defined as the amount of vitreous endosperm expressed in % of total weight or volume of grains, and it is used for evaluation of dry millability of grains. According to the second definition, hardness is defined as a resistance of kernels to breakage i.e. as an indicator of mechanical attributes in transport, handling, and milling. Moreover, it is an attribute of great agronomic relevance, as it may affect the effects and damages of post-harvest handling (Paulsen et al., 2019). Vitreous endosperm makes the kernel stronger and less fragile (García-Lara et al., 2019).

Besides the proportion of glassy versus floury endosperm, hardness is also affected by the endosperm cell structure and pericarp thickness (Szaniel et al., 1984). Cellular compactness, cell sizes, and cell wall thickness all affect the kernel hardness but the most important is the endosperm cell structure: starch granules are surrounded with continuous protein matrix which is thinner and weaker in floury endosperm (Paulsen et al., 2019). In soft endosperm starch granules are larger and loosely packed in contrast to glassy endosperm which consists of densely packed small granules. Hardness is a heritable property which is modified by environmental influences as well as agronomic and processing treatments (García-Lara et al., 2019; Radosavljević et al., 2001). Kernel hardness is strongly directly correlated with test weight and yields of high-quality dry-milling products (Lee et al., 2012). General maize categories (dent, flint, floury, popcorn, and sweet) differ in endosperm structure. Flint maize has the hardest endosperm with the largest and continuous volume of hardest endosperm. The endosperm of floury maize is completely of chalky, soft endosperm. Dent maize has various ratios of vitreous to floury endosperm that range from 0.4 to 1.4 depending on the nitrogen fertility level in soil (Hamilton et al., 1951).

Vitreous endosperm is 1.5–2% higher in proteins compared to floury endosperm (Hinton, 1953).

1.2.2 Density

Density is defined as a ratio of weight and volume and expresses the weight of unit volume i.e. specific weight. It is expressed as bulk density or true density. Density parameters, porosity, and kernel size and shape are important in evaluation of hydrodynamic, aerodynamic, and heat/mass transfer behaviour of grain material, which is necessary in designing storage facilities and handling equipment. Vitreous endosperm increases the density of maize kernels, therefore flint maize is denser than floury maize. Density depends on the moisture content, genetic background, environmental conditions (climate, soil), and agro-technical treatments (fertilization level) (García-Lara et al., 2019). Bulk density or test weight is a very important and frequently used quality feature used in grading and classification of maize grains. It is measured in kilograms per hectolitre. The measured value of test weight depends on the kernel density and the way they are packed in the measurement cylinder. It depends on kernel shape and moisture content but it is less dependent on kernel size (Paulsen et al., 2019). Higher test weights and densities have been recorded in maize hybrids and lower plant populations (Vyn & Tollenaar, 1998). Kernel density is highly correlated with other indicators of hardness, millability, and the ratio of glassy versus floury endosperm (Blandino et al., 2010; Mestres et al., 1991; Wu & Bergquist, 1991). Gustin et al. (2013) reported significant strong correlation (0.80) between bulk test weight and kernel density.

1.2.3 Kernel weight (1000-kernel weight)

Among cereals, maize kernels have the highest weight (around ten times higher in comparison to wheat, sorghum, and rice) (García-Lara et al., 2019). It is closely correlated with the grain size, volume, hardness, and milling yields. The highest kernel weights are characteristic for white maize kernels and those with higher proportion of glassy endosperm (Paulsen et al., 2019). Higher kernel weights are desirable for both dry and wet milling owing to higher yields of prime product (flaking grits, large-particle coarse meal) i.e. higher starch yield per unit of pericarp fibre (Paulsen et al., 2019).

1.2.4 Moisture content

Moisture content is important from an economic point of view as it affects the handling costs (drying in the case of high grain moisture) and maintaining the quality of the grains (excessive moisture leads to fungal development). The standard moisture level is 15%. In trading, higher moisture than the standard reduces the price of the commodity (Paulsen et al., 2019). Most corn traders require moisture content at 14.5% (García-Lara et al., 2019). Moisture content affects several physical properties of maize kernels: bulk density, specific gravity, and kernel weight. Barnwall et al. (2012) studied the physical properties of the bold variety of maize regarding increasing moisture in the range 12.8–29% and found that kernel length, width, thickness, mean diameter, surface area, sphericity, porosity, and 1000-kernels mass increased with increasing moisture levels. A decreasing trend was observed for bulk/true density and rupture/cutting force and energy. The static coefficient of friction also increased. Similar tendencies and a linear relationship between moisture content and physical properties of sweet corn kernels were established in a moisture range of 9.12–17.06% in the study of Karababa and Coşkuner (2007).

The direct indicator of kernel moisture is the actual water content. An indicator known as "moisture shrink" is related to the loss of grain weight due to drying and "handling" loss of solids. During elevation of wet grain mass in storage facilities and transport, grains lose weight partly due to drying and partly due to loss of dust and small kernels. Determination of moisture shrinkage is important for determination of the price of grains in trade.

1.2.5 Other physical parameters

Other important physical parameters of maize grains are their perpendicular geometrical dimensions (length, width, thickness), kernel surface, sphericity, and static coefficient of friction. Their determination is important as they affect different processing phases. Geometrical dimensions are important parameters in the phase of cleaning and classification and are used to calculate kernel volume and sphericity. Kernel sphericity influences seed packaging, thus has an impact on storage capacity. Static coefficient of friction is determined with respect to concrete, wood, and aluminium, and describes moving on an inclined plane. A detailed description of measurement and calculation methods for geometrical dimensions and derived parameters as well as angle of repose and friction coefficients can be found elsewhere (Sangamithra et al., 2016).

1.2.6 Contribution of physical properties to maize grain quality

Maize quality is a term with multiple interpretations and that has different meanings for different users (Radosavljević et al., 2001). The definition of maize quality must be made with reference to its intended end use because each use requires grains of different sets of quality traits best suited for the particular end-use requirement. Maize grain quality is characterized by a range of traits grouped under physical, intrinsic, and sanitary properties (US Grains Council, 2015). Physical properties include test weight, moisture content, kernel size, stress cracking, breakage susceptibility, presence of damaged/broken kernels, and foreign impurities. Intrinsic properties include millability/grinding behaviour, kernel hardness, density, colour, protein, starch and oil content, storability, and feed value. Sanitary properties are odour, microbiological quality, presence of insects and rodent excrements, pesticide residues, and dust.

Different quality grades and classification systems for maize grains exist at national levels. They are used to support maize trading and determine the market value of maize grains but do not relate well to the specific requirements of processors. Maize processors usually have their own specifications and quality requirements in place, based on available varieties, processing equipment, and types of end products demanded by local markets (Anderson & Almeida, 2019). The major grain quality factors in most countries are test weight and amount of damaged kernels, heat damaged kernels, broken kernels, and foreign materials. In countries with defined maize grading standards, moisture is not part of the grades, but its limits are regulated in commercial trade or it is recorded on official certificates (Paulsen et al., 2019). Test weight is the dominant quality criterion used in maize classification which is especially important for dry millers. In wet milling, low test weights are usually associated with low oil yields and higher solubles, which negatively affect wet processing efficiency (Paulsen et al., 2019). In dry milling, test weight is a crucial quality factor which is directly correlated with yields of large grits fractions (flaking, coarse) (Paulsen et al., 2019). More details on maize quality requirements for grains intended for dry and wet milling will be addressed in subsequent sections of this chapter.

1.3 Dry milling

The vast majority of maize is used as animal feed but ample amounts are used in a number of different industrial processes such as dry and wet milling, alkaline nixtamalization, ethanol processing, and seed production. In this and following sections, milling properties and quality requirements for dry and wet milling of maize grains will be discussed. Dry- and wet-milling techniques are very distinct categories and no similarities exist between the processing methods.

1.3.1 Milling and mechanical properties of maize kernels

The processing of maize for food and feed involves various types of mechanical treatments using external forces (Babić et al., 2011). Most of the maize processing techniques require milling or grinding as a first step. It is a very important step that influences the end-use processing performance and power consumption. The main purpose of milling is to mechanically disintegrate the maize kernels by grinding and crushing forces. The millability of maize kernels primarily depends on the physical properties and textural behaviour of the single kernels but also depends on the mechanical properties of grains as affected by friction between the particles, inter-particle contact history, and load history (Lupu et al., 2016). Endosperm hardness is a physical attribute with extreme importance for the milling efficacy (Nikolić et al., 2020).

Maize endosperm makes up over 80% of maize kernels and consists of starch granules embedded and surrounded by a continuous protein matrix (Darrah et al., 2019). Its structure is complex and depends on the interaction between starch and protein such as the intensity of binding forces between them. Endosperm structure also depends on the thickness of protein matrix and cell walls as well as the degree of packaging of cellular compounds (Pereira et al., 2008). However, the correlations between protein content and physical traits of kernels are relatively low, if existing (Dorsey-Redding et al., 1991; Mestres et al., 1991; Abdelrahman & Hoseney, 1984; Manoharkumar et al., 1978; Simmonds, 1974). The endosperm in maize is made up of hard, vitreous and soft, floury endosperm; the proportion of which differs in different types of maize (Wolf et al., 1952). Maize kernel hardness depends on the ratio of hard and soft endosperm (Nikolić et al., 2020).

In industrial processing, milling performance and the ratio of hard to soft endosperm are the most important milling traits (Milašinović, 2005) that actually define the suitability of maize grains for dry and soft milling. Hybrids with larger portions of soft endosperm are more suitable for wet milling due to the weaker protein matrix surrounding starch granules, which enables easier extraction of starch (Milašinović, 2005). In contrast, maize hybrids with a higher ratio of glassy endosperm are more suitable for dry milling because it favours coarser grits formation (Lee et al., 2007). The structure of endosperm in the maize kernels primarily depends on the genetic background and the environmental conditions (Nikolić et al., 2020) but to a lesser extent on grain processing conditions e.g. drying (Dobraszczyk et al., 2002). Several studies were conducted to correlate maize quality and dry-milling properties and efficiencies in order to contribute to the development of standardized testing procedures and correct interpretation of data. Yuan and Flores (1996) revealed significant correlations between dry-milling yields and protein content, true density and ratio of hard to soft endosperm, as well as endosperm to bran ratio. A linear relationship was found between protein content in white corn and the yield of flaking grits (Yuan & Flores, 1996). Macke et al. (2016) investigated the association between dry-milling performance and agronomic and physical traits of maize kernels. They reported that dry-milling performance measured as dry-milling efficiency (DME) negatively correlated with grain yield and positively correlated with test weight. They confirmed that maize millability was moderately heritable and controlled by additive gene action. Flaking-grits yield (FLY) as another attribute of milling performance could be moderately predicted by kernel thickness and 100-kernel volume (Macke et al., 2016). Wu and Bergquist (1991) found a significant positive correlation between grain density and total grits yield. Grain density is a parameter closely associated with endosperm hardness. Chiremba (2012) confirmed that maize kernel hardness was best estimated by test weight and hardness determined on tangential abrasive dehulling device.

1.3.1.1 Mechanical properties of maize kernels

Few studies can be found that investigate the mechanical properties of maize kernels and their susceptibility to grinding. Knowledge on mechanical properties of maize grains is important for equipment construction, definition of the operational parameters, and estimation of energy demand of the equipment. In addition, knowledge on the mechanical properties of maize kernel tissues could contribute to minimizing the mechanical damage of grains during transport and handling. According to anatomical data, in mature maize kernels, endosperm accounts for 82% of the kernel's mass and consists of four types of tissue: the starchy endosperm, the aleurone outer layer, the basal endosperm transfer layer, and the embryo-surrounding region (Garcia-Lara et al., 2019). Endosperm essentially contains two types of structurally different tissues: floury and glassy, where the proportion of each differs according to the type of maize (flint, dent, etc.) (Garcia-Lara et al., 2019). The glassy, floury, and germ tissues respond differently to external forces (Wang & Wang, 2019). The available studies on the mechanic properties of maize kernel tissues differ in testing methodologies, their scopes, used devices, and measuring conditions (Babić et al., 2011) reported on lowering of maize kernel compressive strength properties with increasing moisture content. Lupu et al. (2016) also studied the relationships between the moisture content of maize kernels and grinding behaviour. They concluded that the kernel surface area and deformation increased at higher moisture levels whereas the yield point, force, energy, and modulus of elasticity decreased. Kruszelnicka (2021) determined the mechanical properties of maize grains by means of a compression test, measuring force and displacement during the compression. This allowed the estimation of apparent modulus of elasticity, stresses/energies at point of inflection, bioyield, and rupture points. The authors determined physical properties of the grains (mass, dimensions, sphericity index, volume-equivalent sphere diameter, aspect ratio, geometric volume, and volume correction factors). The results showed that the grain mechanical properties depend on the grain size, more precisely, on the grain thickness. Wang and Wang (2019) argued that compression tests are suitable for determination of mechanical properties of whole samples but are inadequate for determination of the mechanical properties of maize tissues. They used a puncture test to determine the mechanical properties of three tissues (glassy endosperm, floury endosperm, and germ) in several maize varieties differing in kernel structure. Moreover, they studied the effect of moisture on the change of tissue mechanical properties. The study revealed an absence or only weak influence of variety on the mechanical properties of glassy, floury endosperm tissue, and germ tissue. Mechanical properties of tissues of outer endosperm layers (glassy endosperm) were higher than those of floury and germ tissues. These inner tissues showed higher bioyield and rupture deformation. Upon moisture increase, rupture force, ultimate strength, toughness, initial firmness, average firmness, (apparent) modulus of elasticity, bioyield force, bioyield strength, and penetration force of the tissues substantially decreased, whereas bioyield and rupture deformations decreased. Yet, knowledge on the relationships between mechanical properties and internal kernel structure are insufficient.

1.3.2 Grain quality requirements for dry milling

Dry-milling yields refined endosperm fractions of different granulation, ranging from large grits to flour, germ, and dietary fibres. Some of these products can be used directly in human nutrition, but, usually, dry-milling maize fractions are further processed by bakery, snack, breakfast cereal, and brewing industries into flakes, snacks, bread, syrups, beer, and ethanol (Serna-Saldivar, 2016). A general requirement regarding grain suitability for dry milling includes the following quality traits: large and/or uniform kernel size/shape; easy removal of pericarp; low level of stress cracks; hard (vitreous) endosperm; minimum foreign materials; minimum breakage; and absence of odours, moulds, mycotoxins, and excreta of birds and rodents (Anderson & Almeida, 2019). According to Morris (2016), one of the most important quality traits for dry milling is a high proportion of vitreous endosperm in a yellow dent variety. Flint maize, although entirely vitreous, is much less suitable due to its spherical kernel morphology that is not conducive to producing grits.

Some more specific quality requirements for maize in dry milling are presented in Table 1.2. As can be seen, suitability of maize grains for dry milling is estimated using indirect parameters such as test weight, kernel

Table 1.2 Quality attributes of maize grain for dry milling

Important quality traits for maize dry milling		
Parameter	Ranges	Reference
Test weight	Max. 75.9 kg/hl, optimal range 72.1–73.4 kg/hl	Anderson & Almeida (2019)
Moisture	Max. 16%, optimal range 15.5%	
Small kernel	Max. 15%, optimal range 10.0	
Stress cracks	Max. 5%, optimal range: None	
Total damage	2%	
Infestation	None	
Aflatoxin	None	
Soybeans	None	
Odour	None	
Mould	None	
Density	Optimal range 1.27 g/cm³	
Contamination	None	
Heat damage	Max. 3, optimal range: None	
Density	High	Radosavljević et al. (2001)
Hardness	Dent	
Protein content	High	
Oil content	Low	

hardness, breakage susceptibility, and protein content. For dry millers, test weight is a crucial quality factor which is directly correlated with yields of large grit fractions (flaking grits, coarse grits) (Paulsen et al., 2019). Test weight is the dominant quality criterion in maize classification and grading standardization used in maize trading across the world.

Direct indicators of dry-milling performance are the yields of meal fractions obtained after milling and sieving under laboratory conditions. Macke et al. (2016) defined dry-milling efficacy (DME) as a proportion of weight of flaking grits obtained from a 100 g grain sample expressed as a percentage, having in mind that flaking grits are the prime product of the dry-milling process and are highly valued on the market. Another parameter derived from DME is the weight of flaking grits per unit of land (FGY) (Macke et al., 2016). Mestres and Matencio (1996) estimated the dry-milling characteristics of maize grains by measuring the yields of regular and coarse grits (RCG), cornmeal flour (CF), and flaking-grits yields (FG). There is a large variation in the milling characteristics (proportion and composition of grits, flour and hominy feed fractions) among maize hybrids (Rausch et al., 2009). All the aforementioned yields of different milling fractions actually reflect grain millability, which is defined as the ease of separation of anatomic components of grain. Determination of DME implies the development and use of small-scale dry-milling devices able to produce results representative of DME in industrial lines. Moreover, there is the need to develop and standardize new laboratory or pilot-scale procedures for milling that allow quick estimation of fraction yields and their compositions. Critical steps in the development of efficient milling procedures are tempering and degermination (Rausch & Eckhoff, 2016a; Rausch et al., 2009). A lack of widely accepted and standardized methods for measurement of millability of maize for dry milling (Yuan & Flores, 1996) limits the development of a standardized set of criteria to classify and segregate maize grains (Lee et al., 2007).

1.3.3 Maize dry-milling procedure

There are different types of dry millings of maize. Rausch & Eckhoff (2016a) distinguish three different processes:

1. Stone grinding which relates to dry milling of whole maize kernels and does not seek to separate grain components. It is also referred to as full-fat dry milling.
2. Grinding with hammer mill prior to jet cooking in order to produce alcohol.
3. Dry-milling process that aims at separating the maize grain into its components: germ, coarse fibre (pericarp), and several fractions of endosperm differing in particle size.

According to Anderson and Almeid (2019), the following dry-milling processes exist for maize:

1. Full-fat milling
2. Bolted milling
3. Tempering-degerming milling

   a. Beall degerminating system
   b. Buhler degerminating system
   c. Satake degerminating system

Full-fat milling aims at grinding maize kernels to uniform particle size using different mills like hammer, pin, or disc-mills. The oldest way of full-fat milling is grinding on stone grinding plates. This process yields only one fraction which contains all anatomic parts of maize kernels. Due to presence of germ, the obtained products have fuller flavour but shorter shelf-life, as they are highly susceptible to oxidative rancidity (Anderson & Almeida, 2019; Rausch & Eckhoff, 2016a).

Bolted milling involves a step of sieving through a bolting cloth, removing in this way larger particles such as bran, tip cap, and germ fractions from the milled maize. The obtained maize fractions are similar to that after full-fat milling but have some restrictions regarding fibre and fat content as legally defined by the Standard of Identity in the US Code of Federal Regulations (Anderson & Almeida, 2019). According to the identity standard, bolted maize meal has a fat content of 2.25% dry basis (d.b.), whereas full-fat maize has 4% d.b. The fibre content in full-fat meal must be more than 1.2% d.b, while bolted maize meal should be below 1.2% d.b.

Tempering-degerming milling is the most prevalent industrial way of dry maize processing and includes the step of moisture addition to grain mass (tempering) for facilitation of segregation of grain anatomic parts (Figure 1.1).

1.3.3.1 Preparation steps for grain milling

1.3.3.1.1 Cleaning Prior to conditioning (tempering), maize grain is subjected to mechanical cleaning for removal of foreign material and broken kernels. This step is relatively simple and is easier to perform in comparison to wheat grain cleaning. Firstly, ferromagnetic impurities are removed by passing the grain mass through a magnet. Metal-free grain mass is then aspirated to remove coarse, light, and dusty impurities. Coarse impurities are removed on sieves with round openings 16 mm in diameter. Light impurities are aspirated and settled in aspiration chambers or dust filters in cyclones. Light impurities that contain broken kernels and weed seeds are separated as thrus (throughs) on a lower sieve of an aspirator (openings 2.7 × 20 mm). The sieve on the top of the aspirator with round openings of 12 mm in diameter segregates the maize grains into two fractions. The overs on this sieve pass to the tempering phase whereas the thrus move on to stone removal.

1.3.3.1.2 Tempering The tempering step is performed by treating the grains with water and/or steam in one to three steps depending on the processing goals (Žeželj, 1995). There are numerous tempering methods differing in the amount of added moisture, residence time, and temperature (Rausch & Eckhoff, 2016a) that point to absence of sufficient criteria in processing maize by milling as well as to diverse quality of maize as a raw material (Žeželj, 1995). Usually the tempering phase aims at reaching 21–25% moisture in grains (or 16% in the so-called dry method) (Žeželj, 1995). The purpose of tempering is to induce different hydration in the anatomic parts

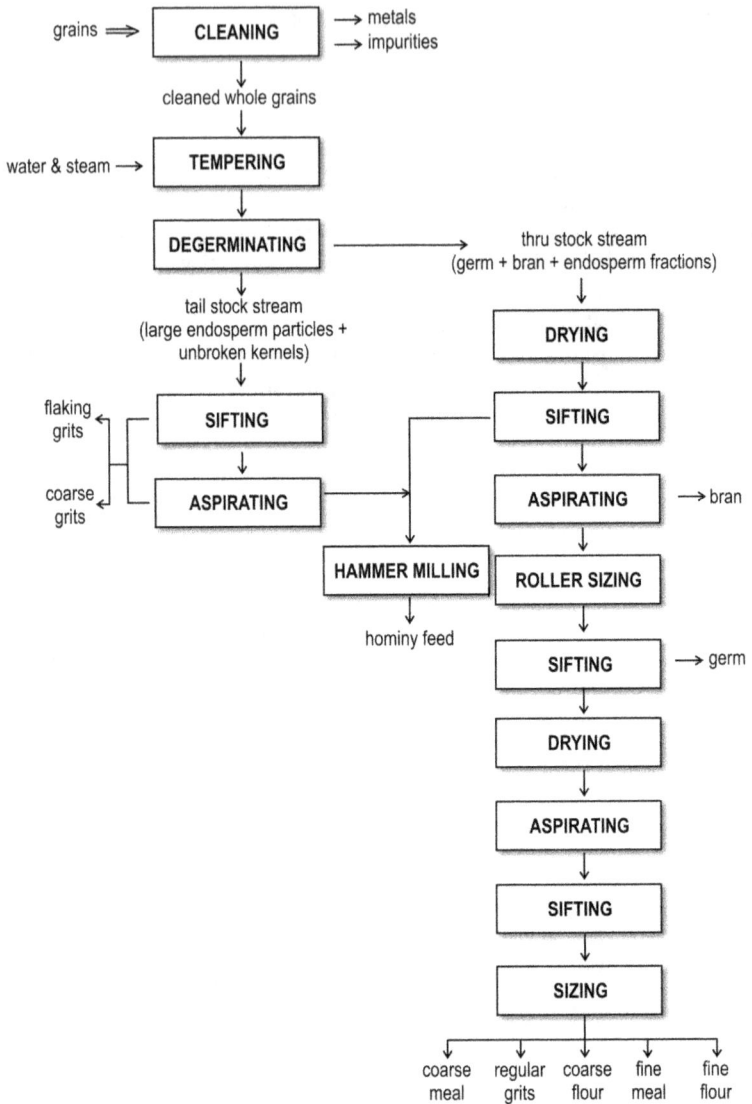

Figure 1.1 Flow chart for maize dry milling

of the grain that facilitate effective segregation: the germ and pericarp swell faster than the endosperm, become resilient against breakage, and the connecting tissue between the pericarp and the aleurone layer of the endosperm and between the germ and the endosperm weakens (Rausch & Eckhoff, 2016a). However, a moisture increase in maize endosperm is not desirable

Table 1.3 Typical conditions applied in maize tempering

Tempering method	Grain temperature (°C)	Residence time (min)	Final grain moisture content (%)	Reference
Multistage tempering				
I step	27–31	65	20.5	Žeželj (1995)
II step	38	75	22.0	
III step	43	5	24.0	
Single-stage tempering	–	10–40	6–8% moisture added in the process	Rausch & Eckhoff (2016a)

because it raises the occurrence of stress cracks that lower the yield of large-granulated grits. In addition, moisture absorbed by the endosperm would need to be removed by drying, which leads to additional costs. Tempering is a vital step in dry-milling processing that determines the effectiveness of the whole procedure. Multi-stage tempering methods are more frequently reported in laboratory studies, whereas industrial practice relies on a single-stage method (Rausch et al., 2009). Table 1.3 presents some of the typical regimes in single- and multistage tempering of maize grains.

1.3.3.1.3 Degermination   After tempering and conditioning with water or steam, the maize grains are subjected to abrasive action in degermina-tors which create vigorous friction between the maize grains, grains and the rotor jacket, and grains and the beaters. This step allows removal of pericarp and partial removal of germ. There are different constructions of degerminators, each of which have their strengths and weaknesses. The oldest and most used one is the Beall-type degerminator. Types of degermi-nators and their effects are listed in Table 1.4.

1.3.3.1.4 Grinding and separation of grits and meals   Depending on the degermination type, the obtained fractions are more or less heterogeneous. According to Rausch & Eckhoff (2016a), the least efficient separation of germ and pericarp is achieved by roller milling. After degermination, the obtained fractions are high in moisture, therefore drying is the necessary step that follows. Drying is performed in dryers of different construction but moisture is partly removed by circulating hot air used for meal trans-port. The material is usually dried to 15–16% moisture (Žeželj, 1995). After drying, the meals are further separated and reduced by a series of milling and sifting operations in which schemes vary depending on the processing goal. According to Žeželj (1995), the dried material is firstly separated into three fractions by sifting. The first fraction consists of large particles, over 8 mm that are returned to the degermination phase. The second fraction

Table 1.4 Type of degerminators and their outputs

Type	Design	Yields	Reference
Beall	Cone-shaped rotor with three surface configurations (auger, knob, and stud-shaped protrusions)	Large flaking grits, effective removal of germ	Rausch & Eckhoff (2016a)
Impact	Horizontal and vertical disc pin-type mills	Less good separation of pericarp and germ, less flaking grits, more whole germ, large bran particles	Anderson & Almeida (2019)
Multiple impact/ shear	Two horizontal or vertical cylinders with flat surfaces to connect impact attachments of different types, shapes, and angles	Less yield of flaking grits, higher prime product yield (maize meal, flour, snack grits, brewer's grits)	Rausch & Eckhoff (2016a)
Roller milling	Series of roller mills with different corrugation followed by sifting and aspiration	Grits, coarse meal and flours, no recovery of large flaking grits. Less effective separation of germ and pericarp	Rausch & Eckhoff (2016a)

contains medium large particles, 3–8 mm in size that contain germ and pericarp. This fraction is further separated into two stocks: one with particles in the range of 5–8 mm and the other with particles 3–5 mm in size. The larger particles are parts of pericarp that are separated by air separators. The smaller-sized fraction is pneumatically transported to specific gravity separators where it is separated into three or four fractions. One fraction contains pure endosperm that is moved on to further size reduction. The second fraction is a blend of endosperm particles and germ that requires further separation on a gravity separator. Flaking grits are produced using air separators on the fraction of large endosperm particles and cylinder sifters with 5–7 mm openings. The fraction of thrus on a 7 mm sifter and overs on a 5 mm sieve are large flaking grits. If grits of smaller granulation are required, the endosperm fraction is ground on a system of four to six successive roller mills and sifters.

1.3.4 Products/by-products of dry milling

The simplest classification of products derived from maize dry milling includes five product categories (Rausch & Eckhoff, 2016a; Okoruwa, 1997;

Žeželj, 1995): grits, meal, flour, germ, and hominy feed. The primary products are categories of endosperm material with different granulation profiles (grits, meal, flour). The smaller the particle sizes of endosperm material, the lower its value on the market (AGRIC, 2016). The fraction of grits has the highest particle size, but different granulation ranges are reported in the literature: 5–7 mm (Žeželj, 1995), 600–1400 µm (Baltenspreger, 1996), 0.6–1.2 mm (Okoruwa, 1997), and 638–5660 µm (Rausch & Eckhoff, 2016a). The particles in the grit fraction mainly come from the glassy endosperm. The grits' yield depends on the proportion of glassy endosperm in the maize kernel and processing conditions. Within this fraction, various distinct product types can be marketed, which each have a specific granulation profile and fat content (flaking grits, large grits, brewers grits, regular grits) (Rausch & Eckhoff, 2016a). Grits have various commercial and home uses (Okoruwa, 1997). The most important commercial use is production of corn flakes and other ready-to-eat (RTE) breakfast cereals upon hydrothermal treatment (Žeželj, 1995).

Maize meal represents an array of products of similar composition and different granulation profiles. According to Okoruwa (1997), maize milling products with particle sizes in the range of 0.2–0.6 mm fall into the category of meal. Raush and Eckhoff (2016a) give a more precise range for meal of 194–638 µm. Baltenspreger (1996) distinguishes meal of 300–600 µm and fine meal of 212–300 µm. Similarly to the grits fraction, meal also contains particles of the glassy endosperm (Žeželj, 1995). Žeželj (1995) distinguishes different categories of meal (brewers' meal, consumers' meal, instant meal, meal for extrusion, and meal for enzymatic degradation). One of the most important quality requirements for meal is low-fat content, where the requirement is especially strict for brewers' meal (max 0.8%) (Žeželj, 1995). Consumers' meal is widely used in the bakery industry (for chemically leavened baked or fried products, batters, breading) and home preparations (porridge, polenta). Maize meal intended for the production of instant flours or extruded products is similar to consumers' meal. After hydrothermal treatment with steam or using an extruder cooker, starch is pregelatinized, dried, and ground into instant flour or meal. Pregelatinized products have higher moisture absorption and retention as well as adhesion (Rausch & Eckhoff, 2016a). Extrusion is also used to produce maize flips, chips, puffs, and sheets for the snack industry and meal for pasta processing.

Maize flour is the finest fraction with particle sizes less than 0.191 mm (Serna-Saldivar & Carillo, 2019), less than 0.200 mm (Okoruwa, 1997) or less than 0.212 mm (Baltenspreger, 1996). Maize flour contains higher amounts of fat (2% d.b.) in comparison to that in grits and meal (0.6–1.2 % d.b.) (Rausch & Eckhoff, 2016a) due to higher concentration of particles from starchy endosperm (Žeželj, 1995). However, maize flour can be produced to contain a higher proportion of glassy endosperm which can be regulated by appropriate recombination of milling streams coming from different anatomic parts of the maize kernel. Maize flour with a higher fat content is more suitable for use in animal food. Maize flour has wide industrial use in the production of RTE breakfast cereals and snacks, and as a binder in meat products (Serna-Saldivar & Carillo, 2019). Maize flour is frequently used in the production of nutritionally enhanced composite flours, in combination with soy flour, in food aid programs (Serna-Saldivar & Carillo, 2019). The germ fraction is further processed by cold-pressing into premium oil which is priced around 5% higher than solvent-extracted wet-milled oil (Rausch & Eckhoff, 2016a). This fraction should contain germ of the highest purity. In well-adjusted mills, around 10% of germ with 18% oil and 15% proteins can be extracted (Žeželj, 1995). After oil removal, a high-quality protein meal remains.

The fraction with the lowest value is hominy feed that contains parts of germ and pericarp, cracked kernels, fines removed from cleaners and degerminator, pressed germ, and foreign material (Rausch & Eckhoff, 2016a; Žeželj, 1995). It is actually the trash fraction that contains the anatomic parts of the kernel that cannot be marketed. Anderson and Almeida (2019) include the fraction of maize bran when characterizing the categories of dry-milling maize products. The bran fraction contains particles from pericarp, testa, and aleurone layer (Tufail et al., 2019). Most of the pericarp particles are present in the thru-stock stream from the Beall degerminator, which contains germ, bran, and fines. It is possible to separate and concentrate raw bran particles from this stream using aspiration, owing to differences in the density between the particles. After additional refining of the recovered bran by mechanic separation of non-bran particles adhered to bran as well as additional sizing, a relatively pure bran product is obtained that is characterized with high total dietary fibre content (88% d.b.) and low oil, starch, ash, and proteins. This fraction can be resized to meet different granulation needs and has numerous food applications in the production of high-fibre RTE cereals, baked products, and low-calorie drinks. Maize bran can be used as a water binder, non-caloric bulking agent, emulsifier, oil retention agent, and stabilizer of oxidative rancidity (Elleuch et al., 2011). Chemically, maize bran consists of hemicellulose, cellulose, and phenolics (Benkő et al., 2007). It is a novel functional and nutraceutical ingredient (Saeed et al., 2021) with appreciable content of dietary fibres, phenolic antioxidants, oil, and phytosterol esters (Rose & Inglett, 2010).

1.3.5 Benefits and deficiencies of dry milling

One of the benefits of maize dry-milling processing is its flexibility regarding the choice of investment levels. It can be set with different levels of complexity depending on market demands: either as a relatively simple but less versatile process, or a more complex process with enhanced quality of obtained fractions that yield high-value food-grade germs, fibres, and prime category endosperm fractions. In comparison to wet-milling processing, dry milling of maize produces a less versatile product range but gives wider particle size range. Dry-milling processing requires high energy consumption.

1.4 Wet milling

The objective of wet milling is to recover constituents that are chemically pure (relatively pure) constituents from maize grain: starch, oil, protein, and fibres that are suitable for food and feed uses and further industrial refining into other solid or liquid products. In wet milling, maize kernels are ground after steeping for a prolonged period of time at elevated temperature and low pH with the aim of recovering starch and germ.

1.4.1 Grain quality requirements for wet milling

Unlike the dry-milling process where hard endosperm is required, in wet milling, soft maize endosperm is a preferred choice. The grains should be free from broken kernels and foreign materials, mould, and mycotoxins. High susceptibility to breakage and damaged kernels are objectionable traits that may generate problems in the phase of steeping, though new enzyme-aided technologies for wet milling have been developed to allow processing of broken kernels and dust (Paulsen et al., 2019). A general requirement of wet millers is yellow dent maize with average chemical composition (Rausch & Eckhoff, 2016b) and large whole kernels, with minimum pericarp damage, stress cracks less than 20%, low susceptibility to breakage, low broken kernels and foreign material, and moderate true density (Paulsen, Singh, & Singh, 2019). According to Radosavljević et al. (2001), desirable quality traits for wet milling include low proteins, high oil and starch content, moderate density, and low endosperm hardness. Some producers require specialty maize varieties with highly-extractable starch due to the larger yield of starch available for specialty starch production and lower coproduct yield (Rausch & Eckhoff, 2016b).

Although maize with a higher than average starch content is believed to have better wet-milling performance, Fox et al. (1992) did not show that greater starch content increased starch yield. This study showed that starch extractability is affected by some other factor(s) as well. Their work exerted that the best models for predicting starch yields in wet milling included protein content and one of the following: test weight, absolute density, kernel hardness, or water absorptivity. According to Rausch and

Eckhoff (2016b), starch yield is affected by kernel composition, test weight, grain millability, level of mechanical damage, grain drying conditions, and grain storage time. According to them, any compositional change in maize grain, like increased protein or oil content, affects the wet-mill balance and does not necessarily produce desirable results in terms of more efficient production. Test weight is associated with endosperm structure and higher weights usually mean higher hard (vitreous) endosperm which is undesirable for wet millers due to longer steeping times and low starch extractability. On the other hand, test weights that are too low lower the capacity of wet mills, which is disadvantageous. Low test weights are usually associated with low oil yields and higher solubles, which negatively affect wet processing efficiency (Paulsen et al., 2019). Maize hybrids with genetically good millability are desirable due to higher yields of components owing to their easier mechanical separation upon grinding. Damaged and broken kernels are unwanted due to problems in the steeping phase and low quality of obtained starch that results as a consequence of direct contact of starch with sulphur dioxide from steepwater and its modifying effect on starch cooking properties. The drying temperature may affect the recovery of starch and oil during wet milling therefore it is recommended to avoid using maize grains dried at temperatures over 60°C (Paulsen et al., 2019) or 70°C (Rausch & Eckhoff, 2016b).

1.4.2 Maize wet-milling procedure

The process of wet milling is complex and involves chemical, biochemical, and mechanical operations in several steps (Rausch et al., 2019):

1. Soaking the grain in warm water and sulphur dioxide under controlled conditions to soften the kernels.
2. The softened grains are wet milled and the ground materials is fractionated by washing, screening, centrifuging, and hydrocycloning, taking the differences in the density and particle size of ground material in water into account. As the consequence of steeping, the germ becomes rubbery and may be separated in intact form while the rest of the kernel is reduced in size during wet milling. This allows the separation of germ, starch, fibre, maize gluten meal, and maize gluten feed.

This process is highly water- and energy-intensive in comparison to other food processing industries. It can be divided into five steps: steeping, germ separation, fibre separation, protein separation, and starch washing (Figure 1.2).

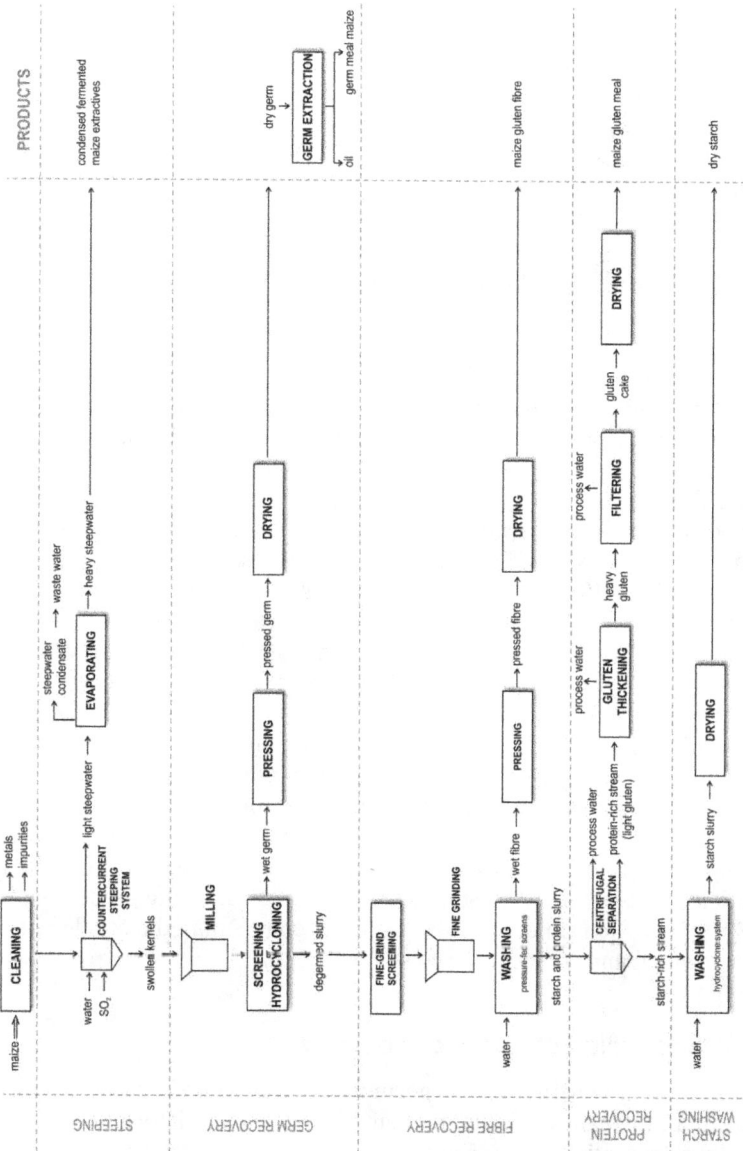

Figure 1.2 Flow chart of maize wet milling

1.4.2.1 Cleaning and steeping

Prior to steeping, maize is mechanically cleaned to remove foreign matter (metal particles, stones, chaff, weed seed, dust, sand, etc.) and small impurities (broken kernels, cobs particles). It is important to remove all impurities because they can cause problems especially during the steeping step (limited water circulation, increased steep-liquor viscosity, impaired quality of recovered starch). Steeping is an essential step in wet milling of maize. The appropriateness of all other steps that follow depends on the proper conduction of steeping, which all effects the yield and quality of starch. The steeping process is a complex phenomenon that involves physical, chemical, and biochemical changes in maize kernels, allowing separation of germ, pericarp, endosperm fibre, protein, and starch in later processing steps.

Steeping is performed in a batch-wise counter-count system of steep tanks interconnected in batteries with recirculating steepwater that contains sulphur dioxide (2000 ppm) at 52°C. During steeping, maize kernels swell and hydrate. Steeping temperature is optimized to favour lactic acid production. During the process, pH is usually in the range 3.5–4.2 (Rausch & Eckhoff, 2016b). In the first stage of steeping, lactic acid is produced by *Lactobacillus* flora until it reaches 1–3% concentration and SO_2 concentrations are reduced to 250 ppm (Rausch & Eckhoff, 2016b) (Rausch et al., 2019). In an acidic environment, maize kernels rapidly hydrate, swell, and soften. Soluble material leaches from the germ into the steepwater. Different levels of swelling weakens the connecting tissues between the kernel's anatomic parts. In the second stage of steeping, SO_2 concentrations are high enough to allow its diffusion into the maize kernel and inhibit the lactic acid fermentation. SO_2 penetrates deep into endosperm and absorbs on the surface of interstitial pores. The leaching from the germ continues. In the third stage, the absorbed SO_2 begins to react with the protein matrix by breaking the disulphide bonds and disrupts the continuity of protein matrix that surrounds starch granules allowing their separation in the downstream stages of milling. The steeping process lasts from 20 to 48 h (Rausch et al., 2019). After its completion, the swollen kernels are drained and subjected to coarse grinding.

1.4.2.2 Separation of kernel components

Firstly, after coarse grinding, the germ is removed in order to prevent its damage, maximize oil yield, and prevent oil leakage and interference with succeeding separation steps. Coarse grinding is performed in disk mills with special teeth-like surface corrugation that breaks the kernels into pieces. During steeping the germ becomes rubbery and does not break during coarse milling if operational parameters of disc-mills are properly adjusted. In this stage, the endosperm is lightly broken and the intact germ is separated using a two-stage hydrocyclone system, owing to the differences in density. The germ has lower density in comparison to the endosperm particles due to high oil content. After separation, the germ fraction contains 42–50% oil and is washed to remove residual starch. Washed germs are pressed and dried to less than 3% moisture. Drying must be carefully conducted at temperatures below 80°C to avoid oil heat damage and maximize oil yields during extraction.

After removal of the germ, the remaining slurry is screened to separate fibre from already released starch and protein. Material retained by the screens consists of fibre, protein, and starch and is subjected to fine grinding on counter-rotating disc-mills. The objective of fine grinding is to liberate as much as possible amounts of starch connected to fibre tissues and surrounded by weakened protein matrix. The finely ground slurry is further processed by washing on pressure-fed concave wedge-wire screens in several stages. The washed fibres contain 10-15% solids and after dewatering on presses or centrifuges achieve around 40% solids (Rausch et al., 2019). The next stage is separation of starch from proteins. The separation is based on density difference: proteins have lower density (1.06 g/cm^3) in comparison to starch (1.6 g/cm^3) (Rausch et al., 2019). Separation is performed on disc-nozzle centrifuges. After separation on the primary centrifuge, two fractions are obtained: 1) the underflow stream which contains mainly starch, 1.5–3% protein and small amounts of soluble and insoluble impurities; 2) the overflow stream with high protein content (around 70%, 1.5–3% solids). The starchy stream is directed towards the washing system where it is washed in several cycles to recover the maintaining fine fibre and protein particles. The protein-rich stream is routed to a centrifuge (gluten thickener) where it is dewatered to nearly 16.5% solids. Final dewatering to 40% solids of proteins is performed on a vacuum belt centrifuge. The washing system for starch-rich fraction consists of clusters of hydrocyclons operated in a counter-current mode, arranged to perform 10–14 separate washing cycles in series. After washing, the starch contains 40% solids and should have less than 0.30% total proteins and 0.01% soluble proteins (Rausch et al., 2019). The wet-milled starch may be directly dried or subjected to chemical modifications. Starch is dried on belt-type dryers at 65–150°C or on flash dryers. After drying it is milled, blended, and packaged.

1.4.3 Benefits and deficiencies of wet milling of maize

Wet-milling processing yields a wide range of highly valuable main products (germ, starch, fibre) which are of higher purity than those obtained from dry milling. They are also suitable for further refinement into multiple value-added, well-priced products (oil, modified starch, etc.). In the wet-milling procedure, starch loss is minimized in comparison to the dry milling. The disadvantages of wet milling are that expensive equipment and high overall capital investment costs are required, there is intensive energy and water consumption, an increased environmental burden due to large water consumption and discharge despite effective water recycling, and use of hazardous gas (sulphur dioxide) in the process.

1.5 Small-scale milling and milling of speciality maize

Small-scale milling is a type of dry milling characteristic for developing countries and rural communities. It typically encompasses household production and custom production i.e. small-capacity merchant mills with outputs ranging between 1–8 t of maize per day (UNIDO, 1986). Industrial processing of maize involves the use of highly efficient roller mills. In small-scale facilities, other milling techniques and equipment is used: mortar and pestle or hand grinders in households or group of households, water-operated stone mills, engine-powered stone mills or hammer mills, or small-diameter roller mills. Usually, untreated, shelled, and dried grains are simply ground into meal or flour which is directly used for feed or food preparation without sifting and separation of anatomic parts. The small-scale milling systems are occasionally extended with rudimentary type reel or centrifugal sifters which enable removal of the coarsest fractions of germ and bran further used as animal feed (Shamrock Milling Systems, 2021). The remaining meal (around 90% extraction) is used for food. In some cultures, grains are pre-processed prior to milling. The pre-processing techniques are alkali cooking and light toasting. Alkali cooking (nixtamalization) involves cooking and steeping of mature maize grains in food-grade lime that facilitates pericarp removal and affects the shelf-life, aroma, taste, and nutritional value (Serna-Saldivar & Chuck-Hernandez, 2019). After washing, the grains are wet milled to obtain tortilla dough masa. Toasting is a thermal treatment which improves the taste of maize, deactivates oxidative enzymes, and improves digestibility by partial starch gelatinization.

Upgrading the basic milling operation on stone, hammer, or small-diameter roller mills with more units of sieving and cleaning equipment leads to the development of small-scale and micro milling plants. This refinement allows more sophisticated separation and yields more variant maize fractions such as special, sifted, and super maize meal (Shamrock Milling Systems, 2021) that are positioned on specialized markets for specialty products.

In industrial processing by dry milling, the most frequently used maize type is yellow dent maize (Serna-Saldivar & Gomez, 2001). In the

Figure 1.3 Specialty maize products exhibited on a fair in Serbia

US, the US No. 2 yellow dent maize dominates the commodity maize market. However, not all commodity maize is suitable for human consumption. The food-grade maize suitable for dry milling has hard endosperm and high density. In colour, it can be yellow, white, or coloured, mostly blue (Scott et al., 2019). In the US, there are no specifications and grades for food-grade maize and the desired quality traits are specified in the identity-preserved contracts between farmers or other suppliers and buyers (Scott et al., 2019). In small-scale milling, it is not rare that types of maize other than commodity maize are used. For example, in Serbia, yellow dent maize is processed industrially using degermination dry milling, while white maize is exclusively processed on a small scale into grits/meal/flour by stone mills and it is considered as specialty maize and sold in health stores (Figure 1.3). Coloured maize varieties are also targeted by small-scale processors because of nutritional advances (Figure 1.3). In addition to high anthocyanins content, blue maize was shown to have lower glycemic index and 20% more proteins than white maize (Nankar et al., 2017). Blue maize is processed by crushing in stone mills followed by sieving to remove the largest pericarp particles and obtain meal or flour (Betrán et al., 2001). Organic maize is also attractive for small-capacity processors owing to the rapidly growing trend of organic market and limited amounts of organic maize available for processing.

1.6 Functional properties of maize grinding fractions

The industrial processing and food applications of maize meal and flours depend on their functional properties such as solubility, absorption, water retention, emulsification, viscoelasticity, gelation, cohesion, adhesion, etc. Typical functional properties for flours include water and oil absorption capacity, swelling capacity, foaming capacity, emulsion capacity, bulk density, particle size distribution, least gelling concentration, pasting properties, etc. Kinsella (1979) defines functional properties as those significant physicochemical properties that reflect the complex interaction among the composition, structure, and molecular conformation together with the nature of environment in which these properties are associated. Pour-ELA (1981) describes functional properties as those intrinsic physicochemical properties that affect the behaviour of food systems and indicate their usefulness and acceptability for consumption. Determination of functional properties is frequently used to evaluate the food application potential of composite flours (Hasmadi et al., 2020).

1.6.1 Water/oil absorption capacity (WAC, OAC) and swelling capacity (SC)

Water and oil absorption capacity is associated with the ability of flour to bind water or physically entrap oil. Water binding includes hydrodynamic water, capillary water, and physically entrapped water. High WAC indicates the presence of hydrophilic constituents such as fibres, proteins, or starch (Hasmadi et al., 2020; Chandra & Samsher, 2013; Celik et al., 2004, Kethireddipalli et al., 2002). WAC is dependent on the flour particle size, starch structure, and composition (amylose to amylopectin ratio, starch granule size, crystallinity) (Adegunwa et al., 2011; Scanlon et al., 1988). Determination of OAC is important since fats positively affect the mouthfeel and flavour of food (Hasmadi et al., 2020). OAC is primarily governed by the capability of proteins to bind with fats i.e. the hydrophobic interaction between the non-polar amino acid chains and hydrocarbon chains in fat (Jitngarmkusol et al., 2008). The swelling capacity (SC) indicates the ability of starch to absorb water and swell, reflecting the intensity of hydrogen bonding and crystalline arrangement within the granule (Biliaderis, 1982). Higher values of SC in flours is usually accompanied with higher pasting viscosity. Composition and structure of starch granules affects SC: high amylose levels tend to decrease SC, leading to highly associated and intensively bonded starch granules that resist to swelling (Buckman et al., 2018).

Table 1.5 illustrates the WAC/OAC/SC of maize flour or meals from different studies. WAC ranged between 1.54 g/g to 3.11 g/g. Bolade et al. (2009) did not show significant differences in WAC of differently granulated maize meals. In contrast, Houssou and Ayernor (2002) and (Shad et al. 2013) reported higher WAC in degermed maize flour in comparison to wholemeal flour. Higher level of damaged starch may increase the WAC in flour (Craig & Stark, 1984). OAC of maize flours ranged between 1 g/g to

Table 1.5 WAC/OAC of different maize grinding/milling fractions

Maize flour, meal	WAC	OAC	SC	Reference
Maize flour (white variety)				
Dry milled, non-tempered	2.31 ml/g	2.31 ml/g	5.82 ml/g	Adedeji & Tadawus (2019)
Dry milled, tempered	2.34 ml/g	2.34 ml/g	5.82 ml/g	
Maize flour fractions				
<75 μm	2 g/g	1.9 g/g	–	Bolade et al. (2009)
75–150 μm	2 g/g	2.1 g/g		
150–300 μm	2 g/g	1.9 g/g		
300–425 μm	1.9 g/g	1.7 g/g		
>425 μm (wholemeal)	2.1 g/g	2 g/g		
Maize flour				
Full fat	1.49 g/g	1.07 g/g	–	Shad et al. (2013)
Defatted	3.11 g/g	2.16 g/g		
Quality protein maize flour	2 g/g	1.8 g/g	2.34–2.84 g/g	Shobha et al. (2014)
Maize flour				
Whole	1.54 g/g	–	Degermed/dehulled flour was 1.5 times higher in SC than the whole maize flour	Houssou & Ayernor (2002)
Degermed and dehulled	1.97 g/g			
Maize flour	1.0 g/g	–	2.47 g/g	Nawaz et al. (2015)
Defatted maize germ flour	3.63 g/g	1.8 g/g		Siddiq et al. (2009)
Maize germ protein isolate (73.4% proteins)	0.6 g/g protein	45.2 g/g protein		Messinger et al. (1987)

2.34 g/g. Bolade et al. (2009) observed an OAC increase with a decrease in particle size, with the exception of the flour fraction with the lowest granulation and explained this finding as a consequence of different protein content in the flour fractions. Low proteins lead to lower OAC due to reduced lipophilicity (Walde et al., 2005). High starch and low fibre content may also lead to reduction in OAC (Roopa & Premavalli, 2008) (Aletor et al., 2002). The SC of Quality Protein Maize (QPM) flour was more than two times lower than that of white maize flour which may be due to lower starch content in QPM. Defatting and decortication of maize grain increases the SC of flour (Table 1.5) which may be due to the changed proportion of starch in flour.

1.6.2 Foaming/emulsifying capacity and stability

It was observed that flours of different origin have an ability to create foam. Suresh et al. (2015) ranked the flours according to their foaming capacity (FC) in the following order: green gram flour (24%), wheat flour (13%), potato flour (7%), and rice flour (3.5%). This trait is associated with the surface properties of the proteins present i.e. their ability to reduce the surface tension at the water-air interface by forming a continuous, cohesive film around the air bubbles in the foam (Kaushal, Kumar, & Sharma, 2012). Foaming stability (FS) is governed by the attractive and repulsive electrostatic forces among the involved polypeptides. Proteins that can be opened and quickly adsorbed present better foaming properties than those slowly adsorbed and whose structures are more difficult to be opened in the air-water interface (Alleoni, 2006).

Defatted maize germ flour (DMGF) has been reported to have foaming and emulsifying properties. FC of DMGF at 19.7% was determined by Siddiq et al. (2009). Increasing levels of DMGF in blends with wheat flour lowered the FC of the blends but increased foam stability (90%). Vani & Zayas (1995) reported that maize germ protein flour had lower foaming capacity than both wheat germ and soy flours at 1% concentration level. Water absorption and foaming properties of maize germ protein isolate (MGPI) were improved by partial hydrolysis with trypsin (Messinger et al., 1987).

Emulsifying capacity (EC) is an important protein functionality trait related to its ability to lower the tension at the interface of water and oil. Maize germ proteins have been found to improve not only EC but also emulsion stability (ES) in food systems by binding excess water (Bhattacharya & Hanna, 1985; Luallen, 1985). Siddiq et al. (2009) reported 63.6% EC and 58.5% ES for DMGF. DMGF contains high quantities of soluble proteins able to build a protective barrier around fat droplets preventing their coalescence (Lawton & Wilson, 2003; Kinsella, 1976). The emulsifying stability of wheat flour increased by raising levels of added DMGF (Siddiq et al., 2009). Earlier, Lin &Zayas (1987) studied the functionality of DMGF and found that its functionality depended on the defatting method. They concluded that better functional properties were exerted by germ flour defatted using supercritical CO_2 in comparison to the hexane treated one, presumably due to protein denaturation caused by hexane. Maize germ protein isolate showed similar emulsifying capacity to that of soy protein isolate (Nielsen et al., 1973). Functional properties (water binding, solubility, EC, ES) of maize germ protein flour improved at more alkaline pH 8 (Wang & Zayas, 1991; Wang & Zayas, 1992). Hojilla-Evangelista (2012) reported that maize germ protein recovered from dry-milled germ had better foaming and emulsifying properties than that obtained from wet milling. Protein extraction combined with ultrafiltration-diafiltration improved the EC of wet-milled maize germ protein but did not improve its ES (Hojilla-Evangelista, 2014). Maize gluten meal is known for its poor functional properties (Bhattacharya & Hanna, 1985). Maize gluten is a by-product of wet milling during which it undergoes severe treatments such as steeping with sulphuric acid and drying at high temperature (Lin & Zayas, 1987). These operations led to modifications in protein structure that affect the functionality of maize gluten and its potential uses in the food industry. Wu (2001) studied the emulsifying properties of commercial maize gluten meals and observed no emulsifying properties at their normal pH (pH=4). At pH above 6.6 and particle size below 44 mm the emulsifying properties of maize gluten meal improved and ranged as follows: EC 49.3–51.5% and ES 39.7–49.5%. Researchers studied various chemical modifications to improve functionality of maize proteins: chemical deamidation (Flores et al., 2010) and microfluidization and pH shifts to alkaline region (Ozturk & Mert, 2019).

References

Abdelrahman, A., & Hoseney, R. (1984). Basis for hardness in pearl millet, sorghum and corn. *Cereal Chemistry, 61*, 232–235.

Adedeji, O., & Tadawus, N. (2019). Functional properties of maize flour (*Zea mays*) and stability of its paste (tuwo) as influenced by processing methods and baobab (*Adansonia digitata*) pulp inclusion. *Ukrainian Journal of Food Science, 7*(1), 49–60.

Adegunwa, M., Sanni, L., & Maziya-Dixon, B. (2011). Effects of fermentation length and varieties on the pasting properties of sour cassava starch. *African Journal of Biotechnology, 42*, 8428–8433.

AGRIC. (2016). Processing and utilization of cereal grains. *Animal Agriculture and Food Science, Module 10-Lecture notes 10*. University of Saskatchewan.

Aletor, O., Oshodi, A., & Ipinmoroti, K. (2002). Chemical composition of common leafy vegetables and functional properties of their leaf protein concentrates. *Food Chemistry, 78*(1), 63–68.

Alleoni, A. (2006). Albumen protein and functional properties of gelation and foaming. *Scientia Agricola, 63*(3), 291–298.

Anderson, B., & Almeida, H. (2019). Corn dry milling: Processes, products, and applications. In S. Serna-Saldivar (Ed.), *Corn: Chemistry and technology* (3rd ed., pp. 405–434). Duxford: Woodhead Publishing, Elsevier Inc.

Babić, L., Radojčin, M., Pavkov, I., Turan, J., Babić, M., & Zoranović, M. (2011). Physical properties and compression loading behaviour of corn (Zea mays L.) seed. *Journal on Processing and Energy in Agriculture, 15*(3), 118–126.

Baltenspreger, W. (1996). Maize milling. In *Technical bulletins* (pp. 6742–6748). Leawood, KS: International Association of Operative Millers.

Barnwal, P., Kadam, D., & Singh, K. (2012). Influence of moisture content on physical properties of maize. *International Agrophysics, 26*(3), 331–334.

Bekrić, V., & Radosavljević, M. (2008). Contemporary approaches to maize utilisation. *Journal of Processing in Energy and Agriculture, 12*, 93–96.

Benkő, Z., Andersson, A., Szengyel, Z., Gáspár, M., Réczey, K., & Stålbrand, H. (2007). Heat extraction of corn fiber hemicellulase. *Applications in Biochedmistry and Biotechnology, 137*, 253–265.

Betrán, F., Bockholt, A., & Rooney, L. (2001). Blue corn. In A. Hallauer (Ed.), *Specialty corns* (2nd ed., pp. 1–9). Boca Raton, FL, London, New York, Washington, DC: CRC Press.

Bhattacharya, M., & Hanna, M. (1985). Extrusion processing of wet corn gluten meal. *Journal of Food Engineering, 50*(5), 1508.

Biliaderis, G. (1982). Physical characteristics, enzymatic digestibility and structure of chemically modified smooth pea and waxy maize starches. *Journal of Agricultural and Food Chemistry, 30*(5), 925–930.

Blandino, M., Mancini, M. C., Peila, A., Rolle, L., Vanara, F., & Reyneria, A. (2010). Determination of maize kernel hardness: A comparison of different laboratory tests to predict dry-milling performance. *Journal of Science of Food and Agriculture, 90*(11), 1870–1878.

Bolade, M., Adeyemi, I., & Ogunsua, A. (2009). Influence of particle size fractions on the physicochemical properties of maize flour and textural characteristics of a maize-based nonfermented food gel. *International Journal of Food Science and Technology, 44*(3), 646–655.

Buckman, E., Oduro, I., Plahar, W., & Tortoe, C. (2018). Determination of the chemical and functional properties of yam bean (*Pachyrhizus erosus* (L.) Urban) flour for food systems. *Food Science and Nutrition, 6*(2), 457–463.

Celik, I., Isik, F., & Gursoy, O. (2004). Couscous, a traditional Turkish food product: Production method and some applications for enrichment of nutritional value. *International Journal of Food Science and Technology, 39*(3), 263–269.

Chandra, S., & Samsher. (2013). Assessment of functional properties of different flours. *African Journal of Agricultural Research, 8*(38), 4849–4852.

Chiremba, C. (2012). *Sorghum and maize grain hardness: Their measurement and factors influencing hardness*. PhD Thesis. Pretoria, South Africa: University of Pretoria.

Craig, S., & Stark, J. (1984). The effect of physical damage on the molecular structure of wheat stach. *Carbohydrate Research, 125*(1), 117–125.

Darrah, L., McMullen, M., & Zuber, M. (2019). Breeding, genetics and seed corn production. In S. Serna-Saldivar (Ed.), *Corn: Chemistry and technology* (pp. 19–42). Duxford: Woodhead Publishing, Elsevier Inc.

Dobraszczyk, B., Whitworth, M., Vincent, J., & Khan, A. (2002). Single kernel wheat hardness and fracture properties in relation to density and the modelling of fracture in wheat endosperm. *Journal of Cereal Science, 35*(3), 245–263.

Dorsey-Redding, C., Hurburgh, C., Johnson, L., & Fox, S. (1991). Relationships among maize quality factors. *Cereal Chemistry, 67*, 602–605.

Dudley, J. (2007). From means to QTL (2007): Illinois long-term selection experiment as a case study in quatitative genetics. *Crop Science, 47*, S20–S31.

Elleuch, M., Bedigian, D., Roiseux, O., Besbes, S., Blecker, C., & Attia, H. (2011). Dietary fibre and fibre-rich by-products of food processing: Characterization, technological functionality and commercial applications: A review. *Food Chemistry, 124*(2), 411–421.

FAOSTAT. (2020). *Food and agricultural data*. Rome: Food and Agriculture Organization of the United Nations. http://www.fao.org/faostat/en/#data

Flores, I., Cabra, M., Quirasco, M., Farres, A., & Galvez, A. (2010). Emulsifying properties of chemically deamidated corn (Zea mays) gluten meal. *Food Science and Technology International, 16*, 241–250.

Fox, S., Johnson, L., Hurburg, J. C., & Dorsey-Redding, C. (1992). Relations of grain proximate composition and physical properties of wet-milling characteristics of maize. *Cereal Chemistry, 69*(2), 191–197.

García-Lara, S., Chuck-Hernandez, C., & Serna-Saldivar, S. (2019). *Development and structure of the corn kernel*. Duxford: Woodhead Publishing, Elsevier Inc.

Gulati, M., Kohlman, K., Ladisch, M., Hespell, R., & Bothast, R. (1996). Assesment of ethanol production options for corn products. *Bioresource Technology, 58*(3), 253–264.

Gustin, J., Jackson, S., Williams, C., Patel, A., Armstrong, P., Peter, G., & Settles, A. M. (2013). Analysis of maize (Zea mays) kernel density and volume using microcomputed tomography and single-kernel near infrared spectroscopy. *Journal of Agricultural and Food Chemistry*, *61*(46), 10872–10880.

Hamilton, H., Hamilton, B., Johnson, B., & Mitchell, H. (1951). The dependence of the physical and chemical composition of the corn kernel on soil fertility and cropping system. *Cereal Chemistry*, *28*, 163–176.

Hasmadi, M., Noorfarahzilah, M., Noraidah, H., Zainol, M., & Jahurul, M. (2020). Functional properties of composite flour: A review. *Food Research*, *4*(6), 1830–1831.

Hinton, J. (1953). The distribution of protein in the maize kernel in comparison with that in wheat. *Cereal Chemistry*, *30*, 441–445.

Hojilla-Evangelista, M. (2012). Extraction and functional properties of non-zein proteins in orn germ from wet-milling. *Journal of American Oil Chemical Society*, *89*(1), 167–174.

Hojilla-Evangelista, M. (2014). Improved solubility and emulsification of wet-milled corn germ protein recovered by ultrafiltration-diafiltration. *Journal of American Oil Chemists Society*, *91*(9), 1623–1631.

Houssou, P., & Ayernor, G. (2002). Appropriate processing and food functional properties of maize flour. *African Journal of Science and Technology*, *3*(1), 126–131.

IGCMR. (2013, November 28). International grains council market report. Retrieved from http://www.igc.int/downloads/gmrsummary/gmrsu mme.pdf

Jitngarmkusol, S., Hongsuwankul, J., & Tananuwong, K. (2008). Chemical composition, functional properties and microstructure of defatted macademice flours. *Food Chemistry*, *110*(1), 23–30.

Karababa, E., & Coşkuner, Y. (2007). Moisture dependent physical properties of dry sweet corn kernels. *International Journal of Food Properties*, *10*(3), 549–560.

Kaushal, P., Kumar, V., & Sharma, H. (2012). Comparative study of physicochemical, functional, anti-nutritional and pasting properties of taro (Colocasia esculanta), rice (Oryza sativa), pegion pea (Cajanus cajan) flour and their blends. *LWT-Food Science and Technology*, *48*(1), 59–68.

Kethireddipalli, P., Hung, Y., McWatters, K., & Phillips, R. (2002). Effect of milling method (wet and dry) on the functional properties of cowpea (Vigna unguiculata) pastes and end-product (akara) quality. *Journal of Food Science*, *67*(1), 48–52.

Kinsella, J. (1979). Functional properties of soy proteins. *Journal of the American Oil Chemists' Society*, *56*(3), 242–258.

Kinsella, J., & Melachouris, N. (1976). Functional properties of protein in foods: A survey. *Critical Reviews in Food Science*, *7*(3), 219–280.

Kruszelnicka, V. (2021). Study of selected physical-mechanical properties of corn grains important from the point of view of mechanical processing systems designing. *Materials*, *14*(6), 1467.

Lawton, W., & Wilson, C. (2003). Proteins of the kernel. In P. White & L. Johnson (Eds.), *Corn: Chemistry and technology* (2nd ed., pp. 314–354). St. Paul, MN: American Association of Cereal Chemists.

Lee, E., Fregeau-Reis, J., & Good, B. (2012). Genetic architecture under-lying kernel quality in food-grade maize. *Crop Science, 52*(4), 1561–1571.

Lee, K.-M., Herrman, T., Rooney, L., Jackson, D., Lingenfelser, J., Rausch, K., ... Fox, S. R. (2007). Corroborative study on maize quality, dry-milling and wet-milling properties of selected maize hybrids. *Journal of Agricultural and Food Chemistry, 55*(26), 10751–10763.

Lin, C., & Zayas, J. (1987). Functionality of defatted corn germ proteins in a model system: Fat binding capacity and water retention. *Journal of Food Science, 52*(5), 1308–1311.

Luallen, T. (1985). Starch as a functional ingredient. *Food Technology, 39*, 59.

Lupu, M. I., Pădureanu, V., Canja, C. M., & Măzărel, A. (2016). The effect of moisture content on grinding process of wheat and maize single kernel. *ModTech International Conference-Modern Technologies in Industrial Engineering IV IOP Conf. Series: Materials Science and Engineering, 145*, 1–7.

Macke, J., Bohn, M., Rausch, K., & Mumm, R. (2016). Genetic factors underlying dry-milling efficiency and flaking-grit yield examined in US maize germplasm. *Crop Science, 56*(5), 2516–2526.

Manoharkumar, B., Gestenkorn, P., Zwingelberg, H., & Bolling, H. (1978). On some correlations between grain composition and physical char-acteristics to the dry milling performance in maize. *Journal of Food Science and Technology, 15*, 1–6.

Messinger, J., Rupnow, J., Zeece, M., & Anderson, R. (1987). Effect of par-tial proteolysis and succinylation on functionality of corn germ protein isolate. *Journal of Food Science, 52*(6), 1620–1624.

Mestres, C., Louis-Alexandre, A., Matencio, F., & Lahlou, A. (1991). Dry-milling properties of maize. *Cereal Chemistry, 68*, 51–56.

Mestres, C., & Matencio, F. (1996). Biochemical basis of kernel milling characteristics and endosperm vitreousness of maize. *Journal of Cereal Science, 24*(3), 283–290.

Milašinović, M. (2005). *Physical, chemical and technological characteristics of novel ZP maize hybrids.* Master Thesis. Novi Sad, Serbia: University of Novi Sad, Faculty of Technology.

Morris, C. (2016). Grain quality attributes for cereals other than wheat. In C. Wrigley, H. Corke, K. Seetharaman, & J. Faubion (Eds.), *Encyclopedia of food grains* (2nd ed., Vol. 3, pp. 257–261). Oxford, GB: Academic Press, Elsevier Inc.

Nankar, A., Holguin, F., Scott, P., & Pratt, R. (2017). Grain and nutritional quality traits southwestern US blue maize landraces. *Cereal Chemistry, 94*(6), 950–955.

Nawaz, H., Shad, M., Mefmood, R., Rehman, T., & Munir, H. (2015). Comparative evaluation of functional properties of commonly used cereal and legume flours with their blends. *International Journal of Food and Allied Sciences, 1*(2), 67–73.

Ni, S., Zhao, W., Zhang, Y., Gasmalla, M., & Yang, R. (2016). Efficient and eco-friendly extraction of corn germ oil using aqueous ethanol solution assisted by steam explosion. *Journal of Food Science and Technology, 53*(4), 2108–2116.

Nielsen, H., Inglett, G., Wall, J., & Donaldson, G. (1973). Corn germ protein isolate-preliminary studies on preparation and properties. *Cereal Chemistry, 50*, 435–443.

Nikolić, V., Žilić, S., Radosavljević, M., Simić, M., Filipović, M., Čamdžija, Z., & Sečanski, M. (2020). Grain properties of new inbred lines in comparison with maize hybrids. *Journal on Processing and Energy in Agriculture, 34*(3–4), 95–99.

Okoruwa, A. (1997). *Utilization and processing of maize.* Ibadan: International Institute of Tropical Agriculture (IITA).

Ortiz, R., Taba, S., Chávez Tovar, V., Mezzalama, M., Xu, Y., Yan, J., & Crouch, J. H. (2010). Conserving and enhancing maize genetic resources as global public goods – A perspective from CIMMYT. *Crop Science, 50*(1), 13–28.

Ozturk, O., & Mert, B. (2019). Characterization and evaluation of emulsifying properties of high pressure microfluidized and pH shifted corn gluten meal. *Innovative Food Science and Emerging Technologies, 52,* 179–188.

Paulsen, M., Singh, M., & Singh, V. (2019). Measurement and maintenance of corn quality. In S. Serna-Saldivar (Ed.), *Corn: Chemistry and technology* (3rd ed., pp. 165–212). Duxford: Woodhead Publishing, Elsevier Inc.

Pereira, R., Davide, L., Pedrozo, C., Carneiro, N., Souza, I. P., & Paiva, E. (2008). Relationship between structural and biochemical characteristics and texture of corn grains. *Genetics and Molecular Research, 7*(2), 498–508.

Pour, E. L. A. (1981). Protein functionality. In J. Cherry (Ed.), *Protein functionality in foods.* ASC Symposium Series, 147.

Radosavljević, M., Božović, I., Bekrić, V., Jakovljević, J., Jovanović, R., Žilić, S., & Terzić, D. (2001). Contemporary evaluation methods of maize quality and technological values. *Journal on Processing and Energy in Agriculture, 5*(3), 85–88.

Rajendran, A., Singh, R., Mahajan, V., Chaudhary, D., & Sapna Kumar, R. (2012). Corn oil: An emerging industrial product. Directorate of Maize Research, New Delhi, India, *Technical Bulletin,* 8, 36 p. https://cupdf.com/document/corn-oil-an-emerging-industrial-product-dmr.html

Rausch, K., & Eckhoff, S. (2016a). Maize: Dry milling. In C. Wriglwy, H. Corke, K. Seetharaman, & J. Faubion (Eds.), *Encyclopedia of food grains* (Vol. 3, pp. 458–466). Oxford, GB: Academic Press, Elsevier Inc.

Rausch, K., & Eckhoff, S. (2016b). Maize: Wet milling. In C. Wriglwy, H. Corke, K. Seetharaman, & J. Faubion (Eds.), *Encyclopedia of food grains* (Vol. 3, pp. 467–481). Oxford, GB: Academic Press, Elsevier Inc.

Rausch, K., Hummel, D., Johnson, L., & May, J. (2019). Wet milling: The basis for corn biorefineries. In S. Serna-Saldivar (Ed.), *Corn: Chemistry and technology* (3rd ed., pp. 501–536). Duxford: Woodhead Publishing, Elsevier Inc.

Rausch, K., Pruiett, L., Wang, P., Xu, L., Belyea, R., & Tumbleson, M. (2009). Laboratory measurement of yield and composition of dry-milled corn fractions using a shortened, single-stage tempering procedure. *Cereal Chemistry, 86*(4), 434–438.

Roopa, S., & Premavalli, K. (2008). Effect of processing on starch fractions in different varieties of finger millet. *Food Chemistry, 106*(3), 875–882.

Rose, D., & Inglett, G. (2010). Production of feruloylated arabinoxylo-oligosaccharides from maize (Zea mays) bran by microwave-assisted autohydrolysis. *Food Chemistry, 119*(4), 1613–1618.

Rosengrant, M., Ringler, C., Msangi, S., Sulser, T., Zhu, T., & Cline, S. (2008). *International model for policy analysis of agricultural commodities and trade (IMPACT): Model description.* Washington, DC: International Food Research Institute.

Saeed, F., Hussain, M., Arshad, M., Afzaal, M., Munir, H., Imran, M., ... Anjum, F. M. (2021). Functional and nutritional properties of maize bran cell wall non-starch polysaccharides. *International Journal of Food Properties, 24*(1), 233–248.

Sangamithra, A., Swamy, G., Sorna, P., Nandini, K., Kannan, K., Sasikala, S., & Suganya, P. (2016). Moisture dependent physical properties of maize kernels. *International Food Research Journal, 23*(1), 109–115.

Scanlon, M., Dexter, J., & Biliadeis, C. (1988). Particle-size related physical properties of flour produced by smooth roll reduction of hard red spring wheat farina. *Cereal Chemistry, 65*, 486–482.

Scott, P., Pratt, R., Hoffman, N., & Montgomery, R. (2019). Specialty corns. In C. Wrigley, H. Corke, K. Seetharaman, J. Faubion (Eds.), *Corn: Chemistry and technology* (3rd ed., pp. 289–304). Duxford: Woodhead Publishing, Elsevier Inc.

Serna-Saldivar, S. (2016). Maize: Foods from maize. In S. Serna-Saldivar (Ed.), *Encyclopedia of food grains-grain-based products and their processing* (2nd ed., Vol. 3, pp. 97–109). Oxford, GB: Academic Press, Elsevier Inc.

Serna-Saldivar, S., & Carillo, E. (2019). Food uses of whole corn and dry-milled fractions. In S. Serna-Saldivar (Ed.), *Corn: Chemistry and technology* (3rd ed., pp. 435–468). Duxford: Woodhead Publishing, Elsevier Inc.

Serna-Saldivar, S., & Chuck-Hernandez, C. (2019). Food uses of lime-cooked corn with emphasis in tortillas and snacks. In S. Serna-Saldivar (Ed.), *Corn: Chemistry and Technology* (3rd ed., pp. 469–500). Duxford: Woodhead Publishing, Elsevier Inc.

Serna-Saldivar, S., & Gomez, M. R. (2001). Food uses of regular and specialty corns and their dry-milled fractions. In A. Hallauer (Ed.),

Specialty corns (2nd ed., pp. 1–35). Boca Raton, FL, London, New York, Washington, DC: CRC Press.

Shad, M., Nawaz, H., Noor, M., Ahmad, H., Hussain, M., & Choudry, M. (2013). Functional properties of maize flour and its blends with wheat flour: Optimization of preparation conditions by response surface methodology. *Pakistan Journal of Botany, 45*(6), 2027–2035.

Shah, T., Prasad, K., & Kumar, P. (2016). Maize – A potential source of human nutrition and health: A review. *Cogen Food & Agriculture, 2,* 1166995. https://doi.org/10.1080/23311932.2016.1166995

Shamrock Milling Systems. (2021). *Maize milling technical advice.* Shamrock Milling Systems. Retrieved August 24, 2021, from https://www.millingsystems.com/cp/36175/technical-advice

Shobha, D., Dileep Kumar, K., Sreeramasetty, T., Puttaramanaik, P. G., & Shivakumar, G. (2014). Storage influence on the functional, sensory and keeping quality of quality protein maize flour. *Journal of Food Science and Technology, 51*(11), 3154–3162.

Siddiq, M., Nasir, M., Ravi, R., Dolan, K., & Butt, M. (2009). Effect of defatted maize germ addition on the functional and textural properties of wheat flour. *International Journal of Food Properties, 12*(4), 860–870.

Simmonds, D. (1974). Chemical basis of hardness and vitreousity in the wheat kernel. *The Bakers Digest,* October, 16–29.

Suresh, C., Samsher, S., & Durvesh, K. (2015). Evaluation and functional properties of composite flours and sensorial attributes of composite flour biscuits. *Journal of Food Science and Technology, 52*(6), 3681–3688.

Szaniel, J., Sagi, F., & Palvolgyi, I. (1984). Hardness determination and quality prediction of maize kernels by a new instrument, the molograph (Zea mays, technological properties). *Maydica, 29,* 9–20.

Tufail, T., Saeed, F., Arshad, M. U., Afzaal, M., Rasheed, R., Bader Ul Ain, H., ... Shahid, M. Z. (2019). Exploring the effect of cereal wall on rheological properties of wheat flour. *Journal of Food Processing and Preservation, 44*(3), e14345.

UNIDO. (1986, January 16). *Small-scale maize milling. UNIDO/ILO Technical Memorandum No 6.* Vienna, Austria: The International Labour Office. Retrieved August 2021.

US Grains Council. (2015, May 13). Chemical and physical properties contribute to corn quality grade. *Council News.* Retrieved from https://grains.org/lta/chemical-and-physical-properties-contribute-to-corn-quality-grade/

Vani, B., & Zayas, F. (1995). Foaming properties of selected plant and animal proteins. *Journal of Food Science, 60*(5), 1025–1028.

Vyn, T., & Tollenaar, M. (1998). Changes in chemical and physical quality parameters of maize grain during decades of yield improvement. *Field Crop Research, 59*(2), 135–140.

Walde, S., Tummala, J., Lakshminarayan, S., & Balaraman, M. (2005). The effect of rice flour on pasting and particle size distribution of green gram (Phaseolus radiata, L. Wilczek) dried batter. *International Journal of Food Science and Technology, 40*(9), 935–942.

Wang, B., & Wang, J. (2019). Mechanical properties of maize kernel horny endosperm, floury endosperm and germ. *International Journal of Food Properties, 22*(1), 863–877.

Wang, C., & Zayas, J. (1991). Water retention and solubility of soy proteins and corn germ proteins in a model system. *Journal of Food Science, 56*(2), 455–458.

Wang, C., & Zayas, J. (1992). Emulsifying capacity and emulsion stability of soy proteins compared with corn germ protein flour. *Journal of Food Science, 57*(3), 726–731.

Wolf, M., Buzan, C., MacMasters, M., & Rist, C. (1952). Structure of the mature corn kernel. 1. Gross anatomy and structural relationships. *Cereal Chemistry, 29*, 321–333.

Wu, V. (2001). Emulsifying activity and emulsion stability of corn gluten meal. *Journal of the Science of Food and Agriculture, 81*(13), 1223–1227.

Wu, Y., & Bergquist, R. (1991). Relation of corn grain density to yields of dry-milling products. *Cereal Chemistry, 68*(5), 54–60.

Yuan, J., & Flores, R. (1996). Laboratory dry-milling performance of white corn: Effect of physical and chemical corn characteristics. *Cereal Chemistry, 73*(5), 574–578.

Žeželj, M. (1995). *Tehnologija žita i brašna.* Novi Sad: Faculty of Technology.

Chapter 2

Chemistry of maize components

Adeleke Omodunbi Ashogbon

DOI: 10.1201/9781003245230-2

2.1 Introduction

Maize or corn (*Zea mays* L.), like wheat and rice, belongs to the grass family *Gramineae* and is widely cultivated throughout the world. Corn has a high yield potential and is cultivated globally; the economic and nutritional value associated with corn is solely due to the high starch content in its seeds (Niu, Ding, Zhang, & Wang, 2019). It is the most produced cereal crop in the world, with over one billion tons produced yearly (USDA, 2017). The major chemical component of the corn kernel is starch, which constitutes up to 73% of the kernel weight (Rocha-Villarreal, Hoffmann, Vanier, Serna Saldivar, & Garcia-Lara, 2018). The next largest chemical component is protein, varying in common varieties from 6 to 12% of the kernel weight (Sinha et al., 2011), followed by lipids and dietary fibre quantitatively. The categorization of corn can be based on either various kernel properties (e.g. flour corn, flint corn, dent corn, sweet corn, and waxy corn) or pericarp pigmentation (white, yellow, blue, purple, red, and black) (Micheletti, 2013). Floury corn (*Zea mays* var. *amylacea*), also known as soft corn, possesses solely white grains with rounded or flat crowns, having mostly soft starch and a small fraction of hard starch. Flint corn (*Zea mays* var. *indurata*), also known as Indian corn, has intermediate soft starch, encircled by a hard shell, and its colour varies from white to red. Dent corn (*Zea mays* var. *indentata*) might have yellow or white pigmentation, with a depressed crown. Sweetcorn (*Zea mays* var. *saccharata* and *Zea mays* var. *rugosa*) possesses higher sugar content than other corn varieties, and it is eaten in various forms (boiled/roasted/frozen/canned). Popcorn (*Zea* mays var. *everta*) is solely utilized for popping and has a greater tendency to pop, which is connected to dense starch filing in the endosperm. Waxy corn (*Zea mays* var. *ceratina*) has starch richly populated with amylopectin (AP) (99%) and an infinitesimal quantity of amylose (AM). The above corn kernel characteristics elucidation is taken from Singh, Singh, and Shevkani (2019).

This chapter is about the chemistry of maize components. In the broadest classification, there are two types of maize: coloured and uncoloured maize. The basic distinction between the two is in their content of phytochemical bioactive compounds. There are more bioactive compounds in the coloured maize compared to the uncoloured maize. The bioactive compounds in maize grains have two main functions – the manifestation of colour and the enhancement of better health via their regular consumption. The uniqueness and individuality of each of the varieties of the corn grain are undeniable due to variations in genetics, environmental factors (both biotic and abiotic), and anthropogenic influences. As previously documented, maize grown in Nigeria and the US will obviously have differences in their physicochemical and functional properties (Ashogbon, 2018a). The excessive mechanization, fertilization, and usage of pesticides will deeply influence corn grown in the US. In contrast, Nigerian maize will be deeply organic in nature, depending mainly on rainfall, sunshine, and the component of the natural soil.

It must be taken into consideration that none of the cereal grains, including maize grain, can be consumed without undergoing some sort of processing, such as heating, boiling, steaming, roasting, extrusion, germination, fermentation, ultrasonication, high-pressure treatment, and microwaving, etc. The natural design of human teeth and the digestive system are not capable of dealing with the raw maize grain; therefore, processing is born. The processing of the maize grain, such as preservation by reducing water content, has so many advantages: it becomes more nutritious and palatable, and it alters starch content and the bioactive constituents of the maize grain. Most consumed maize is eaten in a gelatinized form due to the high content of starch. There has been much emphasis in this chapter on starch because apart from being the most abundant component of the maize grain, it is also the principal polysaccharide in the human diet. Starches also possess plenty of applications in the food and nonfood industries. There is also plenty of discussion about the bioactive phytochemicals, with a focus on the phenolic anthocyanins and the non-aromatic carotenoids. Both are pivotally associated with colour and health-providing features of the maize grain.

2.2 Chemical composition of maize

The chemical constituents of maize can be divided into the following major groups: carbohydrate, protein, lipids, ash, and phytochemicals (phenolics, carotenoids, anthocyanins, phytosterols, and tocopherols). The carbohydrate can be further categorized into monosaccharides (glucose, fructose, galactose, mannose, idose, talose, etc.), disaccharides (maltose, lactose, sucrose, and cellobiose), oligosaccharides (stacyose, raffinose), and polysaccharides (starch, cellulose, and inulin). The carbohydrates are the most abundant constituent of maize; this is closely followed by proteins and lipids. The monomeric or repeat units of the protein are amino acids, and those of the lipids are alkanols and fatty acids. The phytochemicals are significant because they are responsible for the manifestation of colour and for health-promoting processes due to their antioxidant and bioactive activities.

2.2.1 Carbohydrate

For convenience, the carbohydrate of maize can be divided into starch polysaccharides and non-starch polysaccharides (NSPs). The principal NPSs are dietary fibre (DF) such as cellulose and hemicelluloses that are insoluble DF (IDF) and soluble DF (SDF) like mucilages, gum, oligosaccharides, pectins, and inulin.

2.2.1.1 Starch (amylose and amylopectin)

A whole corn kernel consists of four various distinct parts: endosperm (82–84%, dry-weight basis (dwb)), germ (10–12%, dwb), bran (5–6%, dwb), and tip cap (1%, dwb) as depicted in Figure 2.1 (Britannica, 2015). There is a heterogeneous distribution of starch, NSPs, protein, and lipids in the corn kernel. Undisputedly, starch is the principal component (61–78%) of the normal corn kernel and it is mainly located in the endosperm (98–99% of total starch) (Watson, 2003). There is great genetic diversity in maize kernel colour. The colours are diverse and consist mainly of orange, yellow, brown, purple, red, black, blue, and pink, as shown in Figure 2.2 (Zilic, Serpen, Akillioglu, Gokmen, & Vancetovic, 2012).

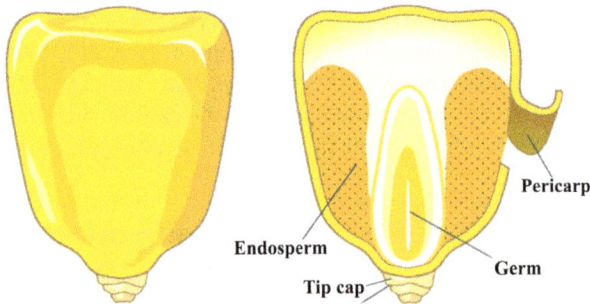

Pericarp

Endosperm

Germ

Tip cap

Maize kernel (monocotyledon cereal)

Figure 2.1 Whole corn kernel indicating endosperm, germ, bran, and tip cap (Britannica, 2015)

white lemon yellow yellow orange red-yellow

red dark red light blue dark blue multicolored

Figure 2.2 Maize kernel colours (Zilic et al., 2012)

The native starch must first be isolated from the maize grain and purified before usage. The following methods below are utilized for the isolation of native maize starch from its grains.

The problems associated with corn starch (CS) isolation are the high content of protein in its grain, especially the presence of disulphide bonds in some of the amino acids constituting the protein. Any method of soaking or steeping involving the sulphur compounds will not be environmentally accepted due to the negative pollutant impacts of sulphur compounds. Due to the high hydrophobicity and disulphide bonds in corn proteins, there is a need for pre-treatment that will disrupt or weaken the bonding forces between the proteins and starch granules (Wang et al., 2015).

The grains contain approximately 8–10% protein content, which are mainly prolamins (Chaidez-Laguna et al., 2016) and can be utilized for CS isolation by soaking the grains in sulphurous acid (H_2SO_3) (0.2–0.4%, w/v) at a temperature of 50⁰C for 24 hour. Sometimes, lactic acid is added to the soaking step to aid its impact (Wang et al., 2015). It should be noted that the functionality of the H_2SO_3 is to split the disulphide bonds of the amino acids that constitute the corn proteins. Additionally, the soaking acid disorganized the structural organization of the proteins and weakens its attachment to the starch granules so that the latter can be easily separated during centrifugation. The wet-milling process, as pointed out in the isolation of legume starches (Ashogbon et al., 2020, produces purer starch with lower protein and ash contents than the dry-milling process.

An escape from the pollutant effect of sulphur compounds leads to the usage of the environmentally friendly method of Li et al. (2015) and the alkaline method. The former method made use of liquid acid and L-cysteine. The L-cysteine is the reductant used to cleave the disulphide bonds in the amino acids, and the physicochemical properties of the obtained corn starch were better than that of the commercial corn starch. The alkaline method was also a fine method, as long as the concentration of the alkali used for soaking the grains is kept as low as possible, although the alkaline conditions could also enhance the multiplication of spoilage microorganisms (El Halal et al., 2019). A high concentration of alkali (especially NaOH) will result in morphological defects of the starch granules and produce alkali gelatinization in the cold.

Other two methods of corn starch isolation are based on Uriarte-Aceves et al. (2018) and Paraginski et al. (2014). The first authors utilized sodium bicarbonate (NaHCO$_3$). The corn grains were wet-milled, and the water pH was modulated to 3.75 with NaHCO$_3$ (0.2 molL^{-1}) solution. The results were 99.32–99.60%, 16.49–26.50%, and 0.44–2.85% for total prime starch, AM content, and resistant starch, respectively. In the second isolation method, the corn grains were soaked in 0.1% sodium bisulphite (NaHSO$_3$) solution and kept at 50⁰C for 20 hour. Subsequently, the following sequential processes followed: water drainage and grains crushing →

double filtration (100–270 mesh sieves) → decantation of supernatant → re-suspension of sediment in distilled water → centrifugation (500 × g for 20 minutes) → drying of wet prime starch (Paraginski et al., 2018; El Halal et al., 2019).

This hydrothermal process results in the production of annealed MS. This changes starch structural characteristics, increases its gelatinization temperature, and facilitates crystalline perfection by interactions between starch chains (Jayakody & Hoover, 2008).

Starch is the most abundant carbohydrate in maize and also the most abundant polysaccharide in the human diet. It is also the second most abundant organic compound in the biosphere after cellulose. Starch is a natural condensation polymer, and the monomeric unit or repeat unit is D-glucose. The endless demand for starch in the food and nonfood industries is because of its cheapness, abundant availability, biodegradability, renewability, and compatibility. Furthermore, starch is easily altered by chemical, physical, biological modification or multiple modifications. The possession of ubiquitous hydroxyl groups by starch facilitates chemical modification. Starch is also hydrophilic due to these hydroxyl groups. The starch granules or particles are constituted of AM and AP. Simplistically, AM is linear (Figure 2.3) with limited random branching. The limited branching associated with AM destroys any symmetry and uniformity of its linearity; therefore AM has nothing to do with the crystallinity of starch. It must be noted that in traditional organic chemistry, a linear structure is more crystalline than a random branched structure. On the other hand, AP is a bigger structure

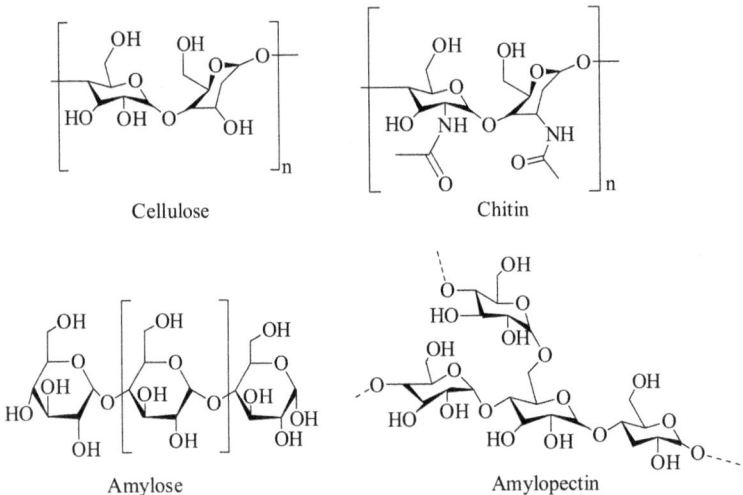

Figure 2.3 Comparative structures of the three main polysaccharides

than AM, but it is highly branched and branching is absolutely non-random (Figure 2.3). This non-randomness associated with AP branching tends to create some sort of symmetry and uniform-ness; therefore AP is almost absolutely responsible for the crystallinity of starch. The differences in the chemistries of AM and AP are responsible for the differences in the physi-cochemical and functional properties of various varieties of maize starches. There is an endeavour to elucidate the complex structure of AP with two theories: the old cluster theory and the new building block theory (Bertoft, 2017). It has been reported that AP is principally responsible for the highly ordered organization of the starch granules, while AM contributes to the defects of the layered structure (Bertoft, 2017).

The structural chemistries of the three main polysaccharides are compared, i.e. starch, cellulose, and chitin (Figure 2.3). The supposed sim-ilarity between starch and cellulose is not attainable due to differences in glycosidic linkages; starch possessed alpha-glycosidic linkage, while that of cellulose is beta (Figure 2.3). On the other hand, the lone pair of elec-trons on the nitrogen of the 2-position of chitin makes it more basic and nucleophilic when compared to starch and cellulose. Therefore, chitin will be more susceptible to electrophilic reaction than starch and cellulose.

The individuality and uniqueness of each of the various starches are undeniable. No two starches are totally identical. A generalization, despite its limitation, can be used in the classification of starches into cereal, tuber, legume, and green fruit starches. Cereal starches, of which corn starch is a member, usually have small granule sizes. The tuber starches have large granule sizes that are weak and therefore possess high swelling power (SP), especially in the case of potato and cassava starches. On the other hand, restricted swelling is associated with legume starches.

The thermal, morphological, crystalline, and pasting properties of other corn starch-like starches are usually evaluated with a differential scanning calorimeter (DSC), scanning electron microscope (SEM), X-ray diffractometer (XRD), and rapid visco-analyser (RVA), respectively.

There are generally three types of corn starches: normal (regular) corn starch, high-amylose corn starch, and waxy corn starch. Normal corn starch usually consists of approximately 25% AM and 75% AP. It is also possible to cultivate corn plants that produce starches with AM to AP con-tents outside the stipulated "normal range"; for instance maize has been cultivated with an AM content as elevated as 85% (high-amylose maize) or as low as almost zero or zero AM (waxy maize) (Wang & Guo, 2020). It has also been documented that the apparent AM content of normal, high-amylose and waxy corn starches were 23.0%, 85.0%, and 2.0%, respectively (Tester, Debon, & Sommerville, 2000).

Morphologically, the granules of normal and waxy starches differ in shapes from small globular to large polyhedral (Figure 2.4) (Jane, 2009) with a 3 to 20 μm range of average particle size. Some granular surfaces

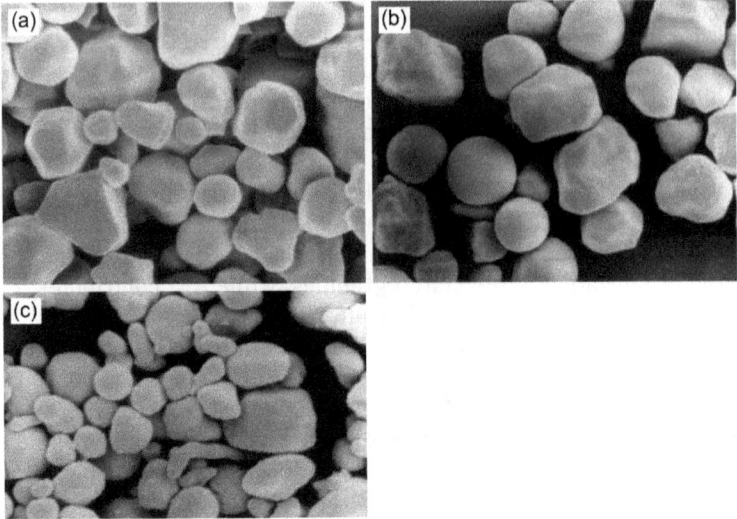

Figure 2.4 Scanning electron micrographs: (a) normal maize starch; (b) waxy maizestarch (Jane, 2009); (c) SEM of high-amylose maizestarch (Jane, 2009) **ANN**-Annealing, **HMT**-Heat moisture treatment, **PEF**-Pulsed electric field, **HHP**-Hydroxypropylation, **MW**-Microwaving, **AA**-α-Amylase, **BA**-β-Amylase, **PUL**-Pullulanare

possess pores of different sizes and depths (Wang, Wang, Yu, & Wang, 2014). In disparity, high-amylose maize starch, unlike regular and waxy starches, has two kinds of granules: small oval granules and large elongated granules (Figure 2.4c) (Jane, 2009; Wang & Guo, 2020). The latter granules, which are around 7–32%, were created by the amalgamation of many small granules (Jiang, Campbell, Blanco, & Jane, 2010). The B-type X-ray diffraction pattern is manifested by high-amylose maize starch; on the other hand, regular and waxy maize starches showed the characteristic typical A-type of all cereals. Obviously, the similarity between normal and waxy starches must be noted and they differ from the high-amylose maize starch. More evidence is that the gelatinization temperature range of waxy and normal starches usually happens at 60 and 80^{0}C. In dissimilarity, the high-amylose maize shows a wide and unspecific endothermic transition, which ends at a high temperature (Wang et al., 2014).

Generally, native starches (NSs) are deficient in their functionalities and therefore not suitable for certain applications in the food and nonfood industries. Some of the shortcomings associated with NSs are easy retrogradation tendency, lower gelatinization transition temperatures, insolubility in cold water, and loss of viscosity after cooking (Goel, Semwal, Khan, Kumar, & Sharma, 2020). Other deficiencies of NSs are undesirable hardness, poor transparency, high hydrophilicity, low tensile strength, and poor freeze-thaw stability. Specifically, corn starch produces weak-bodied, cohesive, rubbery paste and unacceptable gel when cooked (BeMiller, 2016). Therefore, NSs are modified physically (Ashogbon, 2018b), chemically, and biologically or multiple modifications for the betterment of their functionalities.

The multiple or compound starch modifications could be dual (Ashogbon, 2021a), triple (Ashogbon, 2021b), and quadruple (Ashogbon, 2021c). The single physical modification (microwaving, ultrasonication, heat-moisture treatment, annealing, pre-gelatinization, etc.), chemical modification (acetylation, cross-linking, phosphorylation, hydroxypropylation, carboxymethylation, etc.), and enzymatic modification (α-amylolysis, β-amylolysis, amyloglucosylosis, pullulanalysis, etc.) also possess inherent problems associated with them. For instance, succinylation, an esterification process, despite its desirability of producing succinylated starches with high solubility in water, better thickening power, high viscosity, increased paste clarity, inhibited retrogradation, and better freeze-thaw stability (Moin, Ali, & Hasnain, 2016) compared to their NSs; still, succinylated starches are unstable during shearing at high temperature, therefore the need for dual modification for better functionalities. Another problem associated with single chemical modification during acid hydrolysis is the random attack of the inorganic acids (HCl, H_2SO_4, H_3PO_4) on the glycosidic bonds with a tendency to produce undesirable products such as co-products and by-products apart from the net products. Furthermore, the acid is not easily retrievable from the net product.

There are also problems associated with dual and triple modifications of various starches, and on rare occasions quadruple modification is applied. The whole corn starch modification is depicted in Figure 2.5, starting from single physical, chemical, biological modification: the latter including modifications that are genetic and enzymatic. This is followed by dual, triple, and quadruple starch modification. The compound modification could be homogeneous or heterogeneous depending on the various combinations of the single modifications. Some of the dual and triple modifications of corn starch will be briefly discussed, and the rarity of quadruple modification involving corn starches in the literature should be noted.

Briefly, homogeneous dual physical modification involving dry heating and annealing (ANN) of native corn starch. The 7–14% moisture content of the starch was dry heated at 110–200°C for 1–20 hour. The dry-heated corn starch was annealed at 50°C for 24 hour to obtain the dry-heated annealed starch (DHAS) (Chi et al., 2019). It was found that the properties of the DHASs were better than that of the dry-heated corn starch and the annealed starch alone. The DHASs possess higher-order structures when compared to each of the single modified starches, and this was ascribed to the synergistic effect between dry heating and ANN (Chi et al., 2019). Other dual

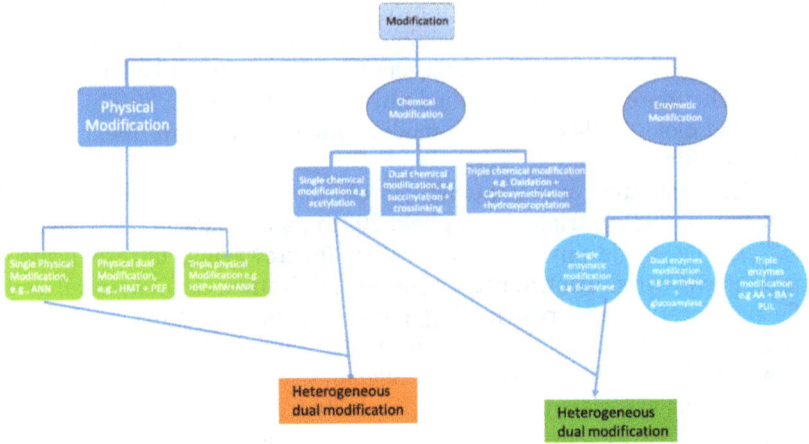

Figure 2.5 Classification of starch modification (Ashogbon, 2021c)

modifications involving corn starch include: ball milling/phosphorylation (Beninca et al., 2019); acid hydrolysis/succinylation (Basilio-Cortes et al., 2019); cross-linking/heat-moisture treatment (HMT) (Park et al., 2018); oxidation/acetylation (Pietrzyk et al., 2018); and debranching/hydroxy-propylation (Hu et al., 2019). It must be noted that in the two stages that constitute the dual modification process, the first stage involves the preparation of the starch structurally and morphologically, which might include the development of granular surface pores and weakening of starch chains for preparation of the second stage (Ashogbon, 2021a).

In the triple modification involving corn starch, three stages constitute the process and this may be homogeneous or heterogeneous. Some of the triple modification processes of corn starch are: extrusion/α-amylase (AA)/glucoamylase (GA) (Wu, Jiao, Xu, Chen, & Jin, 2020); AA/amylo-glucosidase (AG)/pulsed electric field (PEF) (Han, Han, Wang, Liu, & Buckow, 2020); freeze-thawing/AA/AG (Yu et al., 2018); and AA/AG/acid hydrolysis (Shang et al., 2018). Furthermore, an important corn starch triple modification involving the production of resistant starch (RS) using a combination of acid hydrolysis, extrusion, and hydrothermal has also been documented (Neder-Suarez et al., 2018). The RS has a positive impact on health; it escapes digestion in the small intestine and goes directly into the colon, where it is metabolized by microbes into short-chain fatty acids.

2.2.1.2 Non-starch polysaccharide

Generally, cereals' (dietary fibre) DF fraction is composed of NSPs, resistant starch (RS), oligosaccharides (chiefly fructans), and the non-carbohydrate aromatic polyphenolic ether, lignin. The principal NSPs in cereals are arabinoxylan, mixed linkages of β-glucan, and cellulose.

Non-starch polysaccharides are the principal dietary fibre (DF) and are categorized based on their water solubility as insoluble dietary fibre (IDF) and soluble dietary fibre (SDF). Fundamentally, DFs are the eatable parts of plants or comparable carbohydrates that are resistant to digestion and absorption in the human small intestine with absolute or partial fermentation in the large intestine (Guardiola-Marquez, Satana-Galvez & Jacobo-Velazquez, 2020). In short, DFs are capable of escaping digestion in the small intestine of a normal human but easily metabolized in the colon by microbes into short-chain fatty acids with positive health implications.

It has been documented that 50% of daily intake of DF comes from cereals, 30–40% from vegetables, 16% from fruits, and 3% from other sources (Rodriguez, Jimenez, Fernandez-Bolanos, Guillen, & Heredia, 2006). Cellulose and hemicelluloses, together known as holocellulose, are the IDF, and the SDFs consist of NSPs, such as mucilage, gums, pectins, β-glucans, oligosaccharides, and inulin. Briefly, mucilages are complex organic compounds connected to the polysaccharides of plant sources and possess glue-like characteristics. On the other hand, gums are viscous, nonvolatile colloidal plant products which either dissolve or swell up in contact with water. On hydrolysis gums produce complex organic acids in addition to pentoses and hexoses. Pectins are calcium–magnesium salts of polygalacturonic acid that are partially bonded to methanol residues by ether linkage. They are found in the middle lamella of cell walls and are soluble in water and be precipitated from aqueous solutions by excess alcohol. Generally, the glucans are anhydrides of glucose such as cellulose, starch, and glycogen, but specifically cellulose is a beta-glucan and is of interest in the study of NSPs, especially as an example of IDF. In contrast, starch is an alpha-glucan. A particular type of starch called resistant starch is also considered as a DF because it is not digested in the small intestine. The oligosaccharides are carbohydrates constituted of about two to ten monosaccharides. Therefore, the disaccharides, trisaccharides, tetrasaccharides, and decasaccharides are all members of the oligosaccharides.

The best-known NPSs in corn grains are cellulose and hemicelluloses; they are both IDF. Cellulose (Figure 2.3) is the most complex polyose, and it is a homopolymer. In contrast, the hemicelluloses are heteropolymer and less stable thermally when compared to cellulose. In a nutshell, hemicelluloses are more vulnerable to environmental abuses and processing conditions than cellulose.

Among the cereal grains, maize has occupied a pivotal dominant position, and unlike rice and wheat, most corn is used for production of animal feedstuffs and fuel bioethanol. The dry-milling and wet-milling or nixtamalization processes are used for converting raw maize into other desired products. The degermed and refined grits, meals, and flours possessing from 3.6% to 4.6% total DF (dwb) are obtained from whole corn with about 8.2% TDF (dwb) by dry-milling process (Saldivar & Soto, 2020). According to the above authors, these products are utilized in the synthesis of corn breads, corn flakes, cooked grits such as polenta, and snacks like expanded puffs. In contrast, wet milling is mainly applied in the production of pure native maize starch that is lacking in co-products such as DF, germ, and drastically reduced protein, lipid, and ash contents. Most of the refined starch is bioenzymatically transformed into maltodextrin, maltose, glucose, and high-fructose syrups (Serna Saldivar, 2010), and nixtamalization impacts the removal of the pericarp and DF content of masa and tortillas.

The fully grown maize kernel usually possesses from 8.3 to 16% total DF on a dry-weight basis, which is chiefly insoluble and situated in the pericarp and makes up 7% of the kernel weight (Serna Saldivar & Hernandez, 2020). The hemicellulose and β-glucan are the leading DF in maize. The principal hemicellulose content in the kernel is arabino-glucoronoxylans, while xylans chiefly dominate in the endosperm cell wall. The corn bran contains a small amount of SDF that is rich in heteroxylans. The principal constituents of the bran are 50% heteroxylan, 20% cellulose, 9–23% starch, 10–13% protein, 4% phenolic acids, 2–3% lipids, and 2% minerals (Saulnier & Thibault, 1999). About 90% of the corn hemicellulose is B-type (Doner, Johnston, & Singh, 2001), chiefly constituted by heteroxylans (Lapierre, Pollet, Christine, Ralet, & Saulnier, 2001). Chemically, the heteroxylans are made up of 34.3% xylose, 22.8% arabinose, 5.6% galactose, 4.8% glucuronic acid, 22.5% glucose, 2.4% protein, 0.4% p-coumaric acid, 3.2% ferulic, 0.7% dehydrodiferulic acid, and 4.2% acetic acid (Boyer & Shannon, 2003). In a nutshell, raw maize consists of 13.1% IDF and 1.0% SDF.

Processing is inevitable and impacts the amount of DF in processed corn food; for instance, corn transformed into masa resulted in an increase of SDF by 1.75% but lowered the IDF to 6.3% (Serna Saldivar & Hernandez, 2020). Accordingly, simultaneous impact of nixtamalization and thermal treatment hydrolyzes insoluble fibre into soluble fibre constituents. It has been asserted that the DF, i.e. cellulose, arabinoxylan, β-glucan, xyloglucan, and fructan, aids in the prevention and alleviation of non-communicable diseases (NCD) such as cancers, gastrointestinal disorders, diabetes, and cardiovascular diseases, but the processing and refining tend to reduce the DF of cereal grains. As a result of processing, important nutrients, DF, and other phytochemicals are depleted from other grain parts. Therefore, the refined products are of lower nutritional quality than the original whole-grain products (Nirmala & Iris, 2020; Newman, Newman, & Fastnaught, 2019). The refined maize products obtained by processing are sweeter, more nutritious, and more presentable than the whole grain, but this must not be at the expense of the health benefits of the whole grain maize. There is a need to strike a balance between the consumption of refined products and whole-grain products. The processing unit operations and unit processes utilized must be the ones that do not ordinarily unnecessarily degrade much-needed nutrients that are beneficial to health. The bottom line is that processing is inevitable and some raw cereal grains are not edible in raw form. Therefore, there must be cooking, drying, roasting, gelatinization, extrusion, microwaving, etc., in order to reduce moisture content for preservation purposes and most pivotal is that most foods are consumed in the gelatinized form for easy digestion in the gut.

2.3 Proteins

Protein is the second most abundant biopolymer in the corn grains, and it ranges from 6.36% (Bello & Udo, 2018) to 13.13% (MutluArslan-Tontu, Candal, Kilic, & Mustafa, 2017) in the different variety of corn grains. The grains of normal (regular) corn (8.05–8.62%) contained lower protein content than flint dent corn (8.5–8.7%) and waxy corn (11.05%) (Thakur, Kaur, Singh, & Virdi, 2015; Odio, Bera, Beckers, Foucart, & Malumba, 2018). The endosperm housed 75% of the corn protein and consists of 40% zein (prolamin), 30% glutelin, and some quantities of globulins and albumens (Wang, Xu, Qu, & Zhang, 2008). Sometimes corn proteins are classified based on their solubility in some solvents, since their solubility is pivotal to functional properties that impact the application and nutritional value of the grain. Therefore, albumins are water-soluble, globulins are salt-soluble, zeins are alcohol-soluble, and glutelins are alkali-soluble. Zein is the most abundant and the most studied corn protein – it consists of various amino acids such as 21.4% glutamine, 19.3% leucine, 9.0% proline, 8.3% alanine, 6.8% phenylalanine, 6.2% isoleucine, 5.7% serine, and 5.1% tyrosine but unfortunately devoid of the acidic amino acids of tryptophan and lysine. It constitutes about 60% of the maize grain proteins, and based on its solubility in alcohol, four major fractions have been identified: α-zein, β-zein, γ-zein, and δ-zein. The total zein proteins consist mainly of 35% α-zein, the biggest fraction with a large quantity of hydrophobic amino acids including leucine, proline, alanine, and phenylalanine (Singh, Singh, Kaur, & Bakshi, 2012).

α-Zein is found in the large central portion of the protein body with β- and γ-zeins located on its periphery, which could be ascribed to the higher thermo-stability of the zein fraction. There is some sort of similarity between corn zein and wheat gluten, but the former is incapable of forming viscoelastic fibrils at room temperature, though it is functionally achievable at higher temperatures (Schober, Bean, Boyle, & Park, 2008).

The sub-classification of glutelin into three sub-groups is denoted by G1, G2, and G3; they make up the alkali-soluble maize storage protein and the poor solubility of the glutelin is attributed to the presence of disulphide bonds in its structure (Landry & Moureaux, 1970). The deficiency of important acidic amino acids such as tryptophan and lysine in the maize protein is capable of manifesting in harmful effects like the development of anae-mia, pellagra, protein malnutrition, retardation of growth, and free radical damage, etc. (Graham, Lembake, Morales, 1990). The different factors that determine the various protein contents in varieties of corn grains are envi-ronmental factors (biotic and abiotic), genetic factors, and the presence or absence of nitrogen fertilizers in the soil.

2.4 Fats

Fats are naturally occurring substances containing glycerides of higher fatty acids, such as palmitic acid ($C_{15}H_{31}COOH$), stearic acid ($C_{17}H_{35}COOH$), and oleic acid [$CH_3(CH_2)_7CH=CH(CH_2)_7COOH$]. They are essential components of the human diet, mending the wastage of human fat, and via catabolism and oxidation, they provide needed energy. In the different maize varieties, the fat content ranges from 3.725% to 7.43%, but the crude fat content of the endosperm is relatively low, about 1%. The maize endosperm lipids con-sist of more saturated fatty acids than the germ lipids (Singh et al., 2019). Kernels of normal corn consist of 3 to 6% lipids, with larger parts located in the germ (Watson, 2003). Corn oil is very rich in nutrients, especially when extracted from the kernel, and consists of 79% triglycerides, 9% phospholip-ids and glycolipids (polar lipids), 5% sterols, 4% mono- and diglycerides, 3% hydrocarbons-sterol esters, 1% free fatty acids, and infinitesimal quantities of waxes, tocopherols, and carotenoids (Lambert, 2001). The unsaturated fatty acid, Linoleic acid, is rich in triglycerides from corn oil, and this essen-tial polyunsaturated acid is nutritionally necessary for the human diet and health. Nevertheless, the fat content of maize grain depends on a variety of environmental factors (biotic and abiotic), soil pH, and degree of rainfall/sunshine during maturation. That is why corn grown in the USA, under heavy mechanization, fertilization, usage of other agrochemicals, and other anthropogenic influences, cannot be compared to that grown in Nigeria, which mostly depends on the mercy of rainfall and sunshine.

2.5 Bioactive compounds in maize

Generally, whole grains are richly blessed with macronutrients [e.g. car-bohydrates (mostly starchy polysaccharides and non-starch polysaccha-rides)], protein, fat and dietary fibre, micronutrients (e.g. minerals and vitamins), and non-nutrient phytochemicals. The latter are bioactive chemi-cal compounds found in corn grains, which may function in reducing and combating the hazard of chronic diseases (Liu, 2004) and are also involved in impacting the colours of the coloured grains. The corn phytochemicals, like in other cereal grains, are located solely in the kernel and bran (Adom & Liu, 2002; Pang et al., 2018). The bioactive composition differs among the various varieties of corn. The anthocyanins are concentrated in red, blue, purple, and black corn; on the other hand, the carotenoids are accumulated in yellow and red corn, and phytosterols are amassed in the kernel (Luo & Wang, 2012). Phytochemicals are the most pivotal bioactive compounds connected with enhancing health and diminishing hazards of age-related chronic diseases. Generally, the categorization of phytochemicals includes phenolic compounds, terpenoids, sulphur-containing compounds, and nitrogen-containing compounds (e.g. alkaloids). Specifically, corn phytochemicals are phenolic compounds (e.g. phenolic acids, anthocyanins) and carotenoids (e.g. carotenes and xanthophylls).

2.5.1 Phenolic compounds

Phenolic compounds are secondary metabolites of plants, which are categorized into many sub-categories such as phenolic acids, flavonoids, stilbenes, coumarins, lignins, and tannins. The flavonoids are a big class of phenolic compounds that are broadly distributed in plants and consist of more than 8000 metabolites (Zhang et al., 2020).

2.5.1.1 Phenolic acids

Phenolic acids are groups of aromatic acids containing one or more hydroxyl groups attached to the benzene nucleus. The extra stability associated with aromaticity due to delocalized electrons in the benzene also helps in the preservation of the aromatic sextet. This means that despite the highly unsaturated nature of the aromatic compounds, they do not undergo addition reactions in order to preserve their stability, hence the preference for substitution reactions. Most aromatic reactions take place with the attached functional groups to the benzene ring. The 2-hydroxy acids are volatile in steam, soluble in cold chloroform ($CHCl_3$), and give a violet or blue colouration with iron (III) chloride ($FeCl_3$). In contrast, the 3-hydroxy acids are the most stable aromatic acids. The pivotal phenolic acids are salicylic acid (HOC_6H_4COOH), gallic acid [$C_6H_2(OH)_3COOH$], and tannin (tannic acid) – a mixture of derivatives of polyhydroxybenzoic acids.

Another sub-categorization of phenolic acids is their division into derivatives of benzene and cinnamic acid. The phenolic acids have been said to potentially prevent chronic diseases because of the presence of unsaturated carboxylic groups (Pandey & Rizvi, 2009).

2.5.1.2 Flavonoids

Flavonoids are omnipresent phytochemicals with different biological characteristics in plants such as protection from ultraviolet radiation, enhancing pollen potency, and improving resistance to pathogens (Casati &

Walbot, 2005). The categorization of flavonoids is into six major sub-categories: flavones, flavonols, flavanones, flavanonols, anthocyanidins, and condensed tannins (Ferreyra, Rius, & Casati, 2012). Flavonoids are mainly found in cereals, vegetables, fruits, and herbs and serve pivotal functions in plant growth, development, and management of stress, as well as in the protection of human health (Jiang, Doseff, & Grotewold, 2016) through regular consumption. Flavonoids are secondary metabolites that are abundantly available in maize crops and serve inherent protective functions (Cases, Duarte, Doseff, & Grotewold, 2014), such as protection of maize crops from radiation damage when excessively exposed to UV-B radiation.

Anthocyanins belong to flavonoids, a large group of compounds that are subdivided into a larger group of compounds called polyphenolics. The anthocyanins are broad families of water-soluble pigments that are responsible for the manifestation of colours such as blue, purple, red, and orange in cereal grains. Naturally, the anthocyanins are in the form of heterosides (Li, Wang, Luo, Zhao, & Chen, 2017). The sugar-free form or the aglycons are named anthocyanidins and are based on flavylium cation or 2-phenyl-benzopyrilium structure. The structure in Figure 2.6 can give various compounds, based on different positioning of the hydroxyl (OH) and the methoxyl (OCH$_3$) groups, and depending on this, more than 635 anthocyanins have been recognized. There are only about six anthocyanidins (pelargonidin, cyanidin, peonidin, delphinidin, petunidin, and malvidin) (Figure 2.6) that are common in cereal grains. In the study of dyes and colours, the OH group is an auxochrome and it helps in the deepening of the colour of chromogen and maintaining the stability of the colours. When the R1, R2 and R3 in the structure of Figure 2.6 is replaced by OH group, the compound is delphinidin. The latter compound among the compounds in Figure 2.6 is the most stable to environmental abuses and other processing conditions, the colour produced by it will not easily fade away.

Structurally, the anthocyanidin is the basic form of the anthocyanins and the latter is made up of aromatic ring (A) fused to a six-membered oxygen-containing heterocyclic ring (C), which is also linked to another aromatic ring (B) by a carbon-carbon bond (Figure 2.6). There is an increasing attempt to eliminate the rather toxic synthetic colourants and replace them with nontoxic natural pigments such as water-soluble anthocyanins and liposoluble carotenoids. It must be noted that the anthocyanins are less stable than the carotenoids; obviously the former are more reactive than the latter and anthocyanins are more susceptible to environmental abuses and processing conditions than carotenoids. The chemistry behind this could be the many reactive hydroxyl groups attached to the benzene ring and the good-leaving methoxyl groups of the anthocyanins; in contrast the hydroxyl groups attached to some of the cyclohexadiene of carotenoids are not as reactive as the phenolic nature of anthocyanins.

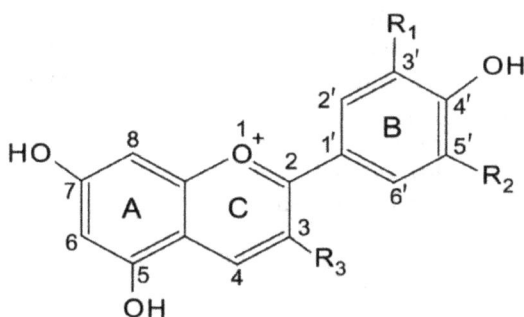

Aglycone	R_1	R_2	R_3
Delphinidin	OH	OH	OH
Cyanidin	OH	H	OH
Petunidin	OCH_3	OH	OH
Peonidin	OCH_3	H	OH
Malvidin	OCH_3	OCH_3	OH
Pelargonidin	H	H	OH

Figure 2.6 Structure of anthocyanidins and replaceable functional groups for naming

The anthocyanins are the most pivotal natural pigments of the vascular plants; they are nontoxic and easily incorporated into aqueous media and are responsible for the shiny orange, pink, red, violet, and blue colours (Castaneda-Ovando, Pacheco-Hernandez, Paez-Hernandez, Rodriguez, & Galan-Vidal, 2009). The colour showed by the anthocyanin molecules is due to the resonant structure of the flavylium ion that caused the intensity of their colours (Wrolstad, Durst, & Lee, 2005). There is enormous genetic heterogeneity in the colour (black, blue, red, and brown) of maize kernels (Paulsmeyer et al., 2017). Anthocyanin pigments tend to concentrate in the pericarp, which is thick in the purple maize kernel. On the other hand, pigments broadly distribute in the aleurone layers of the blue maize kernel, which has a thin and colourless kernel (Paulsmeyer et al., 2017). The maize cob is also rich in anthocyanins (Lao & Giusti, 2016). Enormous genetic diversity in the contents of anthocyanins from coloured maize has been documented (Collison, Yang, Dykes, Murray, & Awika, 2015; Harakotr, Suriharn, Tangwongehai, Scott, & Lertrat, 2014; Harakotr, Suriharn, Scott, & Lertrat, 2015; Nankar et al., 2016; Zilic, Serpen, Akilhoglu, Gokmen, & Vancetovic, 2012). Generally, purple maize kernel has a higher anthocyanin content than blue, red/blue, and red samples (Collison et al., 2015). It has been reported that cyanidin 3-glucoside, peonidin 3-glucoside, and pelargonidin 3-glucoside are the main anthocyanins, although the anthocyanin composition greatly varied among various studies (Camelo-Mendez, Agama-Acevedo, Sanchez-Rivera, & Bello-Perez, 2016; Lao & Giusti, 2016). It has also been asserted that the quantity of anthocyanins in maize kernels increased with increasing duration of maturation (Harakotr et al., 2014).

The blue and purple maize varieties differ from each other, and the main difference between them is that the purple variety predominates in its production of anthocyanins in the pericarp of the kernel (Suriano, Balconi, Valoti, & Redaelli, 2021). On the other hand, anthocyanins in blue maize are mainly located in the aleurone (Li, Somavat, Singh, Chatham, & de Mejia, 2017).

The principal six types of anthocyanidins are found in nature; they are structurally depicted in Figure 2.7. They are cyanidin (50%) – which is the purple-reddish coloured pigment commonly found in purple corn (Cevallos-Casals & Cisneros-Zevallos, 2003); delphinidin (12%) – red-blue pigment (Katsumoto et al., 2007); and pelargonidin (12%) – red-orange colour (Jaakola, 2013). The other three anthocyanidins – peonidin, malvidin, and petunidin – are methoxylated (Figure 2.7). The intensity and type of colour displayed by anthocyanins depend on the number of available hydroxyl and methoxyl groups (He & Giusti, 2010). Plenty of hydroxylation alters the colour to blue, and methoxylation enhances redness (Kahkonen & Heinonen, 2003). Generally, anthocyanins are bluish in basic conditions, pinkish in neutral conditions, and reddish in acidic conditions (Roy & Rhim, 2020).

Structurally, the basic difference between anthocyanidin and anthocyanin is shown in Figure 2.8. During glycosylation, anthocyanidins can be converted to anthocyanins with the help of glucosyltransferase on various hydroxyl moieties of the molecule with 3-OH as the preferred position to produce 3-D-β-glucosides (e.g. chrysanthemin from cyanidin) (He & Giusti, 2010).

There are structural differences between the anthocyanins based on the number of hydroxyl groups, methoxyl groups, and the nature of the sugar moieties attached to the molecule (Figure 2.9) (Khoo, Azlan, Tang, & Lim, 2017). The attached glycosides are commonly known sugars such as glucose, xylose, galactose, rhamnose, arabinose, and rutinose (Figure 2.9) (Miguel, 2017).

Some of the basic properties of anthocyanin need to be stated. Anthocyanins are a family of polyphenol-based flavonoids, which are natural colourants that reflect light from red to blue in the visible spectrum (Khoo, Azlan, Tang, & Lim, 2017; Siguadson, Tang, & Giusti, 2017; Wallace & Giusti, 2019). Anthocyanin's pH reliance on colour alteration

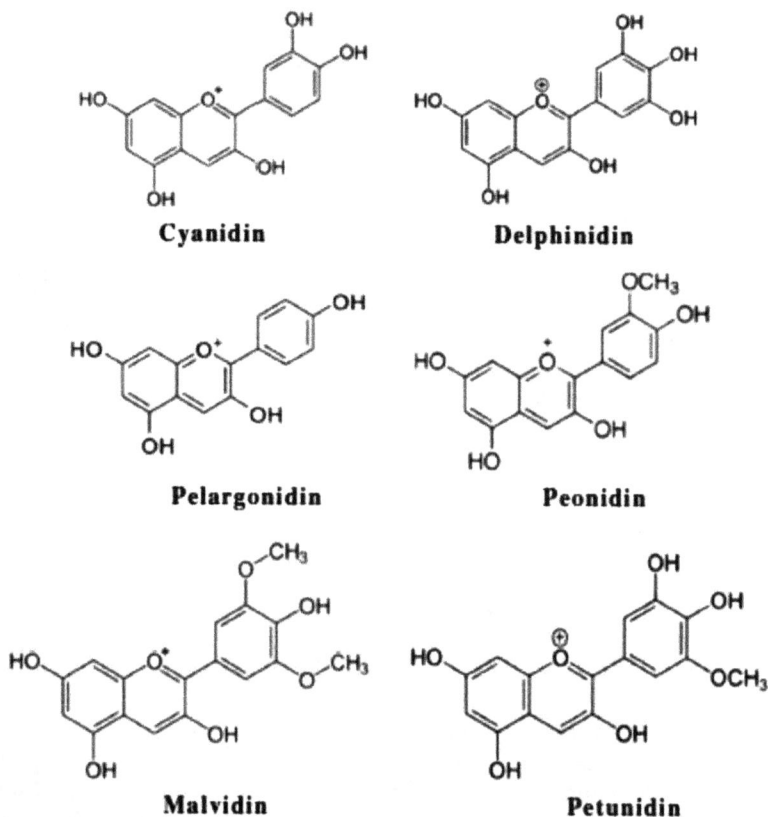

Figure 2.7 Major anthocyanidins in plants

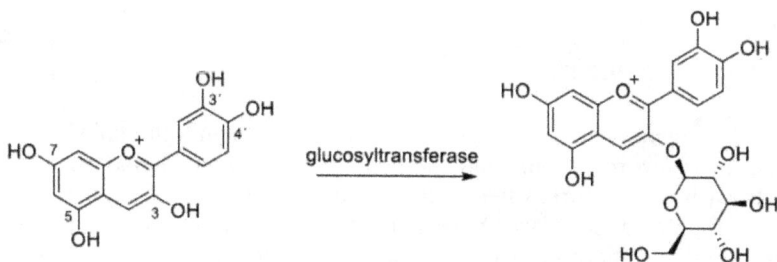

Figure 2.8 Vivid structural differences between anthocyanidin (cyanidin) and anthocyanin (chrysanthemin)

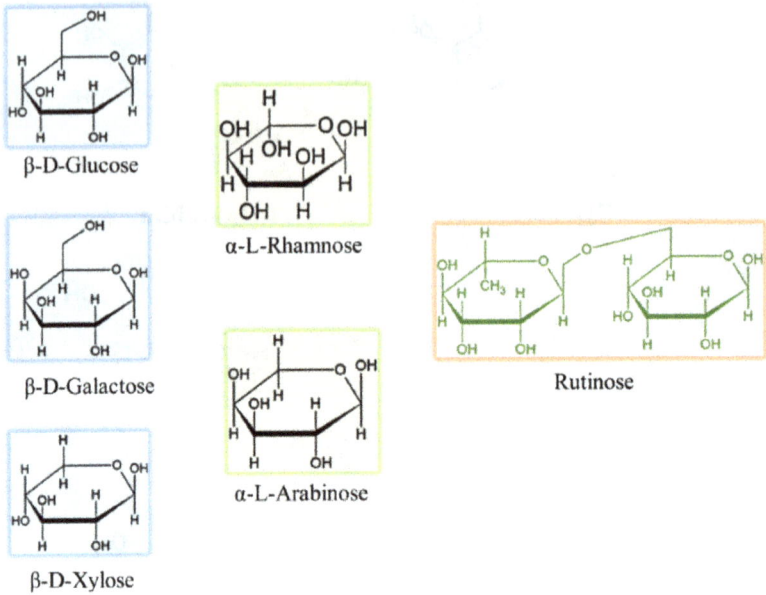

Figure 2.9 The glycosyl units of anthocyanins

characteristics is very pivotal in monitoring food quality, showing the shelf life of food, increasing consumer desire for food, and ultimately being utilized as a colour indicator in food packaging applications (Roy & Rhim, 2020). It is odourless with an astringent taste, and the anthocyanin colour is immensely impacted by the structure, pH, temperature, enzymes, UV radiation co-pigmentation, and the availability of oxygen (Khoo et al., 2017; Singh, Gaikwad, & Lee, 2018). The molecular weight of anthocyanins is in the range of 40–1200 g/mol, and it is polar and, therefore, soluble in polar solvents such as water, methanol, and ethanol.

2.5.2 Carotenoids

The carotenoids can be classified into carotenes and xanthophylls. The former are hydrocarbons, and the latter possess oxygen apart from the hydrocarbon structure. Chemically, due to the presence of oxygen atoms in xanthophylls, they are more polar than the rather inert carotenes. Structurally, at both ends of carotenoids are cyclohexadiene, with some having attached OH groups. In between the two extremes of the cyclohexadiene are extensive conjugated double bonds, which are chromophores responsible for imparting colour to the chromogens (carotenoids)

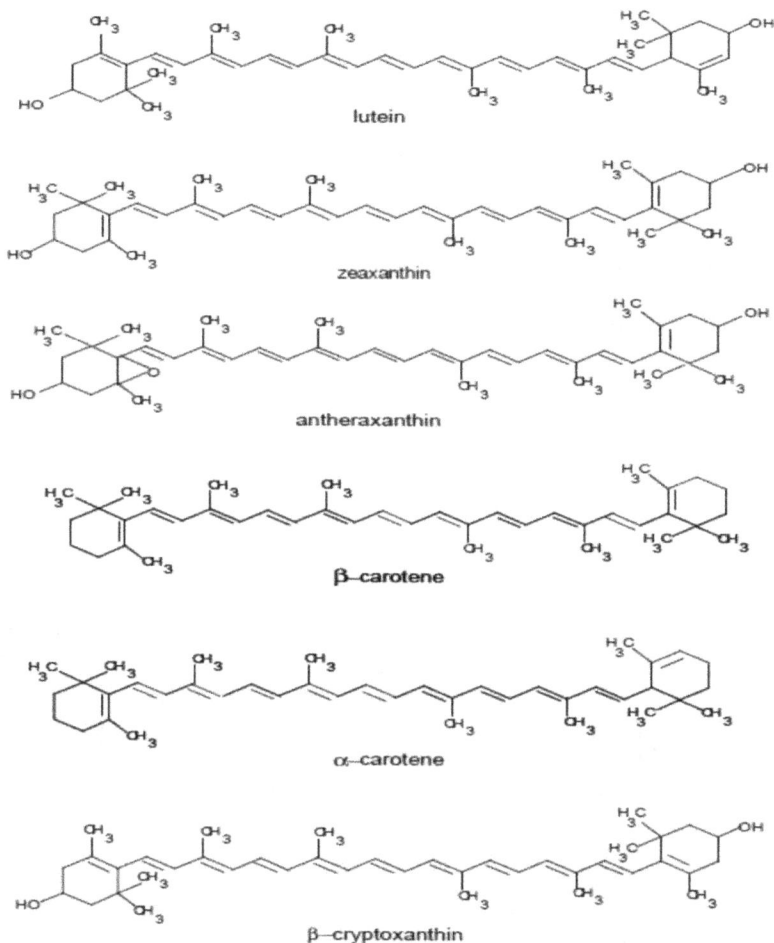

Figure 2.10 Structure of carotenoids identified in yellow and other coloured grains

(Figure 2.10). It must be noted that unlike the phenolic compounds, the carotenoids are non-phenolic and non-aromatic because there is no benzene ring to which the hydroxyl group is directly attached. At least theoretically if the conjugated double bonds on carotenoids are saturated, the coloured compounds become colourless or white. This is achievable by the carotenoids as a result of environmental abuses or processing conditions.

The carotenoids are lipophilic secondary metabolites, and more than 600 of these natural pigments of yellow, orange, and red colours have been identified. They are natural food colourants (Ntrallou, Gika, & Tsochatzis, 2020) in place of toxic synthetic additives.

Carotenoids are C40 isoprenoids synthesized by plants and other organisms such as bacteria, fungi, and algae and consist of a large family with more than 700 yellow, orange, or red fat-soluble pigments (Trono, 2019). These fat-soluble pigments are not synthesized by humans and animals, so they are obtained via dietary sources. There are plenty of acclaimed health benefits associated with carotenoids, for instance the provitamin A activity of α-carotene, β-carotene, and β-cryptoxanthin. After ingestion, these carotenoids are changed into retinol, otherwise known as vitamin A, which serves as the progenitor of the light sensor molecules in the retinol (Trono, 2019). The prevention of degenerative eye damage, like night blindness, xerophthalmis, corneal ulcerations, and lesions, requires sufficient intake of provitamin A (Faustino et al., 2016).

The content, composition, and distribution of carotenoids among the kernels of cereals differ in importance. The zeaxanthin is the dominant carotenoid in maize and is higher than in non-maize cereals (Berman et al., 2017). There is a very low level of carotenoid in raw rice, chiefly β-carotene and lutein (Lamberts & Delcour, 2008). On the other hand, β-carotene and zeaxanthin are the principal carotenoids in sorghum with smaller amounts of lutein (Moreau, Harron, Powell, & Hoyt, 2016). Comparatively, among non-maize cereals, lutein is the principal carotenoid in wheat (Panfili, Fratianni, & Trano, 2004). Obviously, carotenoids are not uniformly distributed throughout the cereal kernel, and in maize, the total carotenoid content (TCC) is derived from the endosperm (Ndolo & Beta, 2013).

Two other components of the maize grain are phytosterols and vitamin E (tocopherols). The former is a terpene alcohol that closely resembles cholesterol. On the hand, vitamin E is a fat-soluble vitamin, and the richest source is the oil of cereal germs. Three closely related tocopherols are known, and the most active is α-tocopherol.

The categorization of the about 250 phytosterols found in nature is based on their methyl group number at the carbon-4 position: 4-desmethylsterols (simple sterols), 4-monomethylsterols, and 4,4-dimethylsterols. The maize oil is abundant in phytosterols, and the most consumed are sitosterol, campesterol, and stigmasterol. There is an uneven distribution of physterols in the endosperm, pericarp, and germ of the maize kernel (Harrabi et al., 2008). The health implication of dietary phytosterols is in a series of physiological and metabolic reactions, which eventually results in the elimination of cholesterol in the stools (Luo & Wang, 2012). There are eight naturally occurring antioxidant components known as vitamin E: four tocopherols (α, β, τ, γ) and four tocotrienols (α-T3, β-T3, τ-T3, γ-T3) (Xu, Li, & Wang, 2019). The amount of tocopherols in corn is determined by environmental factors (biotic and abiotic) and genotypes. The human system is incapable of synthesizing vitamin E (Andelkovic, Massarovic,

Srebric, & Drinic, 2018), and they are obtained from the diet. The two most predominant isomeric tocopherols (γ-tocopherol and α-tocopherol) are more prevalent in corn grains than in other cereals (Rocheford, Wong, Egesel, & Lambert, 2002). Comparatively, γ-tocopherol is naturally more prevalent, but α-tocopherol possessed higher metabolic activity and therefore more desired for human consumption.

2.6 Processing methods for maize components

Processing of coloured cereal grains is absolutely necessary to convert the raw inedible grains into the edible gelatinized form due to the high content of starch. The purposes of processing are numerous and include preservation (minimization of the moisture content in order to reduce spoilage by water-loving microorganisms), increment of health-promoting constitutes such as the anthocyanins and carotenoids, and reduction of the infestation of the cereal grains with insects. For example, corn grain has to be dehulled, cooked, roasted, or steamed before consumption by humans. In the case of corn grain, it has also been asserted and claimed that it is better to cook without de-husking because sometimes the useful nutrient components in the hull will become part of the corn during consumption.

The processing process could be primary, secondary, or tertiary. Sometimes it is enzymatic, especially during fermentation and malting; it could also be chemical treatment (modification), and physical treatment or modification. For the purpose of this write-up, the emphasis will be on physical modification because it is the cheapest method and does not involve the introduction of chemicals or biological agents into the grains. Further, the physical treatment method is in tune with the current mode of green and sustainable chemistry. It is the preferred choice of enlightened consumers and the scientific community. In physical treatment, hot water (hydrothermal process), temperature, high pressure, pulsed electric field (PEF), microwaving, and ultrasonication could be physically utilized to alter the compositions of the grains. A few of these physical modification processes will be discussed here.

Soaking or steeping is one of the cheap processes utilized to improve the quality of raw materials in the food products industry (Rashwan, Yones, Karim, Taha, & Chen, 2021). Soaking enhances the absorption of water and grain moisture, with attendant alterations to physicochemical and structural characteristics (Kaur, Ranawana, & Henry, 2016). During the soaking process, the cereal grains swell and become soft; there is an increase in size, and their constituents hold water and leach into the steep water; the availability of certain constituents improves; and there is a reduction in starch crystallinity (Kale, Jha, Jha, Sinha, & Lal, 2015; Hassani, Procopio, & Becker, 2016). The absorption of water during steeping initiates the activity of enzymes with consequential digestion of food reserve substance; therefore the anti-nutritional contents in steeped grains are reduced (Eltayeb, Mohamed, & Fageer, 2017). It must be noted that whatever happens to the soaked cereal grains depends on the grain types, varieties, and constituents, and also on the soaking time and temperature, and the type of additives in the steep water (Kale et al., 2015; Hassani et al., 2016). There is a direct proportionality relationship between the soaking time or temperature and the amount of water absorbed by the steep grains. During hot soaking, depending on the temperature involved, there could be a weakening of the intra-molecular and intermolecular hydrogen bonding in the starch chains within the grains and this could eventually result in starch gelatinization.

Soaking provides important benefits on the basis of protein digestibility, removal of seed coats, and other anti-nutrients on the surface of the grains (Adeleke, Adiamo, Fawale, & Olamiti, 2017). Steeping also pivotally reduces the molar ratio of phytate to zinc in maize from 41.4% to 40.6% but has no concrete impact on iron availability (Lestienne, Icard-Verniere, Mouquet, Picq, & Treche, 2005). The steeping of corn in 2.5% lactic acid for 48 hours reduces its phytate-phosphorus content by 24.4% (Votterl, Zebeli, Hennig-Pauka, & Metzler-Zebeli, 2019). It has also been documented that soaking yellow maize in water for 48 hours increases the following components: crude protein (10.32 to 12.43%); fat (7.60 to 9.21%); ash (0.66 to 0.94%); and fibre content (0.22 to 0.29%) (Okafor et al., 2018). These authors also reported an elevation in elemental composition (K, P, and Na) with higher contents in Fe, vitamins B1, B2, and B3; the lysine content of amino acids was also enhanced. Pivotally no losses were recorded except the reduction of carbohydrate content from 80.16% to 74.21%.

Germination is a method broadly utilized to enhance the palatability and nutritional properties of cereals for human consumption. These characteristics manifest themselves via the breakdown of anti-nutrient agents such as phytate, tannin, and enzyme inhibitors (Afify, El-Beltagi, Abd El-Salam, & Omran, 2011; Nkhata, Ayua, Kamau, & Shingiro, 2018). Germination enhanced the bioactive compounds and bioactivity of cereal grains such as sorghum (Garzon, Torres, & Drago, 2016), corn (Paucar-Menacho et al., 2016), and millet (Sharma, Saxena, & Riar, 2016). Generally, the improvement of bioactive compounds during germination is ascribed to a release of bound phenolics due to enzymatic action and glycosylation reactions (Saleh et al., 2017).

Germination is a bioprocess that involves four stages: imbibition, onset of enzyme activity, convey of reserve food from endosperm to embryo, and radicle appearance followed by seedling growth. It has been documented that sprouting is a good source of some organic acids, amino acids, and other components such as ascorbic acid, riboflavin, choline, thiamine, tocopherols, and pantothenic acid. Germination of two corn genotypes (Var-113, high phytate content and TL-98B-6225-9.TL617, low phytate content) for six days pivotally reduced 84% of phytic acid, while it increased polyphenol contents by 138.27% (Sokrab, Ahmed, & Babiker, 2012). According to these authors, more extractable calcium and iron were documented for Var-113, while more phosphorus and magnesium were shown for TL-98B-6225-9. TL619. It has also been published that the phenolic and flavonoid content of grains increased in a natural way by germination (Laila & Murtaza, 2014; Paucar-Menacho, Martinez-Villaluenga, Duenas, Frias, & Penas, 2016). The enhancement of protein, ash, energy, Ca, Mg, K, and P content of organic white maize germinated for 72 hours has been mentioned (Bello & Udo, 2018), and reductions were reported for crude fat, fibre, carbohydrate, and Fe.

Lactic acid bacteria and yeast are the main microorganisms utilized in fermentation. It is employed during the preparation of porridge, bread, and other cereal products. Experimental work on fermented grain foods has indicated higher digestibility and an enhanced amino acid profile than unfermented grain foods (Day & Morawicki, 2018; Nkhata et al., 2018). Fermentation is regularly carried out at an ambient temperature and at critical processing conditions such as concentration, temperature, type of microorganisms, and time, as a biotechnological process must be strictly adhered to. In the preparation of cereal-based foods such as breads and cakes, cereal flour dough in the presence of yeast is subjected to a fermentation process.

The improvement or increment in bioactive compounds such as phenolics during the fermentation of cereal flour could be ascribed to the activity of native flour enzymes and microbial enzymes (Saleh et al., 2017). During the fermentative process, the availability of water and oxygen could enhance sequential reactions such as hydrolysis, oxidation, polymerization, and degradation of susceptible sensitive molecules, which impact the structure and solubility of bioactive compounds (Konopka et al., 2014).

Fermentation certainly involves metabolic processes (catabolism and anabolism) and chemical processes that bring about degradation of complex organic compounds into simpler ones such as free sugars, alcohol, vitamins, and carbon dioxide. The degradation of grain compounds like starch and soluble sugars by fermentation is catalyzed by endogenous and exogenous enzymes (Hassan et al., 2006). The grains of two corn flour genotypes (Var-113 and TL-98B-6225-9.TL617) were fermented for 14 days with a resultant increment of extractable calcium (94.73%), phosphorus (76.53%), and iron (84.93%) (Sokrab et al., 2014). In another experimental work, corn flour was lactic acid fermented for 96 hours at 30^0C and resulted in the reduction of the anti-nutrient phytate content by 61.5% (Ejigui et al., 2005).

The fermentation process can also be categorized into two: submerged fermentation and solid-state fermentation. The former involves a liquid medium, and the latter involves the growth of microorganisms on solid particles (Sandhu, Godara, Kaur, & Punia, 2017). The extent to which fermentation has a positive impact on the phenolic content of cereals depends on the microorganism species (Estrada et al., 2014; Sandhu et al., 2017). The improvement of bioactive compounds during fermentation of maize has been ascribed to the activity of native enzymes and the availability of water and oxygen (Auchi & Ukwuru, 2015).

In other processing methods, the wooden ash treatment process is an alkaline process with plenty of cations (K, Na, and Ca) in solution. The method involves steeping cereal grains in wood ash slurry followed by germination of the grains for four days (Claver et al., 2011). The wooden ash treatment method is desirous in reducing the tannin content of high-tannin-containing cereal grains.

The basis of extrusion cooking is the creation of a product under the influence of heat treatment, pressure, and shear forces. Extrusion is a mechanical process where the plasticization of the cereal-based product is impacted by passing via a hole of the die plat, thus developing a three-dimensional form (Krahl, Fuhrmann, & Dimass, 2016). There are various unit operations such as mixing, kneading, cooking, cutting, and drying involved in the extruder's action. In the making of the extruded products, high-temperature short-time processes at temperatures above 100^{0}C are applied. The extrusion process is mostly utilized in making breakfast cereals.

Other cereal processing methods worth mentioning include boiling in water, steaming, roasting, nixtamalization (thermo-alkaline hydrolysis), microwaving, and high hydrostatic pressure.

2.7 Conclusions

The attention given to starch is due to its abundance, and maize grain is cultivated because of its high starch content. Furthermore, the food and nonfood applications of starch – its cheapness, nontoxic nature, renewability, biodegradability, and easy alterations of its structures that bring about changes in its physicochemical and functional properties – are many. Some of the nonfood uses of starches are in the cosmetic, pharmaceutical, petrochemical, chemical, textile, pulp, and paper industries. The phytochemical bioactivity of maize grain is pivotal in the manifestation of colour and the suggested health-promoting functions that could prevent the occurrence of non-communicable chronic diseases.

The bioactive phytochemicals are broadly categorized into phenolic and non-phenolic. The structural chemistries of the phenolic anthocyanins and the non-aromatic carotenoids are given special attention as the main causative agent of colour. The non-starch polysaccharides (NSPs) are dietary fibres; some such as cellulose and hemicelluloses are insoluble and others such as mucilages, gums, inulin, oligosaccharides, and pectins are soluble. Discussions regarding these are in relation to their non-digestibility in the small intestine. The dietary fibres including resistant starch are metabolized in the colon with positive implications for human health.

References

Achi, O. K., & Ukwuru, M. (2015). Cereal-based fermented foods of Africa as functional foods. *International Journal of Applied Microbiology, 2,* 71–83.

Acosta-Estrada, B. A., Guitierrez-Uribe, J. A., & Serna-Saldivar, S. O. (2014). Bound phenolics in foods, a review. *Food Chemistry, 152,* 46–55.

Adeleke, O., Adiamo, O. O., Fawale, O. S., & Olamiti, G. (2017). Effect of soaking and boiling on anti-nutritional factors, oligosaccharide contents and protein digestibility of newly developed Bambara groundnut cultivars. *Turkish Journal of Agriculture: Food Science and Technology, 5*(9), 1006. https://doi.org/10.24925/turjaf.v5i9.1006-1014.949

Adom, K. K., & Liu, R. H. (2002). Antioxidant activity of grains. *Journal of Agriculture and Food Chemistry, 50*(21), 6182–6187.

Afify, A. E. M., El-Beltagi, H., Abd El-Salam, S., & Omran, A. (2011). Bioavailability of iron, zinc, phytate and phytase activity during soaking and germination of white sorghum varieties. *PLOS ONE, 6*(10), Article e25512. 1–7. https://doi.org/10.1371/journal.pone.0025512.

Andelkovic, V., Masarovic, J., Srebric, M., & Drinic, S. M. (2018). Pigmented maize—A potential source of β-carotene and α-tocopherol. *Journal of Engineering and Processing Management, 10*(2), 1–7.

Ashogbon, A. O. (2018a). Contradictions in the study of some compositional and physicochemical properties of starches from various botanical sources. *Starch/Starke, 70,* 1600372. https://doi.org/10.1002/star.201600372

Ashogbon, A. O. (2018b). Current research addressing physical modification of starch from various botanical sources. *Global Nutrition and Dietetics, 1*(1), 1–7.

Ashogbon, A. O. (2021a). Dual modification of various starches: Synthesis, properties and applications. *Food Chemistry, 342,* 128325. https://doi.org/10.1016/j.foodchem.2020.128325

Ashogbon, A. O. (2021b). The recent development in the syntheses, properties, and applications of triple modification of various starches. *Starch/Starke, 73*(3–4), 2000125. https://doi.org/10.1002/star. 202000125

Ashogbon, A. O. (2021c). Limited quadruple modification of various starches in the literature: Why? *Starch/Starke, 73*(3–4), 2000126. https://doi.org/10.1002/star.202000126

Ashogbon, A. O., Akintayo, E. T., Oladebeye, A. O., Oluwafemi, A. D., Akinsola, A. F., & Imanah, O. E. (2020). Developments in the isolation, composition, and physicochemical properties of legume starches. *Critical Reviews in Food Science and Nutrition.* https://doi.org/10.1080/10408398.2020.1791048

Basilio-Cortes, U. A., Gonzalez-Cruz, L., Velazquez, G., Teniente-Martinez, G., Gomez-Aldapa, C. A., Castro-Rosas, J., & Bernardino-Nicanor, A. (2019). Effect of dual modification on the spectroscopic, calorimetric, viscosimetric and morphological characteristics of corn starch. *Polymers, 11*(2), 333.

Bello, F. A., & Udo, J. A. (2018). Effect of processing methods on the nutritional composition and functional properties of flours from white and yellow local maize varieties. *Journal of Advances in Food Science and Technology, 5*(1), 1–7.

BeMiller, J. N. (2016). Starch: Modification. In *Encyclopedia of food grains* (2nd ed., pp. 282–286). https://doi.org/10.1016/b978-0-12-394437-5.00147-9

Beninca, C., de Oliveira, C. S., Bet, C. D., Bisinella, B., Gaglier, C., & Schnitzler, E. (2019). Effect of ball milling treatment on thermal, structural and morphological properties of phosphate starches from corn and pinhao. *Starch/Starke.* https://doi.org/10.1002/star.201900233

Berman, J., Zorrilla-Lopez, U., Sandmann, G., Capell, T., Christou, P., & Zhu, C. (2017). The silencing of carotenoid-hydroxylases by RNA interference in different maize genetic backgrounds increases the–carotene content of the endosperm. *International Journal of Molecular Science, 18*(12), 2515.

Bertoft, E. (2017). Understanding starch structure: Recent progress. *Agronomy, 7*(3), 56. https://doi.org/10.3390/agronomy7030056

Boyer, C. D., & Shannon, J. C. (2003). Carbohydrates in the kernel. In R. J. White & L. A. Johnson (Eds.), *Corn: Chemistry and technology* (pp. 289–311). St. Paul, MN: American Association of Cereal Chemists.

Britannica, E. (2015). *Corn: Composition of corn kernel.* Retrieved July 24, 2015, from http://kids.britannica.com/comptons/art-181448

Camelo-Mendez, G. A., Agama-Acevedo, E., Sanchez-Rivera, M. M., & Bell-Perez, L. A. (2016). Effect on in vitro starch digestibility of Mexican blue maize anthocyanins. *Food Chemistry, 214*, 281–284.

Casati, P., & Walbot, V. (2005). Differential accumulation of maysin and rhamnosylisoorientin in leaves of high-altitude landraces of maize after UV-B exposure. *Plant, Cell and Environment, 28*(6), 788–799.

Cases, M. T., Duarte, S. M., Doseff, A. I., & Grotewold, E. (2014). Flavone-rich maize: An opportunity to improve the nutritional value of an important commodity crop. *Frontier in Plant Science, 5*, 1–11.

Castaneda-Ovando, A., Pacheco-Hernandez, M., Paez-Hernandez, L., Rodríguez, J. A., & Galán-Vidal, C. A. (2009). Chemical studies of anthocyanins: A review. *Food Chemistry, 113*(4), 859–871.

Cevallos-Casals, B. A., & Cisneros-Zevallos, L. (2003). Stoichiometric and kinetic studies of phenolic antioxidants from Andean Purple corn and Red-fleshed sweet potato. *Journal of Agricultural and Food Chemistry, 51*(11), 3313–3319.

Chaidez-Laguna, L. D., Torres-Chavez, P., Ramirez-Wong, B., Marquez-Rois, E., Isas-Rubio, A. R., & Carvajal-Millan, E. (2016). Corn proteins solubility changes during extrusion and traditional nixtamalization for tortilla processing: A study using size exclusion chromatography. *Journal of Cereal Science, 69,* 351–347. https://doi.org/10.1016/j.jcs .2016.04.004.

Chi, C., Li, X., Lu, P., Miao, S., Zhang, Y., & Chen, L. (2019). Dry heating and annealing treatment synergistically modulate starch structure and digestibility. *International Journal of Biological Macromolecules, 137,* 554–561.

Claver, I. P., Zhou, H.-M., Zhang, H.-H., Zhu, K.-X., Li, Q., & Murekatete, N. (2011). The effect of soaking with wooden ash and malting upon some nutritional properties of sorghum flour used for impeke, a traditional Burundian malt-based sorghum beverage. *Agricultural Sciences in China, 10*(11), 1801–1811.

Collison, A., Yang, L., Dykes, L., Murray, S., & Awika, J. M. (2015). Influence of genetic background on anthocyanin and copigment composition and behavior during thermoalkaline processing of maize. *Journal of Agricultural and Food Chemistry, 63*(22), 5528–5538.

Day, C. N., & Morawicki, R. O. (2018). Effects of fermentation by yeast and amylolytic lactic acid bacteria on grain sorghum protein content and digestibility. *Journal of Food Quality, 29,* 1–8. https://doi.org/10.1155 /2018/3964392.

Donner, L. W., Johnston, D., & Singh, V. (2001). Analysis and properties of arabinoxylans from discrete corn wet-milling fibre fractions. *Journal of Agriculture and Food Chemistry, 49*(3), 1266–1269.

Ejigui, J., Savoie, L., Marin, J., & Desrosiers, T. (2005). Beneficial changes and drawbacks of a traditional fermentation process on chemical composition and anti-nutritional factors of yellow maize (*Zea mays*). *Journal of Biological Sciences, 5*(5), 590–596.

El Halal, S. L. M., Kringel, D. H., da Rosa Zavareze, E., & Dias, A. R. G. (2019). Methods for extracting cereal starches from different sources: A review. *Starch/Starke.* https://doi.org/10.1002/star.201900128

Eltayeb, L. F. E. F., Mohammed, M. A., & Fageer, A. S. M. (2017). Effect of soaking on nutritional value of sorghum (*Sorghum bicolour* L.). *International Journal of Scientific and Engineering and Research, 6,* 1360–1365.

Faustino, J. F., Ribeiro-Silva, A., Dalto, R. F., Souza, M. M., Furtado, J. M., dc Melo Rocha, G., ... Rocha, E. M. (2016). Vitamin A and the eye: An old tale for modern times. *Arquivos brasileiros de oftalmologia, 79*(1), 56–61.

Ferreyra, M. L. F., Rius, S. P., & Casati, P. (2012). Flavonoids: Biosynthesis, biological functions, and biotechnological applications. *Frontier in Plant Science, 3,* 1–15.

Garzon, A. G., Torres, R. L., & Drago, S. R. (2016). Effects of malting conditions on enzyme activities, chemical, and bioactives of sorghum starchy products as raw material for brewery. *Starch/Starke, 68*(11–12), 1048–1054.

Goel, C., Semwal, A. D., Khan, A., Kumar, S., & Sharma, G. K. (2020). Physical modification of starch: Changes in glycemic index, starch fractions, physicochemical and functional properties of heat-moisture treated buckwheat starch. *Journal of Food Science and Technology, 57*(8), 2941–2948.

Graham, G. G., Lembake, J., & Morales, E. (1990). Quality-protein maize as the sole source of dietary protein and fat in rapidly growing young children. *Pediatrics, 85*(1), 85–91.

Guardiola-Marquez, C. E., Santan-Galvez, J., & Jacobo-Velazquez, D. A. (2020). Association of dietary fibre to food components. In J. Welti-Chanes et al. (Eds.), *Science and technology of fibres in food systems.* Switzerland: Springer Nature. Food Engineering Series. https://doi .org/10.1007/978-3-030-38654-2_3

Han, Z., Han, Y., Wang, J., Liu, Z., Buckow, R., & Cheng, J. (2020). Effects of pulsed electric field treatment on the preparation and physico-chemical properties of porous corn starch derived from enzymolysis. *Journal of Food Processing and Preservation, 44*(2), e14353. https://doi .org/10.1111/jfpp.14353.

Harakotr, B., Suriharn, B., Scott, M. P., & Lertrat, K. (2015). Genotypic variability in anthocyains, total phenonics, and antioxidant activity among diverse waxy corn germplasm. *Euphytica, 203*(2), 237–248.

Harakotro, B., Suriharn, B., Tangwongchai, R., Scott, M. P., & Lertrat, K. (2014). Anthocyanin and antioxidant activity in coloured waxy corn at different maturation stages. *Journal of Functional Foods, 9,* 109–118.

Harrabi, S., St-Amand, A., Sakouhi, F., Sebei, K., Kallel, H., Mayer, P. M., & Boukhchina, S. (2008). Phytostanols and phytosterols distribution in corn kernel. *Food Chemistry, 111*(1), 115–120.

Hassan, A. B., Ahmad, I. A. M., Osman, N. M., Eltayeb, M. M., Osman, G. A., & Babiker, E. E. (2006). Effect of processing treatments followed by fermentation on protein content and digestibility of pearl millet (Pennisetumtyphoideum) cultivars. *Pakistan Journal of Nutrition, 5*(1), 86–89.

Hassani, A., Procopio, S., & Becker, T. (2016). Influence of malting and lactic acid fermentation on functional bioactive components in cereal-based raw materials: A review paper. *International Journal of Food Science and Technology, 51*(1), 14–22.

He, J., & Giusti, M. M. (2010). Anthocyanins: Natural colourants with health-promoting properties. *Annul Reviews in Food Science and Technology, 1,* 163–187.

Hu, X., Jia, X., Zhi, C., Jin, Z., & Miao, M. (2019). Improving properties of normal maize starch films using dual modification: Combination treatment of debranching and hydroxypropylation. *International Journal of Biological Macromolecules, 130,* 197–202.

Jaakola, L. (2013). New insights into the regulation of anthocyanin biosynthesis in fruits. *Trends in Plant Science, 18*(9), 477–483.

Jane, J.-I. (2009). Chapter 6 - Structural features of starch granules II. In J. BeMiller & R. Whistler (Eds.), *Starch: Chemistry and technology* (3rd ed., pp. 193–236). New York, USA: Elsevier.

Jayakody, I., & Hoover, R. (2008). Effect of annealing on the molecular structure and physicochemical properties of starches from different botanical sources. A review. *Carbohydrate Polymers, 74,* 691–703.

Jiang, H. X., Campbell, M., Blanco, M., & Jane, J. L. (2010). Characterization of maize amylose-extender (ae) mutant starches: Part II. Structure and properties of starch residues remaining after enzymatic hydrolysis at boiling-water temperature. *Carbohydrate Polymers, 80*(1), 1–12.

Jiang, N., Doseff, A. I., & Gratewold, E. (2016). Flavones: From biosynthesis to health benefits. *Plants, 5*(27), 1–16.

Kahkonen, M. P., & Heinonen, M. (2003). Antioxidant activity of anthocyanins and their aglycons. *Journal of Agricultural and Food Chemistry, 51*(3), 628–633.

Kale, S. J., Jha, S. K., Jha, G. K., Sinha, J. P., & Lal, S. B. (2015). Soaking induced changes in chemical composition, glycemic index and starch characteristics of basmati rice. *Rice Science, 22*(5), 227–236.

Katsumoto, Y., Fukuchi-Mizutani, M., Fukui, Y., Brugliera, F., Holton, T. A., Karan, M., ... Tanaka, Y. (2007). Engineering of rose flavonoid biosynthetic pathway successfully generated blue-hued flowers accumulating delphinidin. *Plant and Cell Physiology, 48*(11), 1589–1600.

Kaur, B., Ranawana, V., & Henry, J. (2016). The glycemic of rice and rice products: A review, and table of GI values. *Critical Review in Food Science and Nutrition, 56*(2), 215–236.

Khoo, H. E., Azlan, A., Tang, S. T., & Lim, S. M. (2017). Anthocyanidins and anthocyanins: Coloured pigments as food, pharmaceutical ingredients, and the potential health benefits. *Food and Nutrition Research, 61*(1), 1361779. https://doi.org/10.1080/16546628.2017.1361779

Konopka, I., Tanska, M., Faron, A., & Czaplicki, S. (2014). Release of free ferulic acid and changes in antioxidant properties during the wheat and rye bread making process. *Food Science and Biotechnology, 23*(3), 831–840.

Krahl, T., Fuhrmman, H., & Dimassi, S. (2016). Colouration of cereal-based products. In *Handbook of natural pigments in food and beverages.* https://doi.org/10.1016/b978-0-08-100371-8.00011-7

Laila, O., & Murtaza, I. (2014). Seed sprouting: A way to health promoting treasure. *International Journal of Current Research and Review, 6*(23), 70–74.

Lambert, R. J. (2001). High-oil corn hybrids. In A. R. Hallauer (Ed.), *Specialty corn* (2nd ed., pp. 131–154). Boca Raton, FL: CRC Press, LLC.

Lamberts, L., & Delcour, J. A. (2008). Carotenoids in raw and parboiled brown and milled rice. *Journal of Agricultural and Food Chemistry, 56*(24), 11914–11919.

Landry, J., & Moureaux, T. (1970). Heterogeneity of corn glutelin: Selective extraction and amino acid composition of the three isolated fractions. *Bulletin of Chemical and Biological Society, 52,* 1021–1037.

Lao, F., & Giusti, M. M. (2016). Quantification of purple corn (*Zea mays* L.) anthocyanins using spectrophotometric and HPLC approaches: Method comparison and correlation. *Food Analytical Methods, 9*(5), 1367–1380.

Lapierre, C., Pollet, B., Christine, M., Ralet, M., & Saulnier, L. (2001). The phenolic fraction of maize bran: Evidence for lignin-heteroxylan association. *Phytochemistry, 57*(5), 765–772.

Lestienne, I., Ieard-Verniere, C., Mouquet, C., Picq, C., & Treche, S. (2005). Effect of soaking whole cereal and legume seeds on iron, zinc and phytate contents. *Food Chemistry, 89*(3), 421–425.

Li, D., Wang, P., Luo, Y., Zhao, M., & Chen, F. (2017). Health benefits of anthocyanins and molecular mechanisms: Update from recent decade. *Critical Reviews in Food Science and Nutrition, 57*(8), 1729–1741.

Li, Q., Somavat, P., Singh, V., Chatham, L., & de Mejia, E. G. (2017). A comparative study of anthocyanin distribution in purple and blue corn coproducts from three conventional fractionation process. *Food Chemistry, 231*, 332–339.

Li, X., Wang, C., Lu, F., Zhang, L., Yang, Q., Mu, J., & Li, X. (2015). Physicochemical properties of corn starch isolated by acid liquid and L-cysteine. *Food Hydrocolloids, 44*, 353–359. https://doi.org/10.1016/j.foodhyd.2014.09.003.

Liu, R. H. (2004). Potential synergy of phytochemicals in cancer prevention: Mechanism of action. *Journal of Nutrition, 134*(2), 3479–3485.

Luo, Y., & Wang, Q. (2012). Bioactive compounds in corn. In L. Yu, R. Tsao, & F. Shahidi (Eds.), *Cereals and pulses* (pp. 85–103). Chichester: Wiley-Blackwell.

Micheletti, A. M. S. (2013). Carotenoids, phenolic compounds and antioxidant capacity of five local Italian corns (*Zea mays* L.) kernels. *Journal of Nutrition and Food Science, 3*, 6.

Miguel, M. G. (2017). Anthocyanins: Antioxidant and/or anti-inflammatory activities. *Jounal of Applied Pharmaceutical Science, 1*(6), 7–15.

Moin, A., Ali, T. M., & Hasnain, A. (2016). Effect of succinylation on functional and morphological properties of starches from broken kernels of Pakistani Basmati and Irri rice cultivars. *Food Chemistry, 191*, 52–58.

Moreau, R. A., Harron, A. F., Powell, M. J., & Hoyt, J. L. (2016). A comparison of the levels of oil, carotenoids, lipolytic enzyme activities in modern lines and hybrids of grain sorghum. *Jounal of American Oil and Chemical Society, 93*(4), 560–573.

Mutlu, C., Arslan-Tontul, S., Candal, C., Kilic, O., & Mustafa, E. (2017). Physicochemical, thermal, and sensory properties of blue corn (*Zea mays* L.). *Journal of Food Science.* https://doi.org/10.1111/1750-3841.14014

Nankar, A. N., Dungan, B., Paz, N., Sudasingle, N., Schaub, T., Holguin, F. O., & Pratt, R. C. (2016). Quantitative and qualitative evaluation of kernel anthocyanins from southwestern United States blue corn. *Journal of the Science of Food and Agriculture, 96*(13), 4542–4552.

Ndolo, V. U., & Beta, T. (2013). Distribution of carotenoids in endosperm, germ, and aleurone fractions of cereal grain kernels. *Food Chemistry, 139*(1–4), 663–671.

Neder-Suarez, D., Amaya-Guerra, C. A., Baez-Gonzalez, J. G., Quintero-Ramos, A., Aguilar-Palazueles, E., Galicia-García, T., ... de Jesús Zazueta-Morales, J. (2018). Resistant starch formation from corn starch by combining acid hydrolysis with extrusion cooking and hydrothermal storage. *Starch: Starke, 70*(5–6), 1700118. https://doi.org/10.1002/star.201700118.

Newman, C. W., Newman, R. K., & Fastnaught, C. E. (2019). Barley. In J. Johnson & T. Wallace (Eds.), *Whole grains and their bioactives: Composition and health* (pp. 135–167). Hoboken, NJ: John Wiley & Sons.

Nirmala, P. V. P., & Iris, J. J. (2020). Dietary fibre from whole grains and their benefits on metabolic health. *Nutrients, 12*(10), 3045. https://doi.org/10.3390/nu12103045

Niu, L., Ding, H., Zhang, J., & Wang, W. (2019). Proteomic analysis of starch biosynthesis in maize seeds. *Starch/Starke.* https://doi.org/10.1002/star.201800294

Nkhata, S. G., Ayua, E., Kamau, E. H., & Shingiro, J.-B. (2018). Fermentation and germination improve nutritional value of cereals and legumes through activation of endogenous enzymes. *Food Science and Nutrition, 6*(8), 2446–2458.

Ntrallou, K., Gike, H., & Tsochatzis, E. (2020). Analytical and sample preparation techniques for the determination of food colourants in food matrices. *Foods, 9*(58), 1–24.

Odio, S., Bera, F., Beckers, Y., Foucart, G., & Malumba, P. (2018). Influence of variety, harvesting date and drying temperature on the composition and the in vitro digestibility of corn grain. *Journal of Cereal Science, 79*, 218–225.

Okafor, U. I., Omemu, A. M., Obadina, A. O., Bankole, M. O., & Adeyeye, A. O. (2018). Nutritional composition and anti-nutritional properties of maize ogi co-fermentated with pigeon pea. *Food Science and Nutrition, 6*(2), 424–439.

Pandez, K. B., & Rizvi, S. I. (2009). Plant polyphenols as dietary antioxidant in human health and disease. *Oxidative Medicinal Cell Longevity, 2*, 270–278.

Panfili, G., Fratianni, A., & Irano, M. (2004). Improved normal-phase high-performance liquid chromatography procedure for the determination of carotenoids in cereals. *Journal of Agricultural and Food Chemistry, 52*(21), 6373–6377.

Pang, Y., Ahmad, S., Xu, Y., Beta, T., Zhu, Z., Shao, Y., & Bao, J. (2018). Bound phenolic compounds and antioxidant properties of whole grain and bran of white, red and black rice. *Food Chemistry, 240*, 212–221.

Paraginski, R. T., Vanier, N. L., Moomand, K., Oliveira, D. M., Zavareze, E. D. R., Marques e Silva, R., ... Elias, M. C. (2014). Characteristics of starch isolated from maize as a function of grain storage temperature. *Carbohydrate Polymers, 102*, 88–94.

Paraginski, R. T., Colussi, R., Dias, A. R. G., da Rosa Zavareze, E., Elias, M. C., & Vanier, N. L. (2018). Physicochemical, pasting, crystallinity, and morphological properties of starches isolated from maize kernels exhibiting different types of defects. *Food Chemistry, 274*, 330–336. https://doi.org/10.1016/j.foodchem.2018.09.026.

Park, E. Y., Ma, J.-G., Kim, J., Lee, D. H., Kim, S. Y., Kwon, D. J., & Kim, J.-Y. (2018). Effect of dual modification of HMT and crosslinking on physicochemical properties and digestibility of waxy maize starch. *Food Hydrocolloid, 75*, 33–40.

Paucar-Menacho, L. M., Martinez-Villaluenga, C., Duenas, M., Frias, J., & Penas, E. (2016). Optimization of germination time and temperature to maximize the content of bioactive compounds and the antioxidant activity of purple corn (*Zea mays* L.) by response surface methodology. *LWT-Food Science and Technology, 76*, 236–244.

Paulsmeyer, M., Chatham, L., Becker, T., West, M., West, L., & Juvik, J. (2017). Survey of anthocyanin composition and concentration in diverse maize germplasms. *Journal of Agricultural and Food Chemistry, 65*(21), 4341–4350.

Pietrzyk, S., Fortuna, T., Labanowska, M., Juszczak, I., Galkowska, D., Baczkowics, M., & Kundziel, M. (2018). The effect of amylose content and level of oxidation on the structural changes of acetylated corn starch and generation of free radicals. *Food Chemistry, 240*, 259–267.

Rashwan, A. K., Yones, H. A., Karim, N., Taha, E. M., & Chen, W. (2021). Potential processing technologies for developing sorghum-based food products: An update and comprehensive review. *Trends in Food Science and Technology, 110*, 168–182.

Rocha-Villarreal, V., Hoffmann, J. F., Vanier, N. L., Serna-Saldivar, S. O., & Garcia-Lara, S. (2018). Hydrothermal treatment of maize: Changes in physical, chemical, and functional properties. *Food Chemistry, 263*, 225–231.

Rocheford, T. R., Wang, J. C., Egesel, C. O., & Lambert, R. J. (2002). Enhance of vitamin E levels in corn. *Journal of American College of Nutrition, 21*(Suppl.), 191–198.

Rodriguez, R., Jimenez, A., Fernandez-Bolanos, J., Guillen, R., & Heredia, A. (2006). Dietary fibre from vegetable products as source of functional ingredients. *Trends in Food Science and Technology, 17*(1), 3–15.

Roy, S., & Rhim, J.-W. (2020). Anthocyanin food colourant and its application in pH-responsive colour change indicator films. *Critical Reviews in Food and Nutrition.* https://doi.org/10.1080/ 10408398.2020.1776211

Saldivar, S. O. S., & Soto, F. E. A. (2020). Chemical composition and biosynthesis of dietary fibre components. In J. Welti-Chances et al. (Eds.), *Science and technology of fibres in food systems* (pp. 15–43). Springer Nature Switzerland AG. Food Engineering Series. https://doi.org/10 .1007/978-3-030-38654-2_2

Saleh, A. S. M., Wang, P., Wang, N., Yang, S., & Xiao, Z. (2017). Technologies for enhancement of bioactive components and potential health benefits of cereal and cereal-based foods: Research advances and application challenges. *Critical Reviews in Food Science and Nutrition.* https://doi .org/10.1080/10408398.2017.1363711

Sandhu, K. S., Godara, P., Kaur, M., & Punia, S. (2017). Effect of toasting on physical functional and anti-oxidant properties of flour from oat (*Avena sativa* L.) cultivars. *Journal of the Saudi Society of Agricultural Sciences*, *16*(2), 197–203.

Saulnier, L., & Thibault, J. F. (1999). Ferulic acid diferulic acids as components of sugar-beet pectins and maize bran heteroxylans. *Journal of Science of Food and Agriculture*, *76*(3), 396–402.

Schober, T. J., Bean, S. R., Boyle, D. L., & Park, S. H. (2008). Improved viscoelastic zein-starch doughs for leavened gluten-free breads: Their rheology and microstructure. *Journal of Cereal Science*, *48*(3), 755–767.

Serna Saldivar, S. O. (2010). *Cereal grains: Properties, processing and nutritional attributes*. Boca Raton, FL: CRC Press, Taylor & Francis Group.

Serna Saldivar, S. O., & Hernandez, D. S. (2020). Dietary fibre in cereals, legumes, pseudocereals and other seeds. In J. Welti-Chanes et al. (Eds.), *Science and technology of fibres in food systems* (pp. 87–122). Springer Nature Switzerland AG. Food Engineering Series. https://doi.org/10.1007/978-3-030-38654-2_5

Shang, Y., Chao, C., Yu, J., Copeland, L., Wang, S., & Wang, S. (2018). Starch spherulites prepared by a combination of enzymatic and acid hydrolysis of normal corn starch. *Journal of Agricultural and Food Chemistry*, *66*(25), 6357–6363.

Sharma, S., Saxena, D. C., & Riar, C. S. (2016). Analyzing the effect of germination on phenolics, dietary fibres, minerals, and γ-aminobutyric acid contents of barnyard millet (*Echinochloafrumentaceae*). *Food Bioscience*, *13*, 60–68.

Sigurdson, G. T., Tang, P., & Giusti, M. M. (2017). Natural colourants: Food colourants from natural sources. *Annual Review of Food Science and Technology*, *8*, 261–280.

Singh, N., Singh, S., Kaur, A., & Bakshi, M. S. (2012). Zein: Structure, production, film properties and applications. In M. J. John & S. Thomas (Eds.), *Natural polymers* (Vol. 1, pp. 204–218). London: RSC.

Singh, N., Singh, S., & Shevkani, K. (2019). Maize: Composition, bioactive constituents, and unleavened bread. In *Flour and breads and their fortification in health and disease prevention*. https://doi.org/10.1016/b978-0-12-814639-2.00009-5

Singh, S., Gaikwad, K. K., & Lee, Y. S. (2018). Anthocyanin – A natural dye for smart food packaging systems. *Korean Journal of Packaging Science and Technology*, *24*(3), 167–180.

Sinha, A. K., Kumar, V., Makkar, H. P. S., de Boeck, G., & Becker, K. (2011). Non-starch polysaccharides and their role in fish nutrition-A review. *Food Chemistry*, *127*(4), 1409–1426. https://doi.org/10.10161/j.foodchem. 2011.02.042.

Sokrab, A. M., Ahmed, I. A. M., & Babiker, E. E. (2012). Effect of malting and fermentation on anti-nutrients, and total and extractable minerals of high and low phytate corn genotypes. *International Journal of Food Science and Technology*, *47*(5), 1037–1043.

Sokrab, A. M., Mohamed Ahmad, I. A., Babiker, E. E., Ahmed, I. A. M., & Babiker, E. E. (2014). Effect of fermentation on anti-nutrients, and total and extractable minerals of high and low phytate corn genotypes. *Journal of Food Science and Technology, 51*(10), 2608–2615.

Suriano, S., Balconi, C., Valoti, P., & Redaelli, R. (2021). Comparison of total polyphenols, profile anthocyanins, colour analysis, carotenoids and tocols in pigmented maize. *LWT-Food Science and Technology, 144*, 111257. https://doi.org/10.1016/j.lwt.2021.111257

Tester, R. F., Debon, S. J. J., & Sommerville, M. D. (2000). Annealing of maize starch. *Carbohydrate Polymers, 42*(3), 287–299.

Thakur, S., Kaur, A., Singh, N., & Virdi, A. S. (2015). Successive reduction of dry milling of normal and waxy corn: Grain, grit, and flour properties. *Journal of Food Science, 80*, 1144–1155.

Trono, D. (2019). Carotenoids in cereal food crops: Composition and retention throughout grain storage and food processing. *Plants, 8*(12), 551. https://doi.org/10.3390/plants8120551

Uriarte-Aceves, P. M., Milan-Carrillo, J., Cuevas-Rodriguez, E. O., Gutierrez-Dorado, R., Reye-Moreno, C., & Milan-Noris, E. M. (2018). In vitro digestion properties of native isolated starches from Mexican blue maize (*Zea mays* L.) landrace. *LWT-Food Science and Technology, 93*, 384–389. https://doi.org/10.1016 /j.lwt.2018.03.015.

USDA. (2017). *Grain: World markets and trade stiff competition among rice exporters as global trade expands.* Reported by United States Department of Agriculture.

Votterl, J. C., Zebeli, Q., Hennig-Pauka, I., & Metzler-Zebeli, B. U. (2019). Soaking in lactic acid lowers the phytate-phosphorus content and increases the resistant starch in wheat and corn grains. *Animal Feed Science and Technology.* https://doi.org/10.1016/j.anifeedsci .2019.04.013

Wallace, T. C., & Giusti, M. M. (2019). Anthocyanins-nature's bold, beautiful, and health-promoting colours. *Foods, 8*(11), 550. https://doi.org /10.3390/foods8110550

Wang, L., Xu, C., Qu, M., & Zhang, J. (2008). Kernel amino acid composition and protein content of introgression lines from Zea mays ssp. Mexicana into cultivated maize. *Journal of Cereal Science, 48*(2), 387–393.

Wang, S., & Guo, P. (2020). Botanical sources of starch. In S. Wang (Ed.), *Starch structure: Functionality and application in foods* (pp. 9–27). Springer Nature Singapore Pte Ltd. https://doi.org/10.1007/978-981 -15-0622-2_2

Wang, S., Wang, J., Yu, J., & Wang, S. (2014). A comparative study of annealing of waxy, normal and high-amylose maize starches: The role of amylose molecules. *Food Chemistry, 164*, 332–338.

Wang, Y., Zhou, Y. L., Cheng, Y. K., Jiang, Z. Y., Jin, Y., Zhang, H. S., … Zang, G. R. (2015). Enzymo-chemical preparation, physicochemical characterization and hypolipidemic activity of granular corn bran dietary fibre. *Journal of Food Science and Technology, 52*(3), 1718–1723.

Watson, S. A. (2003). Description, development, structure, and composition of corn kernel. In P. J. White & L. A. Johnson (Eds.), *Corn: Chemistry and technology* (2nd ed., pp. 69–106). St. Paul, MN: American Association of Cereal Chemists, Inc.

Wrolstad, R. E., Durst, R. W., & Lee, J. (2005). Tracking colour and pigment changes in anthocyanin products. *Trend in Food Science and Technology, 16*(9), 423–428.

Wu, W., Jiao, A., Xu, E., Chen, Y., & Jin, Z. (2020). Effects of extrusion technology combined with enzymatic hydrolysis on the structural and physicochemical properties of porous corn starch. *Food and Bioprocess Technology, 13*(1–2), 1–10. https://doi.org/10.1007/s11947 -020-02404-1.

Xu, J., Li, Y., & Wang, W. (2019). Corn. In J. Wang et al. (Ed.), *Bioactive factors and processing technology for cereal foods* (pp. 33–53). Singapore: Springer. https://doi.org/10.1007/978-981-13-6167-8_3

Zhang, S., Ji, J., Zhang, S., Xiao, W., Guan, C., Wang, G., & Wang, Y. (2020). Changes in the phenolic compound content and antioxidant activity in developmental maize kernels and expression profiles of phenolic biosynthesis-related genes. *Journal of Cereal Science, 96*, 103113. https:// doi.org/10.1016/j.jcs.2020.103113

Yu, L., Zhao, A., Yang, M., Wang, C., Wang, M., & Bai, X. (2018). Effects of the combination of freeze-thawing and enzymatic hydrolysis on the microstructure and physicochemical properties of porous corn starch. *Food Hydrocolloids, 83*, 465–472. https://doi.org/10.1016/j .foodhyd.2018.04.041.

Zilic, S., Serpen, A. I., Akilhoglu, G., Gokmen, V., & Vancetovic, J. (2012). Phenolic compounds, carotenoids, anthocyanins, and antioxidant capacity of coloured maize (*Zea mays* L.) kernels. *Journal of Agricultural and Food Chemistry, 60*(5), 1224–1231.

Chapter 3

Nutritional profile of maize and effect of processing methods

*Sukhvinder Singh Purewal, Pinderpal Kaur,
Kawaljit Singh Sandhu, Sneh Punia Bangar,
Anil Kumar Siroha, Surender Kumar
Singh, Maninder Kaur, Raj Kumar Salar,
and Dilip Kumar Markandey*

DOI: 10.1201/9781003245230-3

3.1 Introduction

Maize (*Zea mays*) is gaining interest worldwide as a major energy source of food and feed (Salar et al., 2012). Maize is an important member of *Poaceae* family, which is grown throughout the world as a staple food. Major types of maize that are utilized by food industries and other sectors are pod corn (*Zea mays* var. *tunicata*); amylomaize (*Zea mays*); waxy corn (*Zea mays* var. *ceratina*); striped maize (*Zea mays* var. *japonica*); flour maize (*Zea mays* var. *amylacea*); sweet maize (*Zea mays* var. *rugosa* and *Zea mays* var. *saccharata*); flint maize (*Zea mays* var. *indurata*); popcorn (*Zea mays* var. *everta*), and dent corn (*Zea mays* var. *indentata*). Different varieties of maize are in use in industry for the production of oil, feed, and bourbon (Arnold et al., 2019; Eckert et al., 2018; Green et al., 2015). Generally, for human consumption breeders grow specific cultivars (sweet corn) of maize that are rich in sugar content. Maize grains and flour are widely being used for the preparation of various delicious and health-benefiting food products (Figure 3.1). Extensive research is being carried out on maize and its milling fractions to improve the bioactive compounds and nutritional profile. Various methodologies had been adapted by food scientists/researchers/industries to modulate the nutritional content and bioactive profile of grains (Kaur et al., 2019; Singh et al., 2019; Sandhu and Punia, 2017; Siroha et al., 2016; Sandhu et al., 2016). In natural resources,

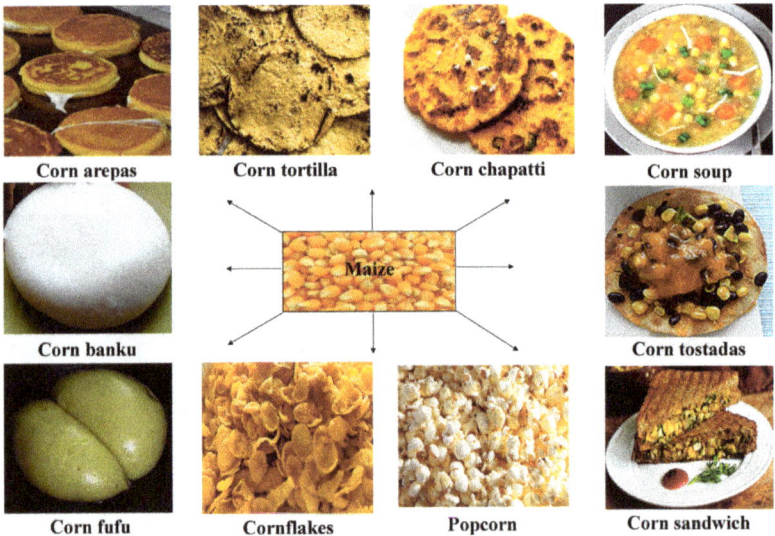

Corn arepas Corn tortilla Corn chapatti Corn soup

Maize

Corn banku Corn tostadas

Corn fufu Cornflakes Popcorn Corn sandwich

Figure 3.1 Various maize-based food products

bioactive compounds with antioxidant properties are not readily available due to their complex nature (Salar et al., 2015). For the purpose of increasing their bioavailability they need specific processing so as to convert them from their bound to free forms (Ming-Zhu et al., 2020; Ciric et al., 2020; Molina et al., 2020; Salar et al., 2016). Processing methods have been applied to grains to meet consumer demand and obtain desirable changes in food products. Processing of cereal grains could be carried out using different methods such as (1) solid-state fermentation (SSF), (2) germination/steeping, and (3) thermal processing. Processing significantly affects the nutritional profile of cereal grains either by enhancement of specific compounds or improvement of minerals and flavour of processed products (Kaur et al., 2019). Solid-state fermentation has been used for human benefit since ancient times. In the fermentation process, a starter culture (microbial strain) is employed to modulate the nutritional profile of specific substrates. The selection of starter culture depends on the need of the experiment and the characteristics required in fermented products (Dulf et al., 2020; Klempova et al., 2020; Mansor et al., 2019; Naik et al., 2019; Liu et al., 2019). The increment in bioactive compounds after the fermentation process depends on the nutrients present in steam-sterilized substrates as they are the only source for the starter culture to obtain moisture and other nutrients to sustain metabolic processes (Purewal et al., 2020; Postemsky et al., 2019; Postemsky and Curvetto, 2015). Sprouting or germination of grains is essential as it results in the enhancement of specific nutrients and bioactive phytochemicals that possess health-benefiting properties. Keeping in mind the effects of processing on the nutritional profile, the present chapter was designed to explore the nutrient content and bioactive profile of maize.

3.2 Maize production in India and worldwide

Production of maize in India and worldwide in terms of production and yield between 2008 and 2018 is presented in Tables 3.1 and 3.2. As observed from the data collected from FAO (2020), maize production in India was highest during the year 2017 (28750000 tonnes) whereas the lowest production was observed during the year 2009 (16719500 tonnes). Data obtained regarding maize production across the world also indicates the highest production during the year 2017 (1164466612 tonnes) and lowest during 2009 (820072448 tonnes). The top maize producers in the world are the United States (392450840 tonnes) followed by mainland China (257173900 tonnes); Brazil (82288298 tonnes); Argentina (43462323 tonnes); Ukraine (35801050 tonnes); Indonesia (30253938 tonnes); India (27820000 tonnes); Mexico (27169977 tonnes); Romania (18663939 tonnes); and Canada (13884800) (Table 3.2). The main maize-growing Indian states are Madhya

Table 3.1 Maize production in India 2008–2018 (FAO, 2020)

Year	2008	2009	2010	2011	2012	2013	2014	2015	2016	2017	2018
Area harvested (ha)	81,73,800	82,61,600	85,53,200	87,80,000	87,10,000	94,30,000	92,58,000	86,90,000	99,00,000	92,20,000	92,00,000
Yield (kg/ha)	24,140	20,238	25,401	24,784	25,557	25,726	26,107	25,972	26,162	31,182	30,239
Production (tonnes)	1,97,31,400	1,67,19,500	2,17,25,800	2,17,60,000	2,22,60,000	2,42,59,510	2,41,70,000	2,25,70,000	2,59,00,000	2,87,50,000	2,78,20,000

Table 3.2 Maize productions worldwide 2008–2018 (FAO, 2020)

Year	2008	2009	2010	2011	2012	2013	2014	2015	2016	2017	2018
Area harvested (ha)	16,31,42,954	15,88,19,581	16,40,20,015	17,12,02,475	17,97,91,974	18,69,57,444	18,57,36,210	19,05,78,754	19,56,09,280	19,74,74,622	19,37,43,247
Yield (kg/ha)	50,829	51,635	51,925	51,791	48,670	54,355	55,952	55,208	57,617	58,968	59,238
Production (tonnes)	82,92,40,208	82,00,72,448	85,16,79,519	88,66,80,581	87,50,39,160	101,62,07,182	103,92,26,655	105,21,39,015	112,70,42,534	116,44,66,612	114,76,89,084

Pradesh (MP); Uttar Pradesh (UP); Himachal Pradesh (HP); Jammu and Kashmir (J&K); and Bihar and Punjab.

3.3 Nutritional profile of maize

Being a rich source of specific nutrients and bioactive compounds, maize is reported to be a distinctive crop (Tabasum et al., 2019). The nutritional profile of maize is reported in Table 3.3. Camelo-Méndez et al. (2017) studied the proximate composition of blue and white maize flour. Blue maize flour indicates the presence of starch (70.7 g/100 g) as a major nutrient, followed by fibre (10.9 g/100 g⁻), protein (9.1 g/100 g), and lipids (5.2 g/100 g). White maize flour showed the presence of different nutrients such as starch (74.2 g/100 g), dietary fibre (11.2 g/100 g), protein (8.4 g/100 g), and lipids (4.7 g/100 g). Other scientific studies on maize flour indicate the presence of protein (8–13 g/100 g), ash (0.8–1.3 g/100 g), moisture (9–10 g/100 g), crude fibre (0.7–2.7 g/100 g), and crude fat (3–7 g/100 g) (Dei, 2017; Qamar et al., 2016). Starch is the major fraction of carbohydrate present in maize grains. Because of the high potential to use sunlight, maize crops can synthesize starch easily. The properties of maize polysaccharides are significantly affected by the concentration of amylose and amylopectin. Starch is used in food industries for the preparation of many important food products (Punia et al., 2020; Siroha et al., 2019). As well as many food products, starch is also used as an important substrate for the production of biodegradable films (Sandhu et al., 2020). The specific functional properties of starch can be explained based on physicochemical, gelatinization, retrogradation, pasting, and rheological properties. These properties significantly affect the texture as well as other important properties of starch-based food products. Twelve different maize cultivars were studied for the evaluation of specific properties present in starch (Sandhu and Singh, 2007; Sandhu et al., 2004). The details of the specific properties of maize starch are presented in Table 3.4. Fibre is an important non-starch carbohydrate and plays an important role in maintaining various metabolic pathways for sustaining a healthy life. Crude fibre occupies 87% of the maize seed coat and includes hemicelluloses (67%), cellulose (23%), and lignin (0.1%) (Burge and Duensing, 1989). Different parts of the maize kernel contain different amounts of fibre: in the pericarp, the fibre content is 86.7%; in the germ, 8.8%; and in the endosperm, 2.7%. The protein content in grains significantly varies with the grain fraction. For instance, a major percentage of protein (glutelin and prolamin) occurs in the endosperm followed by the germ and aleurone layers which possess globulin and albumins. Maize grains possess a high proportion (50–60%) of prolamin proteins, which contain a low amount of lysine; however, the amount of lysine is higher in globulin and albumin. Maize grains are rich in specific minerals such as potassium

Table 3.3 Nutritional profile of maize

Nutrients	Amount (per 100 g)	Reference
Carbohydrates	71.88 g	Shah et al. (2015); Zilic et al. (2011);
Starch	54.59–69.92%	Whelan et al. (2011); Nuss and
Total sugars	0.64 g	Tanumihardjo (2010); Gopalan et al. (2007); Knudsen (1997)
Sugar profile		
Xylose	1.84% (ZP 531su)	
Fructose	1.93%(ZP 531su)	
Glucose	1.59–1.96%	
Sucrose	2.39–4.25%	
Total fibre	2.15–7.3 g	Zilic et al. (2011); Knudsen (1997)
Fibre profile		
Cellulose	3.11–4.15%	
Hemicellulose	7.07–10.29%	
Lignin	0.29–0.80%	
Fructan	0.6 g	
Total protein	8.84–9.42%	Shah et al. (2015); Gopalan et al.
Protein profile		(2007); Serna-Saldivar and Rooney (1994)
Albumin+globulein	6.2%	
Prolamin	39.2%	
Glutelin	22.7%	
Amino acids profile		
Arginine	5.60 g	
Histidine	3.07 g	
Isoleucine	3.76 g	
Leucine	12.52 g	
Lysine	3.40 g	
Methionine	1.73 g	
Phenylalanine	5.16 g	
Threonine	3.84 g	
Tryptophan	0.59 g	
Valine	5.05 g	
Fat	4.57–4.74 g	Ray et al. (2019); Shah et al. (2015); Gopalan et al. (2007)
Fatty acid profile		
Palmitic acid	12.45–13.86%	
Stearic acid	1.64–2.26%	
Oleic acid	26.31–30.76%	

(Continued)

Table 3.3 (Continued)   Nutritional profile of maize

Nutrients	Amount (per 100 g)	Reference
Linoleic acid	48.36–57.20%	
Linolenic acid	0.95–1.85%	
Arachidic acid	0.02–0.76%	
Total Minerals	1.5 g	Shah et al. (2015); Nuss & Tanumihardjo
Mineral profile		(2010); Gopalan et al. (2007)
Phosphorus	210–348 mg	
Sodium	15.9–35 mg	
Sulphur	114 mg	
Calcium	7–10 mg	
Iron	2.3–2.71 mg	
Potassium	286–287 mg	
Magnesium	127–139 mg	
Copper	0.14–0.31 mg	
Zinc	2.21 mg	
Manganese	0.48 mg	
Selenium	15.5 µg	
Vitamins profile		
Riboflavin	0.10–0.20 mg	
Thiamine	0.38–0.42 mg	
Pantothenic acid	0.424 mg	
Pyridoxine	0.62 mg	
Niacin	3.63 mg	
Vitamin C	0.12 mg	
Folate	19 µg	
Vitamin A	11 µg	
Carotenoids		Moros et al. (2002); Zhao et al. (2005);
Carotene	2.20–45.8 mg	Salinas-Moreno et al. (1999); Singh
Xanthophylls	2.07 mg	et al. (2013)
Lutein	1.50–406.2 mg	
Zeaxanthin	0.57 mg	
Phenolic compounds	215–551 mg	
Phenolic profile		Locatelli and Berardo (2014)
Ferulic acid (FA)	102–174 mg	
Anthocyanins	0.57–141.7 mg	
Phytosterols	14.83 mg	
Sitosterol	9.91 mg	
Stigmasterol	1.52 mg	
Campesterol	3.40 mg	

Table 3.4 Specific properties of maize starch

Cultivars	Physicochemical properties		References
African tall, Ageti, early composite, Girja, Navjot, Parbhat, Partab, PbSathi, Vijay, dent corn (bold, long), popcorn (small, medium, large), baby corn	Amylose content (%)	15.3–25.1	Sandhu and Singh (2007); Sandhu et al. (2004)
	Swelling power (g/g)	13.7–20.7	
	Solubility (%)	9.7–20.3	
	Water binding capacity (%)	82.1–107	
	Gelatinization properties		
	T_o (°C)	65.6–69.3	
	T_p (°C)	69.9–74	
	T_c (°C)	75.1–79.7	
	ΔH_{gel} (J/g)	8.9–12.7	
	PHI	1.7–2.98	
	Retrogradation properties		
	T_o (°C)	41.5–43.1	
	T_p (°C)	52.4–54.5	
	T_c (°C)	62–64.3	
	ΔH_{ret} (J/g)	4.4–6.9	
	Pasting properties		
	PV (cP)	804–1252	
	TV (cP)	594–727	
	BV (cP)	113–590	
	FV (cP)	824–1388	
	SV (cP)	141–726	
	$P_{temp.}$ (°C)	75.9–83.8	
	Rheological properties		
	TG' (°C)	73–73.7	
	Peak G' (Pa)	2172–5354	
	Peak G'' (Pa)	383–920	
	Breakdown in G' (Pa)	1102–3184	
	Peak tan δ	0.122–0.181	

T_o: onset temperature; T_p: peak temperature; T_c: conclusion temperature; ΔH_{gel}: enthalpy of gelatinization (dwb, based on starch weight); PHI: peak height index.

(4 mg g^{-1}), followed by phosphorus (3 mg g^{-1}), magnesium (1.6 mg g^{-1}), sulphur (1.4 mg g^{-1}), chlorine (0.7 mg g^{-1}), sodium (0.5 mg g^{-1}), manganese (0.068 mg g^{-1}), calcium (0.06 mg g^{-1}), copper (0.045 mg g^{-1}), and iron (0.025 mg g^{-1}). From a nutritional point of view, oil extracted from maize is of superior quality as it can be used for cooking purposes and is with a low risk of rancidity. The nutritional profile of maize oil indicates a high concentration of bioactive compounds with antioxidant properties and a low concentration of linolenic acid (Martinez et al., 1996). Watson and Ramstad (1987) reported the presence of oil, starch, sugar,

protein, and ash in different milling fractions of maize grains. Their study indicates the presence of oil varying from 0.8–82.6%, followed by starch (0.1–97.8%), sugar (0.8–69.3%), protein (0.9–73.8%), and ash (1–78.4%). Maize endosperm possesses a high amount of starch (97.8%) and protein (73.8%); however the concentration of oil (82.6%), sugar (69.3%), and ash (78.4%) was highest in maize germ. Pillay et al. (2014) studied different varieties of maize (dark orange 10 MAK 7–8; medium orange 10 MAK 7–7; and light orange 10 MAK 7–5) in terms of carotenoids' composition. The results from their study indicate the presence of Zeaxanthin in the range of 14.1–18.7 µg g^{-1}; followed by β-cryptoxanthin (3.7–4.8 µg g^{-1}); β-carotene (3.4–3.6 µg g^{-1}); provitamin A carotenoids (7.3–8.3 µg g^{-1}); and total carotenoids (22.3–26.4 µg g^{-1}).

3.4 Bioactive compounds

Bioactive compounds are naturally occurring secondary metabolites in natural resources (Purewal et al., 2019; Dhull et al., 2016; Kaur et al., 2018; Salar and Purewal, 2016; Salar et al., 2017a, 2017b). They are naturally produced in plants in response to certain stresses to help them in sustaining their routine metabolism. Secondary metabolites could be useful as medicines as their use in specific doses helps to overcome oxidative stress conditions, free radicals, and symptoms of chronic disorders (Salar and Purewal, 2017). Secondary metabolites are responsible for the specific flavour, colour, and aroma in natural extracts. These are solvent-specific compounds which could be extracted using optimal conditions of extractions (organic and aqueous solvents; temperature and time durations) (Salar et al., 2016; Purewal et al., 2020). Extraction conditions play an important role in the recovery of phenolic compounds from cereal grains as well as other natural resources. A list of cereal grains along with details of extraction conditions are reported in Table 3.5. Beta and Hwang (2018) studied the effect of moisture and heat on phenolic compounds and antioxidant properties of orange maize flour. The amount of phenolic compounds present in orange maize flour was 1.66 mg g^{-1}. Camelo-Méndez et al. (2017) demonstrated that phenolic compounds may vary with the type of maize selected for the experimentations. They elucidate that blue and white maize differ significantly from each other in terms of secondary metabolites. White maize possesses 127 mg GAE g^{-1} phenolic compounds, which were significantly lower than the amount present in blue maize (164 mg GAE g^{-1}). Lopez-Martinez et al. (2009) studied 18 different maize cultivars for the presence of phenolic compounds with antioxidant properties. The total phenolics in selected maize cultivars were in the range of 1.7–34 mg GAE g^{-1} whereas the amount reported for free and bound phenolics was in the range of 0.33–1.24 mg GAE g^{-1} and 1.36–27.2 mg GAE g^{-1}. Anthocyanin in

Table 3.5 Extraction conditions for recovery of specific bioactive compounds from maize

Maize cultivar/ plant part	Extraction phase	Temperature	Extraction time	Compounds studied	References
Corn	Ethanol 80%	25°C	24 h	Phenolic compounds	Xu et al. (2018)
Purple corn cob	Ethanol, water	55°C 65°C	90–360 min	Anthocyanins, phenolic compounds	Monroy et al. (2016)
Corn silk	Methanol, ethyl acetate	40°C	96 h	Phenolic compounds, flavonoids	Laeliocattleya (2018)
Corn grains	Ethanol, methanol, acetone	40°C	20–60 min	Free and bound phenolics	Fuentealba et al. (2016)
Corn silage	Aqueous ethanol	90–180°C	40–120 min	Phenolic compounds	Kuzmanovic et al. (2015)
Maize filaments	Ethanol	–	–	Phenolic compounds (benzoic acid, gentisic acid, epicatechin, syringic acid, quercetin, rutin, and chlorogenic acid)	Lingzhu et al. (2015)
Corn silk	Water, ethanol 70%, ethanol 96%, ethyl acetate	–	51 h	Phenolic compounds	Irawaty et al. (2018)
Maize	Ethanol 80%	25°C	15 min	Phenolic compounds	Zavala-Lopez and Garcia-Lara (2017)
Maize kernels	Acetone, methanol, and water	Room temp.	4 h	Phenolic compounds	Zilic et al. (2012)
Corn cob	Aqueous ethanol	25°C	60–120 min	Phenolic compounds	Hernandez et al. (2018)
Corn silk	Ethyl acetate: water (85:15 v/v)	70°C	1.5 h	Phenolic compounds	Haslina and Eva (2017)
Purple corn	Aqueous ethanol	60–120°C	15–60 min	Anthocyanin	Muangrat and Saengcharoenrat (2018)

selected cultivars was in the range of 0.015–8.5 mg g^{-1}. Zilic et al. (2013) reported that maize flour consists of 5.77 mg g^{-1} phenolic compounds and 29.38 mg Kg^{-1} flavonoids. The phenolic acid level in maize flour signifies the presence of high amounts of ferulic acid (4.15 mg g^{-1}) followed by p-coumaric acid (0.62 mg g^{-1}).

3.5 Effect of processing on maize phenolics

Processing of cereal grains and other natural resources is carried out to improve the bioactive compounds, texture, taste, aroma, and shelf life. Processing imparts modifications in processed material via changing its anatomical features, and modulating its nutritional and bioactive compounds. Processing also significantly affects the bioactive compounds and other nutrients in both positive as well as negative ways. Food industries adopt processing methods that are safe, cost-effective, and have little effect on the nutritional profile of substrates. The processing of maize and its milling fraction can be carried out in different ways: (1) solid-state fermentation, (2) germination/steeping, and (3) thermal processing.

3.5.1 Solid-state fermentation (SSF)

SSF is an industrial process which is carried out in limited amounts of moisture i.e. just enough support growth in the starter culture. Natural resources are fermented using different strains of fungi so as to improve their bioactive compounds within a short span of time. The effect of fermentation on the bioactive profile of specific substrates may vary, as it depends on the nutritional profile of the substrate. The substrate being fermented should have specific nutrients so that it can sustain metabolic reactions within the starter culture during the fermentation time. Cui et al. (2012) reported the effect of the fermentation process on amino acids, proteins, in-vitro digestibility of proteins, bioactive phenolics, and phytic acid of maize. The outcome of their study indicated that the fermentation process results in a significant enhancement of protein content (43.5%), followed by an increment of 131.5% in lysine content, and an enhancement of 23.4% in phenolic compounds. However, a decrease in phytic acid content (24%) was observed after the fermentation process. Salar et al. (2012) studied the capability of *Thamnidium elegans* regarding the modulation of maize phenolics. They reported that fungal fermentation results in the enhancement of maize phenolics from 327 to 409 µmol gallic acid equivalent g^{-1}. The enhancement of maize phenolics continued until the fifth day of fungal fermentation; thereafter a decrease in phenolics was observed. The whole process of phenolic enhancement was under the control of enzymatic actions (α-amylase, xylanase, and β-glucosidase). Under certain specific conditions even microbial strains could act as important sources of bioactive phenolics (Salar et al., 2013; Salar et al., 2017c). Depending on the starter strains, the effect on individual phenolics during the fermentation process may vary accordingly (Salar and Purewal, 2016; Adebo and Medina-Meza, 2020; Starzynska-Janiszewska et al., 2019). The genomic status of specific fungal strains might be responsible for encoding genes, which results in phenolic degradation/modulation during the fermentation process (Ripari et al., 2019; Hole et al., 2012). Acosta-Estrada et al. (2019) demonstrated the effect of fermentation technology on phenolic compounds, antioxidant potential, and fibre composition of lime-cooked maize by-products (LCMP). In LCMP, a portion of approximately 85% is occupied by bound phenolic compounds. In their study, the capabilities of four fungal strains (*Aspergillus oryzae*; *Hericium erinaceus*; blue oyster; and pearl oyster) were compared in terms of the enhancement of phenolic compounds in LCMP. Enhancement in LCMP phenolics was observed from the first to the third day of fermentation. Fermentation of LCMP with *A. oryzae* and pearl oyster resulted in a change in phenolic compounds from 132.37 to 159.63 mg ferulic acid equivalent (FAE) 100 g^{-1} and 173.94 mg FAE 100 g^{-1}. An enhancement from 132.37 to 162.20 mg FAE 100 g^{-1} on the third day of fermentation was observed in *Hericium erinaceus*-fermented LCMP. However, the enhancement of LCMP phenolics was limited from 132.37 to 151.20 mg FAE 100 g^{-1} on the first day of fermentation with blue oysters. Xu et al. (2018) reported the effect of *Agaricus bisporus* on corn phenolics, and they observed an increase in the phenolic content from 3.19 to 3.28 after solid-state fermentation (28th day).

3.5.2 Germination

Germination is a processing technique in which sprouting of seeds/grains can be performed under a specific set of conditions. Cereal grains may be germinated for the purpose of modulating their bioactive as well as nutritional profile. Chalorcharoenying et al. (2017) studied the effect of the germination process on corn phytochemicals. They observed an increase in nutrients and health-benefiting phytochemicals in sprouts and seedlings. A significant increase in gamma amino butyric acid, total phenolics, anthocyanin, and carotenoids was observed in seedlings and sprouts. Oluwalana and Babatunde (2014) reported the effect of sprouting on functional properties, mineral content, and proximate composition of yellow and white maize. Significant changes were observed in the proximate composition of yellow and white maize. Sprouted white maize showed a change in white maize, with moisture content falling from 6.29 to 6.13%, as well as fat (4.21 to 3.89%), fibre (2.90 to 1.89%), and carbohydrates (79.80 to 77.90%), while increases were found in protein content (10.23–12.34%), ash (2.16–2.24%), and energy (400.87–402.94 kcal). The authors further reported that sprouted yellow maize showed changes in protein content (10.23–12.13%), followed by fat (4.20–3.88%), ash (2.18–2.27%), fibre (2.89–2.69%), carbohydrates (80.73–79%), and energy (400.72–399.56 kcal). After sprouting, the mineral profile of white maize showed changes in iron content from 0.180–0.210 mg/100 g^{-1}; sodium (10.42–11.56 mg/100 g^{-1}); potassium (113.70–132 mg/100 g^{-1}); calcium (0.150–0.360 mg/100 g^{-1}); and magnesium (24.12–27.11 mg/100 g^{-1}). However, yellow maize after sprouting indicated changes in iron content (0.190–0.270 mg/100 g^{-1}), followed by sodium (10.44–11.65 mg/100 g^{-1}), potassium (114–134.70 mg/100 g^{-1}), calcium (0.210–0.350 mg/100 g^{-1}), and magnesium (24.13–28 mg/100 g^{-1}). Adedeji et al. (2014) demonstrated the effect of germination duration on the functional properties of maize flour. They observed that germination up to a period of 72 hours results in modulation of loose bulk density (LBD) from 0.58 to 0.50 g/ml^{-1}, followed by PBD (packed bulk density) from 0.79 to 0.70 g/ml^{-1}, WAC (water absorption capacity) 0.94–2.79 ml/g^{-1}, OAC (oil absorption capacity) 1.03–2.57 mg/g^{-1}, and FC (foaming capacity) 3.10–2.50%. Imran (2015) studied the effect of germination on the proximate composition of two maize cultivars. In his study two cultivars, viz. Azam and Jalal, were evaluated in terms of moisture content, fat, and protein. After germination, moisture content changed from 7.35 to 40%, followed by protein (12.25–14.88%), and fat (4.5–6.5%).

3.5.3 Thermal processing

Thermal processing of natural resources can be carried out using roasting, steaming, and microwave cooking. Roasting is an important processing method which imparts specific changes in texture, flavour, and nutritional profile under the influence of hot air by Maillard browning and caramelization. For the roasting process, open flame/direct heat source/oven can be used. Steaming includes the cooking of food materials under the effect of steam. The processing is often carried out in food steamer/kitchen appliances to produce desirable features in food being steamed. In microwave cooking, an electric oven is used for the purpose of heating and cooking food under the influence of electromagnetic radiation. During microwave-based processing, the polar molecules present in food rotates and starts producing thermal energy which affects texture, nutrients, and flavour of processed food materials. Ayatse et al. (1983) reported the effect of roasting on the nutritional profile of maize. They observed that roasting results in a change in elementary composition by decreasing the content of potassium by 13.8% and calcium by 41%. Loss in amino acid amount was also observed as indicated by the decrease in lysine (26.7%), leucine by 23%, and isoleucine (20.8%). Prasanthi et al. (2017) studied processed corn in terms of bioactive phytochemicals and nutrients. They found that processing has significant effects on bioactive compounds and specific nutrients of corn. Total phenolic content changes after processing in the following are: baby corn (10.7–6.57 mg GAE 100 g^{-1}); sweetcorn (8.88–7.46 mg GAE 100 g^{-1}); dent corn (14.53–12.61 mg GAE 100 g^{-1}); popcorn with oil (13.47–7.38 mg GAE 100 g^{-1}), and popcorn without oil (11.4–6.88 mg GAE 100 g^{-1}). Steaming modifies vanillic acid, syringic acid, caffeic acid, and ferulic acid from 83.36–57.56 µg GAE 100 g^{-1}, 11.66–8.19 µg GAE 100 g^{-1}, 2.34–1.15 µg GAE 100 g^{-1}, and 10.34–2.47 µg GAE 100 g^{-1}, respectively, in baby corn. In sweetcorn, steaming results in changes in vanillic acid, syringic acid, caffeic acid, and ferulic acid, and was 32.64–20.23 µg GAE 100 g^{-1}, 15.48–13.31 µg GAE 100 g^{-1}, 6.46–3.62 µg GAE 100 g^{-1}, and 3.45–1.67 µg GAE 100 g^{-1}, respectively. Steamed dent corn showed changes in vanillic acid, syringic acid, caffeic acid, and ferulic acid from 63.46 to 40.68 µg GAE 100 g^{-1}, 10.25 to 5.57 µg GAE 100 g^{-1}, 3.72 to 2.53 µg GAE 100 g^{-1}, and 1.50 to 1.16 µg GAE 100 g^{-1}, respectively. Oboh et al. (2010) reported changes in proximate composition of yellow and white maize after the roasting process. Roasted yellow maize showed changes in protein (8.45–7.85%); fat (6.21–7.34%); fibre (1.46–1.26%); carbohydrates (68.23–71.77%); and moisture (13.67–10.93%). White maize shows the effects of roasting on the following nutrients: protein (12.97–10.86%); fat (5.32–6.39%); fibre (1.32–1.24%); and carbohydrates (61.54–65.20%).

3.6 Conclusions

Maize grains are a rich source of specific minerals, bioactive constituents, and other health-benefiting nutrients. They can be used at a household level as a source of food/feed and at an industrial level for the production of functional food products. Processing of food materials is intended to meet the need and fulfil rising customer demands. Processing imparts morphological changes, improves flavours and enhances the shelf life of the product. Adaption of processing methods by industries can depend on consumer choice. For the purposes of enhancing bioactive compounds, solid-state fermentation processes may be preferred; however, to increase the amount of protein and minerals, a germination process may also be integrated. Thermal processing has its own advantages as it improve texture, flavour, and shelf life. Maize and products based on maize flour could be a boon for food and pharmacological industries.

Abbreviations

GAE: Gallic acid equivalent; FAO: Food and Agriculture Organization; SSF: solid-state fermentation; LCMP: lime-cooked maize products; FAE:

ferulic acid equivalent; LBD: loose bulk density; PBD: packed bulk density; WAC: water absorption capacity; OAC: oil absorption capacity; FC: foaming capacity; T_o: onset temperature; T_p: peak temperature; T_c: conclusion temperature; ΔH_{gel}: enthalpy of gelatinization (dwb, based on starch weight); PHI: peak height index.

References

Acosta-Estrada, B. A., Villela-Castrejon, J., Perez-Carrillo, E., Gomez-Sanchez, C. E., & Gutierrez-Uribe, J. A. (2019). Effects of solid-state fungi fermentation on phenolic content, antioxidant properties and fibre composition of lime cooked maize by-product (nejayote). *Journal of Cereal Science, 90*, 102–837. https://doi.org/10.1016/j.jcs.2019.102837

Adebo, O. A., & Medina-Meza, I. G. (2020). Impact of fermentation on the phenolic compounds and antioxidant activity of whole cereal grains: A mini review. *Molecules, 25*(4), 1–19. https://doi.org/10.3390/molecules25040927

Adedeji, O. E., Oyinloye, O. D., & Ocheme, O. B. (2014). Effects of germination time on the functional properties of maize flour and the degree of gelatinization of its cookies. *African Journal of Food Science, 8*(1), 42–47. https://doi.org/10.5897/ajfs2013.1106

Arnold, R. J., Ochoa, A., Kerth, C. R., Miller, R. K., & Murray, S. C. (2019). Assessing the impact of corn variety and Texas terroir on flavor and alcohol yield in new-make bourbon whiskey. *PLOS ONE, 14*(8), e0220787. https://doi.org/10.1371/journal.pone.0220787

Ayatse, J. O., Eka, O. U., & Ifon, E. T. (1983). Chemical evaluation of the effect of roasting on the nutritive value of maize (*Zea mays*, Linn.). *Food Chemistry, 12*(2), 135–147. https://doi.org/10.1016/0308-8146(83)90024-9

Beta, T., & Hwang, T. (2018). Influence of heat and moisture treatment on carotenoids, phenolic content, and antioxidant capacity of orange maize flour. *Food Chemistry, 246*, 58–64. https://doi.org/10.1016/j.foodchem.2017.10.150

Burge, R. M., & Duensing, W. J. (1989). Processing and dietary fibre ingredient applications of corn bran. *Cereal Foods World, 34*, 535–538.

Camelo-Méndez, G. A., Agama-Acevedo, E., Tovar, J., & Luis, A. (2017). Functional study of raw and cooked blue maize flour: Starch digestibility, total phenolic content and antioxidant activity. *Journal of Cereal Science, 76*, 179–185. https://doi.org/10.1016/j.jcs.2017.06.009

Chalorcharoenying, W., Lomthaisong, K., Suriharn, B., & Lertrat, K. (2017). Germination process increases phytochemicals in corn. *International Food Research Journal, 24*, 552–558.

Ciric, A., Krajnc, B., Heath, D., & Ogrinc, N. (2020). Response surface methodology and artificial neural network approach for the optimization of ultrasound-assisted extraction of polyphenols from garlic. *Food and Chemical Toxicology, 135*, 110–976. https://doi.org/10.1016/j.fct.2019

Cui, L., Li, D., & Liu, C. (2012). Effect of fermentation on the nutritive value of maize. *International Journal of Food Science and Technology, 47*(4), 755–760.

Dei, H. K. (2017). *Assessment of maize (Zea mays) as feed resource for poultry* (pp. 1–31). http://doi.org/10.5772/65363

Dhull, S. B., Kaur, P., & Purewal, S. S. (2016). Phytochemical analysis, phenolic compounds, condensed tannin content and antioxidant potential in Marwa (*Origanum majorana*) seed extracts. *Resource-Efficient Technologies, 2*(4), 168–174. https://doi.org/10.1016/j.reffit.2016.09.003

Dulf, F. V., Vodnar, D. C., Tosa, M. L., & Dulf, E. H. (2020). Simultaneous enrichment of grape pomace with γ-linolenic acid and carotenoids by solid-state fermentation with *Zygomycetes* fungi and antioxidant potential of the bioprocessed substrates. *Food Chemistry, 310*, 125–927. https://doi.org/10.1016/j.foodchem.2019

Eckert, C. T., Frigo, E. P., Albrecht, L. P., Albrecht, A. J. P., Christ, D., Santos, W. G., ... Egewarth, V. A. (2018). Maize ethanol production in Brazil: Characteristics and perspectives. *Renewable and Sustainable Energy Reviews, 82*, 3907–3912. https://doi.org/10.1016/j.rser.2017 .10.082

FAO (Food and Agricultural Organization of United Nations). (2018). Retrieved June 10, 2020, from http://www.fao.org/faost at/en/#data/QC

Fuentealba, C., Quesille-Villalobos, A. M., González-Muñoz, A., Saavedra Torrico, J., Shetty, K., & Gálvez Ranilla, L. (2016). Optimized methodology for the extraction of free and bound phenolic acids from Chilean Cristalino corn (*Zea mays* L.) accession. *CyTA – Journal of Food*, 1–8. https://doi.org/10.1080/19476337.2016.1217048

Gopalan, C., Rama Sastri, B. V., & Balasubramanian, S. (2007). *Nutritive value of Indian foods*. Hyderabad: National Institute of Nutrition (NIN), ICMR.

Green, D. I. G., Agu, R. C., Bringhurst, T. A., Brosnan, J. M., Jack, F. R., & Walker, G. M. (2015). Maximizing alcohol yields from wheat and maize and their co-products for distilling or bioethanol production. *Journal of the Institute of Brewing, 121*(3), 332–337. https://doi.org/10 .1002/jib.236

Haslina, H., & Eva, M. (2017). Extract corn silk with variation of solvents on yield, total phenolics, total flavonoids and antioxidant activity. *Indonesian Food and Nutrition Progress, 14*(1), 21–28.

Hernandez, M., Ventura, J., Castro, C., Boone, V., Rojas, R., Ascacio-Valdes, J., & Martinez-Avila, G. (2018). UPLC-ESI-QTOF-MS2-Based identification and antioxidant activity assessment of phenolic compounds from red corn cob (*Zea mays* L.). *Molecules, 23*(6), 1425. https://doi .org/10.3390/molecules23061425

Hole, A. S., Rud, I., Grimmer, S., Sigl, S., Narvhus, J., & Sahlstrøm, S. (2012). Improved bioavailability of dietary phenolic acids in whole grain barley and oat groat following fermentation with probiotic *Lactobacillus acidophilus, Lactobacillus johnsonii, and Lactobacillus reuteri. Journal of Agricultural and Food Chemistry, 60*(25), 6369–6375.

Imran. (2015). Effect of germination on proximate composition of two maize cultivars. *Journal of Biology, Agriculture and Healthcare, 5*, 123–128.

Irawaty, W., Ayucitra, A., & Indraswati, N. (2018). Radical scavenging activity of various extracts and varieties of corn silk. *ARPN Journal of Engineering and Applied Sciences, 13*, 10–16.

Kaur, M., Purewal, S. S., Sandhu, K. S., Kaur, M., & Salar, R. K. (2019). Millets: A cereal grain with potent antioxidants and health benefits. *Journal of Food Measurement and Characterization, 13*(1), 793–806. https://doi.org/10.1007/s11694-018-9992-0

Kaur, P., Dhull, S. B., Sandhu, K. S., Salar, R. K., & Purewal, S. S. (2018). Tulsi (*Ocimum tenuiflorum*) seeds: In vitro DNA damage protection, bioactive compounds and antioxidant potential. *Journal of Food Measurement and Characterization, 12*(3), 1530–1538. https://doi.org/10.1007/s11694-018-9768-6

Klempova, T., Slany, O., Sismis, M., Marcincak, S., & Certik, M. (2020). Dual production of polyunsaturated fatty acids and beta-carotene with *Mucor wosnessenskii* by the process of solid-state fermentation using agro-industrial waste. *Journal of Biotechnology, 311*, 1–11. https://doi.org/10.1016/j.jbiotec.2020.02.006

Knudsen, K. E. (1997). Carbohydrate and lignin contents of plant materials. *Animal Feed Science and Technology, 67*(4), 319–338.

Kuzmanović, M., Tisma, M., Bucić-Kojić, A., Casazza, A. A., Paini, M., Aliakbarian, B., & Perego, P. (2015). High–pressure and temperature extraction of phenolic compounds from corn silage. *Chemical Engineering Transactions, 43*, 133–138. https://doi.org/10.3303/cet1543023

Laeliocattleya, R. A. (2018). The potential of methanol and ethyl acetate extracts of corn silk (*Zea mays* L.) as Sunscreen. 1st International conference on Bioinformatics, *Biotechnology and Biomedical Engineering* (BioMIC-2018) https://doi.org/10.1063/1.5098417

Lingzhu, L., Lu, W., Dongyan, C., Jingbo, L., Songyi, L., Haiqing, Y., & Yuan, Y. (2015). Optimization of ultrasound-assisted extraction of polyphenols from maize filaments by response surface methodology and its identification. *Journal of Applied Botany and Food Quality, 88*, 152–163. https://doi.org/10.5073/jabfq.2015.088.022

Liu, X., Yan, Y., Zhao, P., Song, J., Yu, X., Wang, Z., ... Wang, X. (2019). Oil crop wastes as substrate candidates for enhancing erythritol production by modified *Yarrowia lipolytica* via one-step solid state fermentation. *Bioresource Technology, 294*, 122–194. https://doi.org/10.1016/j.biortech.2019.122194

Locatelli, S., & Berardo, N. (2014). Chemical composition and phytosterols profile of degermed maize products derived from wet and dry milling. Consiglio per la Ricerca e la sperimentazione in Agricoltura, Unità di Ricerca per la Maiscoltura (CRA-MAC), via Stezzano 24, 24126 Bergamo, Italy. Maydica, 59, 261–266

Lopez-Martinez, L. X., Oliart-Ros, R. M., Valerio-Alfaro, G., Lee, C., Parkin, K. L., & Garcia, H. S. (2009). Antioxidant activity, phenolic compounds and anthocyanins content of eighteen strains of Mexican maize. *LWT – Food Science and Technology, 42*(6), 1187–1192. https://doi.org/10.1016/j.lwt.2008.10.010

Mansor, A., Ramli, M. S., Abdul-Rashid, N. Y., Samat, N., Lani, M. N., Sharifudin, S. A., & Raseetha, S. (2019). Evaluation of selected agri-industrial residues as potential substrates for enhanced tannase production via solid-state fermentation. *Biocatalysis and Agricultural Biotechnology, 20*, 101–216. https://doi.org/10.1016/j.bcab.2019.101216

Martinez, B. F., Sevilla, P. E., & Bjarnason, M. (1996). Wet milling comparison of quality protein maize and normal maize. *Journal of the Science of Food and Agriculture, 71*(2), 156–162.

Ming-Zhu, G., Qi, C., Li-Tao, W., Yao, M., Lian, Y., Yan-Yan, L., & Yu-Jie, F. (2020). A green and integrated strategy for enhanced phenolic compounds extraction from mulberry (*Morus alba* L.) leaves by deep eutectic solvent. *Microchemical Journal, 154*, 104–598. https://doi.org/10.1016/j.microc.2020.104598

Molina, G. A., Gonzalez-Fuentes, F., Loske, A. M., Fernandez, F., & Estevez, M. (2020). Shock wave-assisted extraction of phenolic acids and flavonoids from Eysenhardtia polystachya heartwood: A novel method and its comparison with conventional methodologies. *Untrasonics Sonochemistry, 61*, 104–809. https://doi.org/10.1016/j.ultsonch.2019

Monroy, Y. M., Rodrigues, R. A. F., Sartoratto, A., & Cabral, F. A. (2016). Optimization of the extraction of phenolic compounds from purple corn cob (*Zea mays* L.) by sequential extraction using supercritical carbon dioxide, ethanol and water as solvents. *Journal of Supercritical Fluids, 116*, 10–19. https://doi.org/10.1016/j.supflu.2016.04.011

Moros, E. E., Darnoko, D., Cheryan, M., Perkins, E. G., & Jerrell, J. (2002). Analysis of Xanthophylls in corn by HPLC. *Journal of Agricultural and Food Chemistry, 50*(21), 5787–5790. http://doi.org/10.1021/jf020109

Muangrat, R., & Saengcharoenrat, P. (2018). Effect of processing conditions of hot pressurized solvent extraction in batch reactor on anthocyanins of purple field corn. *Agricultural Engineering International: CIGR Journal, 20*, 173–182.

Naik, B., Goyal, S. K., Tripathi, A. D., & Kumar, V. (2019). Screening of agro-industrial waste and physical factors for the optimum production of pullulanase in solid-state fermentation from endophytic *Aspergillus* sp. *Biocatalysis and Agricultural Biotechnology, 22*, 101–423. https://doi.org/10.1016/j.bcab.2019.101423

Nuss, E. T., & Tanumihardjo, S. A. (2010). Maize: A paramount staple crop in the context of global nutrition. *Comprehensive Reviews in Food Science and Food Safety, 9*(4), 417–436. https://doi.org/10.1111/j.1541-4337.2010.00117.x

Oboh, G., Ademiluyi, A. O., & Akindahunsi, A. A. (2010). The effect of roasting on the nutritional and antioxidant properties of yellow and white maize varieties. *International Journal of Food Science and Technology, 45*(6), 1236–1242.

Oluwalana, B. I. (2014). Comparative effects of sprouting on proximate, mineral composition and functional properties of white and yellow sweet maize *(Zea mays var saccharata)*. *Journal of Emerging Trends in Engineering and Applied Sciences, 5*, 111–115.

Pillay, K., Siwela, M., Derera, J., & Veldman, F. J. (2014). Provitamin A carotenoids in biofortified maize and their retention during processing and preparation of South African maize foods. *Journal of Food Science and Technology, 51*(4), 634–644. https://doi.org/10.1007/s13197-011-0559-x

Postemsky, P. D., Bidegain, M. A., Lluberas, G., Lopretti, M. I., Bonifacino, S., Landache, M. I., … Omarini, A. B. (2019). Biorefining via solid-state fermentation of rice and sunflower by-products employing novel monosporic strains from *Pleurotus sapidus*. *Bioresource Technology, 289*, 121–692. https://doi.org/10.1016/j.biortech.2019.121692

Postemsky, P. D., & Curvetto, N. R. (2015). Solid-state fermentation of cereal grains and sunflower seed hulls by *Grifola gargal* and *Grifola sordulenta*. *International Biodeterioration and Biodegradation, 100*, 52–61. https://doi.org/10.1016/j.ibiod.2015.02.016

Prasanthi, P. S., Naveena, N., Rao, M. V., & Bhaskarachary, K. (2017). Compositional variability of nutrients and phytochemicals in corn after processing. *Journal of Food Science and Technology, 54*(5), 1080–1090. https://doi.org/10.1007/s13197-017-2547-2

Punia, S., Sandhu, K. S., Dhull, S. G., Siroha, A. K., Purewal, S. S., Kaur, M., & Kidwai, M. K. (2020). Oat starch: Physico-chemical, morphological, rheological characteristics and its application-A review. *International Journal of Biological Macromolecule, 154*, 493–498. https://doi.org/10.1016/j.ijbiomac.2020.03.083

Purewal, S. S., Salar, R. K., Bhatti, M. S., Sandhu, K. S., Singh, S. K., & Kaur, P. (2020). Solid-state fermentation of pearl millet with *Aspergillus oryzae* and *Rhizopus azygosporus*: Effects on bioactive profile and DNA damage protection activity. *Journal of Food Measurement and Characterization, 14*(1), 150–162. https://doi.org/10.1007/s11694-019-00277-3

Purewal, S. S., Sandhu, K. S., Salar, R. K., & Kaur, P. (2019). Fermented pearl millet: A product with enhanced bioactive compounds and DNA damage protection activity. *Journal of Food Measurement and Characterization, 12*, 1530–1538.

Qamar, S., Aslam, M., & Javed, M. A. (2016). Determination of proximate chemical composition and detection of inorganic nutrients in maize *(Zea mays L.)*. *Materials Today: Proceedings, 3*(2), 715–718.

Ray, K., Banerjee, H., Dutta, S., Hazra, A. K., & Majumdar, K. (2019). Macronutrients influence yield and oil quality of hybrid maize *(Zea mays L.)*. *PLOS ONE, 14*(5), e0216939. https://doi.org/10.1371/journal.pone.0216939

Ripari, V., Bai, Y., & Ganzle, M. G. (2019). Metabolism of phenolic acids in whole wheat and rye malt sourdoughs. *Food Microbiology, 77*, 43–51.

Salar, R. K., Certik, M., & Brezova, V. (2012). Modulation of phenolic content and antioxidant activity of maize by solid state fermentation with *Thamnidium elegans* CCF 1456. *Biotechnology and Bioprocess Engineering, 17*(1), 109–116. https://doi.org/10.1007/s12257-011-0455-2

Salar, R. K., Certik, M., Brezova, V., Brlejova, M., Hanusova, V., & Breierová, E. (2013). Stress influenced increase in phenolic content and radical scavenging capacity of Rhodotorula glutinis CCY 20-2-26. *3 Biotech, 3*(1), 53–60. https://doi.org/10.1007/s13205-012-0069-1

Salar, R. K., & Purewal, S. S. (2016). Improvement of DNA damage protection and antioxidant activity of bio-transformed pearl millet (*Pennisetum glaucum*) cultivar PUSA-415 using *Aspergillus oryzae* MTCC 3107. *Biocatalysis and Agricultural Biotechnology, 8*, 221–227. https ://doi.org/10.1016/j.bcab.2016.10.005

Salar, R. K., & Purewal, S. S. (2017). Phenolic content, antioxidant potential and DNA damage protection of pearl millet (*Pennisetum glaucum*) cultivars of North Indian region. *Journal of Food Measurement and Characterization, 11*(1), 126–133. https://doi.org/10.1007/s11694-016-9379-z

Salar, R. K., Purewal, S. S., & Bhatti, M. S. (2016). Optimization of extraction condition and enhancement of phenolic content and antioxidant activity of pearl millet fermented with *Aspergillus awamori* MTCC-548. *Resource-Efficient Technology, 2*(3), 148–157. https://doi.org/10 .1016/j.reffit.2016.08.002

Salar, R. K., Purewal, S. S., & Sandhu, K. S. (2017a). Relationships between DNA damage protection activity, total phenolic content, condensed tannin content and antioxidant potential among Indian barley cultivars. *Biocatalysis and Agricultural Biotechnology, 11*, 201–206. https ://doi.org/10.1016/j.bcab.2017.07.006

Salar, R. K., Purewal, S. S., & Sandhu, K. S. (2017b). Fermented pearl millet (*Pennisetum glaucum*) with in vitro DNA damage protection activity, bioactive compounds and antioxidant potential. *Food Research International, 100*(2), 204–210. https://doi.org/10.1016/j.foodres.2017.08.045

Salar, R. K., Purewal, S. S., & Sandhu, K. S. (2017c). Bioactive profile, free-radical scavenging potential, DNA damage protection activity, and mycochemicals in *Aspergillus awamori* (MTCC 548) extracts: A novel report on filamentous fungi. *3 Biotech, 7*(3), 164. https://doi.org/10 .1007/s13205-017-0834-2

Salar, R. K., Sharma, P., & Purewal, S. S. (2015). In vitro antioxidant and free radical scavenging activities of stem extract of *Euphorbia trigona* Miller. *CELLMED, 5*(2), 1–6. https://doi.org/10.5667/tang.2015.0004

Salinas-Moreno, Y., Soto-Hernández, M., Martínez-Bustos, F., González-Hernández, V., & Ortega-Paczka, R. (1999). Análisis de antocianinas en maíces de grano azul y rojo provenientes de cuatro razas [Analysis of anthocyanins in four races from blue and Red grain maize]. *Revista Fitotecnia Mexicana, 22*, 161–174.

Sandhu, K. S., & Punia, S. (2017). Enhancement of bioactive compounds in Barley cultivars by solid substrate fermentation. *Journal of Food Measurement and Characterization, 11*(3), 1355–1361. https://doi.org/10.1007/s11694-017-9513-6

Sandhu, K. S., Punia, S., & Kaur, M. (2016). Effect of duration of solid state fermentation by *Aspergillus awamori nakazawa* on antioxidant properties of wheat cultivars. *LWT – Food Science and Technology, 71*, 323–328. https://doi.org/10.1016/j.lwt.2016.04.008

Sandhu, K. S., Sharma, L., Kaur, M., & Kaur, R. (2020). Physical, structural and thermal properties of composite edible films prepared from pearl millet starch and carrageenan gum: Process optimization using response surface methodology. *International Journal of Biological Macromolecule, 143*, 704–713. https://doi.org/10.1016/j.ijbiomac.2019.09.111

Sandhu, K. S., & Singh, N. (2007). Some properties of corn starches II: Physicochemical, gelatinization, retrogradation, pasting and gel textural properties. *Food Chemistry, 101*(4), 1499–1507. https://doi.org/10.1016/j.foodchem.2006.01.060

Sandhu, K. S., Singh, N., & Kaur, M. (2004). Characteristics of the different corn types and their grain fractions: Physicochemical, thermal, morphological, and rheological properties of starches. *Journal of Food Engineering, 64*(1), 119–127. https://doi.org/10.1016/j.jfoodeng.2003.09.023

Serna-Saldivar, S. O., & Rooney, L. W. (1994). Quality protein maize processing and perspectives for industrial utilization. In B. A. Larkins & E. T. Mertz (Eds.), *Quality protein maize 1964–1994: Proceedings of the international symposium on quality protein Maize; 1995* (pp. 89–120). Sete Lagoas: EMBRAPA/CNPMS.

Shah, T. R., Prasad, K., & Kumar, P. (2015). Studies on physicochemical and functional characteristics of asparagus bean flour and maize flour. In G. C. Mishra (Ed.), *Conceptual frame work & innovations in agroecology and food sciences* (1st ed., pp. 103–105). New Delhi: Krishi Sanskriti Publications.

Singh, N., Kaur, A., & Shevkani, K. (2013). Maize: Grain structure, composition, milling, and starch characteristics. In *Maize: Nutrition dynamics and novel uses* (pp. 65–76). https://doi.org/10.1007/978-81-322-1623-0_5

Singh, S., Kaur, M., Sogi, D. S., & Purewal, S. S. (2019). A comparative study of phytochemicals, antioxidant potential and in-vitro DNA damage protection activity of different oat (*Avena sativa*) cultivars from India. *Journal of Food Measurement and Characterization, 13*(1), 347–356. https://doi.org/10.1007/s11694-018-9950-x

Siroha, A. K., Sandhu, K. S., & Kaur, M. (2016). Physicochemical, functional and antioxidant properties of flour from pearl millet varieties grown in India. *Journal of Food Measurement and Characterization, 10*(2), 311–318. https://doi.org/10.1007/s11694-016-9308-1

Siroha, A. K., Sandhu, K. S., Kaur, M., & Kaur, V. (2019). Physicochemical, rheological, morphological and in vitro digestibility properties of pearl millet starch modified at varying levels of acetylation. *International Journal of Biological Macromolecule, 131*, 1077–1083. https://doi.org/10.1016/j.ijbiomac.2019.03.179

Starzynska-Janiszewska, A., Stodolak, B., Socha, R., Mickowska, B., & Wywrocka-Gurgul, A. (2019). Spelt wheat tempe as a value-added whole-grain food product. *LWT – Food Science and Technology, 113*, 108–250.

Tabasum, S., Younas, M., Zaeem, M. A., Majeed, I., Majeed, M., Noreen, A., … Zia, K. M. (2019). A review on blending of corn starch with natural and synthetic polymers, and inorganic nanoparticles with mathematical modeling. *International Journal of Biological Macromolecules, 122*, 969–996. https://doi.org/10.1016/j.ijbiomac.2018.10.092

Watson, S. A., & Ramstad, P. R. (1987). *Corn chemistry and technology.* St Paul, MN: American Association of Cereal Chemists. p. 605.

Whelan, K., Abrahmsohn, O., David, G. J. P., Staudacher, H., Irving, P., Lomer, M. C. E., & Ellis, P. R. (2011). Fructan content of commonly consumed wheat, rye and gluten-free breads. *International Journal of Food Sciences and Nutrition, 62*(5), 498–503. https://doi.org/10.3109/09637486.2011.553588

Xu, L., Guo, S., & Zhang, S. (2018). Effects of solid-state fermentation with three higher fungi on the total phenol contents and antioxidant properties of diverse cereal grains. *FEMS Microbiology Letters, 365*(16), 1–8. https://doi.org/10.1093/femsle/fny163

ZavalaLópez, M., & Garcia-Lara, S. (2017). An improved microscale method for extraction of phenolic acids from maize. *Plant Methods, 13*, 81. https://doi.org/10.1186/s13007-017-0235-x

Zhao, Z., Egashira, Y., & Sanada, H. (2005). Phenolic antioxidants richly contained in corn bran are slightly bioavailable in rats. *Journal of Agricultural and Food Chemistry, 53*(12), 5030–5035. http://doi.org/10.1021/jf050111n

Zilic, S., Milasinovic, M., Terzic, D., Barac, M., & Ignjatovic-Micic, D. (2011). Grain characteristics and composition of maize specialty hybrids. *Spanish Journal of Agricultural Research, 9*(1), 230–241. ISSN: 1695-971-X.

Zilic, S., Mogol, B. A., Akıllıoglu, G., Serpen, A., Babic, M., & Gokmen, V. (2013). Effects of infrared heating on phenolic compounds and Maillard reaction products in maize flour. *Journal of Cereal Science, 58*(1), 1–7.

Žilić, S., Serpen, A., Akıllıoğlu, G., Gökmen, V., & Vančetović, J. (2012). Phenolic compounds, carotenoids, anthocyanins, and antioxidant capacity of colored maize (*Zea mays* L.) kernels. *Journal of Agricultural and Food Chemistry, 60*(5), 1224–1231. https://doi.org/10.1021/jf204367z

Chapter 4

Maize starch

Granules and technological properties and applications trends

Vania Zanella Pinto and Ricardo Tadeu Paraginski

DOI: 10.1201/9781003245230-4

4.1 Introduction

Starch is widely distributed in vegetables as a storage carbohydrate. The cereals are the principal starch source, with a content of around 40%–90% of dry weight (dw). Among cereals, maize starch content is 60–78% (dw) (García-Lara et al., 2018), and is of outstanding economic and technological importance. The maize grains are used in feed and food, consumed *in natura* or from derivatives production, highlighting starch and starch products. There is great genetic variability in maize grains, with multicoloured kernels – white, yellow, red, purple, blue, and black – and related nutritional composition (Žilić et al., 2012). Starch concentration from endosperm in the maize kernel is around 98% of total starch, stored as insoluble granules with 20 μm of average diameter (El Halal et al., 2019; Ranum et al., 2014).

Maize is the primary starch source, providing over 85% of the starch produced worldwide, around 1.2 billion tons (FAO, 2018). It is responsible for thickening, as a colloid stabilizer, as a gelling and bulking agent, affecting stickiness and water (Sujka et al., 2018; S. Wang et al., 2015; Zámostný et al., 2012) in food, drugs, and cosmetics. Also, around 70% of maize starch is converted to syrup, which is often used in food and beverages (Ranum et al., 2014). Normal or regular maize native starch is characterized by the formation of a consistent gel, widely used in dehydrated soups and sauces that require hot viscosity. On the other hand, its high retrogradation and syneresis (Jacobson et al., 1997; Tian et al., 2011; Yuan et al., 1993) are not appropriate for refrigerated stored products. Waxy maize starches are suitable for refrigerated products due to their amylose-free standard (Zhou et al., 2010). This starch has good stability at low temperatures, is weak, highly viscous in cooking, and has clear and cohesive gels. This chapter provides details of the main characteristics and granules' morphology, its isolation, processing, food and non-food uses, and principal trends applications.

4.2 Maize starch

4.2.1 Main characteristics and starch granules morphology

Carbohydrates are typically composed of carbon, hydrogen, and oxygen atoms. They are produced by vegetable cells via photokinesis through carbon dioxide (CO_2) and water conversion into glycide molecules using light energy. Starch is polymerized into plastids from the leaves, chloroplasts, and amyloplasts from storage organs (Taiz et al., 2017). The carbohydrate on maize kernels is synthesized at the differentiation stage from the development and flowering stages and in the filling of maize grains (Bahaji et al., 2014).

Maize kernels are composed of endosperm (82%–84%, dw), germ (10%–12%, dw), pericarp (5%–6%, dw), and tip cap (1%) (R. Zhang et al., 2021). However, due to the great diversity between cultivars, structures vary, with different nutrient compositions and component sizes (Figure 4.1). The chemical composition is related to genetic breeding, soil fertilization, climatic conditions, luminosity, kernel maturation, storage conditions, and defects incidences (Paraginski et al., 2019; Ranum et al., 2014; R. Zhang et al., 2021). Post-harvesting processing can also influence starch chemical composition and properties (Paraginski et al., 2014).

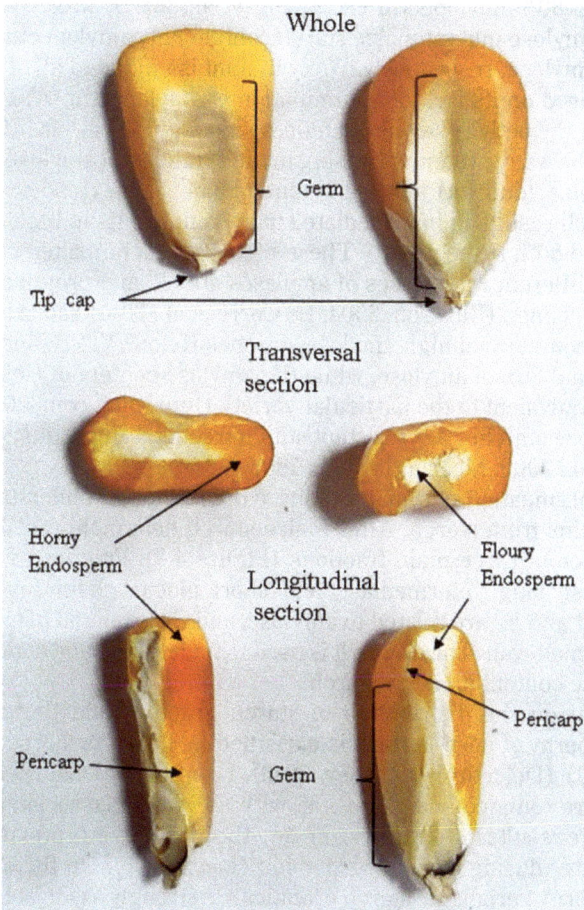

Figure 4.1 Whole, transversal, and longitudinal sections of the mature maize kernel (author's unpublished data)

Maize endosperm (floury and horny) has around 98% of total starch from the kernel (Figure 4.1). Maize endosperm mutants have been reported to show varying carbohydrate compositions (X. Yu et al., 2015). Starch is stored in the amyloplasts as insoluble granules, which comprise amylose and amylopectin macromolecules. Amylose is a linear polymer linked by α-1,4 glucose units, whereas amylopectin is a α-1,4 glucose chain with frequent branches due to α-1-6 bonds. The different ratios of amylose and amylopectin are a consequence of the species' genetic diversity according to the degree of maturation of each plant (Eliasson, 2004; Hamaker et al., 2018; Tester et al., 2004; X. Yu et al., 2015).

By weight, maize starch has around 20–30% amylose and approximately 75%–80% amylopectin (R. Zhang et al., 2021); waxy (wx) mutant and high-amylose mutant maize starch contain only amylopectin and up to 50%–90% amylose, respectively. Waxy mutant is unique to all other known mutants based on its lack of accumulation of amylopectin. Waxy starch is deficient in the activity of one granule-bound starch synthase (GBSS) isoform, and the waxy protein is missing in the synthesis of amylose (Eliasson, 2004). Double (dull:wx) and triple mutant (amylose-extender:dull:wx) of maize starch results in intermediate chain components in high concentration (40 and 80%, respectively). These materials contain altered chain distribution, different resistances of amylases attack, and prominent pasting properties change (Eliasson, 2004; Le Corre et al., 2011; Yao et al., 2004).

The commercial high-amylose starches Hylon® V, VII, and VIII have 50%, 70%, and 80% of amylose, while the amylose content of Gelose 50, 70, and 80 is equivalent to the particular variety (Ingredion.com, 2021). These high-amylose starches are resistant starch fractions (Ingredion.com, 2021; Le Leu et al., 2009; McNaught et al., 1998).

Gel-permeation chromatography (GPC) helps in understanding the glucan chains from starch. After maize starch debranching, GPC reveals the presence of three main fractions (Figure 4.2). Fractions F1, F2, and F3 comprise long, intermediate, and short glucan chains, respectively. Fraction F1 and F2 are related to amylose, and F3 is related to amylopectin chains (Vamadevan et al., 2014). It is used to understand and confirm different amylose content in maize starch.

Based on the distribution of starch granules and protein matrix, the endosperm of corn kernels is classified as floury or horny (vitreous) (Figure 4.1) (Delcour & Hoseney, 2010). From floury endosperm, starch granules are rounded and dispersed, with no granule-associated proteins. This feature results in void spaces during the grain drying, previously occupied by water during grain development (Paes, 2006). On the other hand, from the horny kernel, the starch granules are strongly associated with proteins, which results in polyhedral granules by non-voiding spaces between the structures (Svihus et al., 2005) (Figure 4.3). The size, shape, and structure of corn starch granules vary according to their genetic origin. Starch

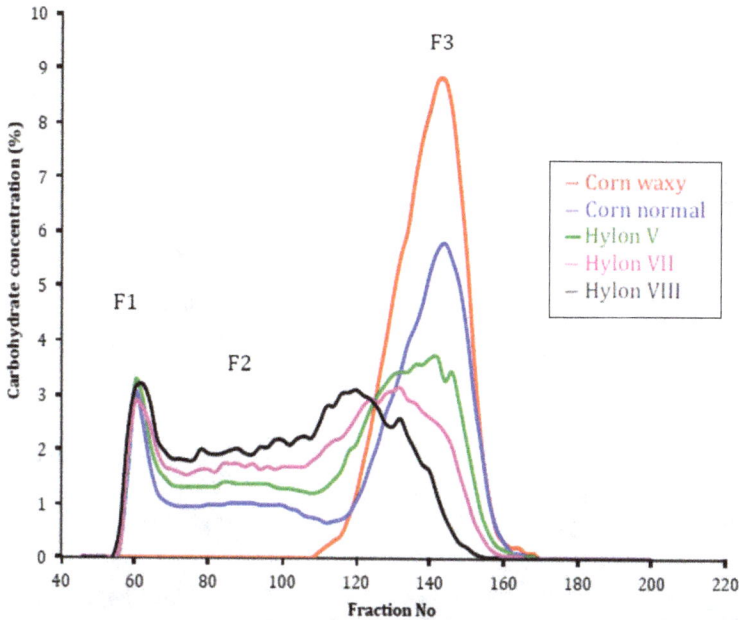

Figure 4.2 Fractionations of debranched starches by size exclusion chromatography (Sepharose CL 6B). The division into long chains (F1), intermediate chains (F2), and short chains (F3) are indicated. Source: Vamadevan et al. (2014)

granules are usually found in sizes between 1 and 100 μm, with regular or irregular spherical or polyhedral shapes, with an average diameter of 7–25 μm (El Halal et al., 2019; Pan & Jane, 2000; Vamadevan et al., 2014). There were changes in starch granule size and amylose percentage during kernel development in all *Zea mays* L. genotypes. Normal and waxy kernels produce large granules, although ae-Ref, ae-il, and ae wx have medium to small granules (Boyer et al., 1976).

Regular maize starch has polyhedral and mostly spherical or ovoid granules that range from 5 to 16 μm in diameter. Waxy maize has irregular shape granules, ranging from 4 to 17 μm in diameter. When the kernel is cut with a knife, the waxy lack of amylopectin makes the cut surface appear shiny and waxy (Eliasson, 2004). Also, pores have been observed in high-resolution images on the regular and waxy maize granule surface and central cavities in normal and waxy maize starch granules (Figure 4.3) (Huber & BeMiller, 1997). Pores develop during a late maturation stage, i.e. 30 days after pollination (L. Li et al., 2007). However, it is still unclear how pores and channels are formed only in specific starch granules.

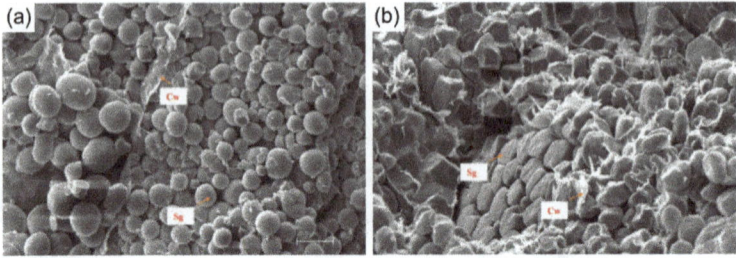

Figure 4.3 Scanning electron micrograph of yellow dent corn endosperm, showing round (A), thigh-packed starch granules (B), cell walls (Cw), and starch granules. Source: García-Lara et al. (2018)

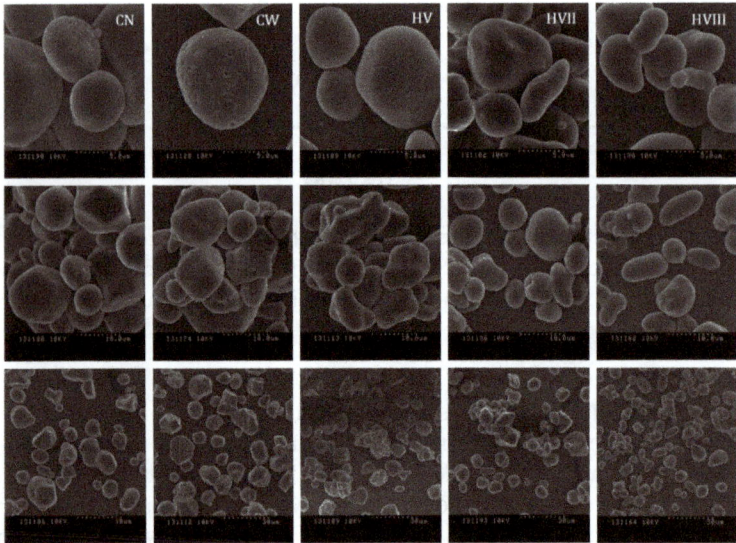

Figure 4.4 Scanning electron microscopy (SEM) of native regular/normal (CN), waxy (CW), Hylon V (HV), Hylon VII (HVII), and Hylon VIII (HVIII) corn starches at different magnifications. Source: Vamadevan et al. (2014)

High-amylose starch has irregular granules, some being elongated, others with large protrusions on their surfaces, or even regular (Figures 4.4 and 4.5) (Imam et al., 2012; S. Pérez & Bertoft, 2010; Zeeman et al., 2010). Apparent amylose content may increase with granule size (Dhital et al., 2011).

A birefringence pattern in the central hilum in the maize starch granules, under cross-polarized light microscopy, is interfered by a Maltese cross

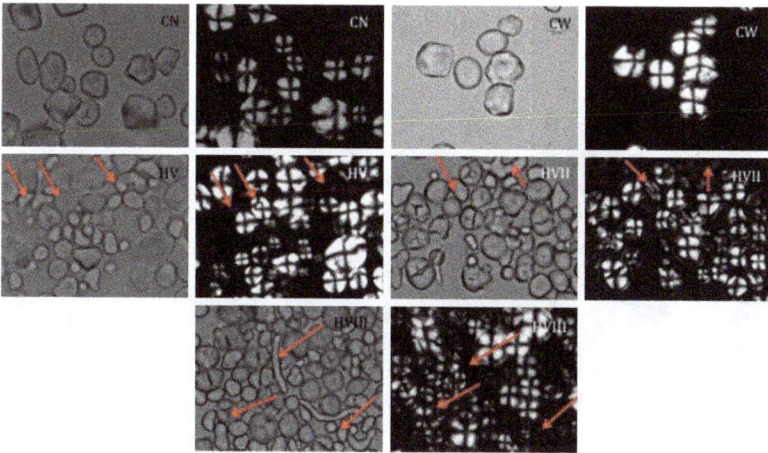

Figure 4.5 Light microscopy of native regular/normal, waxy, Hylon V, Hylon VII, and Hylon VIII corn starches under bright field (left frame) and cross-polarized light (right frame). The arrows depict the different patterns of birefringence. Source: Vamadevan et al. (2014)

phenomenon (Figure 4.5). The radial disposition of amylopectin within the granules is thought to be responsible for the formation of optical polarization. It results in visible optical polarization in the order of the visible light wavelength (100–1000 nm) (S. Pérez & Bertoft, 2010). The gelatinization heating temperature causes an irreversible transition in the crystal structure and consequently loss of granules birefringence (Jayakody & Hoover, 2008).

Some changes may occur in the starch granules' morphology during the storage of maize kernels (Figure 4.6c and d). Kernel storage at 5 °C results in pits on the surface of the starch granules. At this temperature, the enzymes remain while the kernels are in storage and are activated at steeping for starch isolation (50 °C), promoting the structure change. However, at 35 °C there is some residual protein after starch isolation (Figure 4.6i and j) (Paraginski et al., 2014). The temperature of storage is critical for improving the quality of kernel and maize derivates.

Starch granules change when maize kernels possess defects, from incomplete kernels due to adverse effects of climate or even from storage damage (Kim & Kim, 2021; W. Zhang et al., 2021; Ziegler et al., 2021). Elevated temperatures reduce starch deposition in endosperm by reducing the activity of soluble starch synthase (Keeling et al., 1993). Thus, the control of cleaning and drying processes and storage conditions over time allows the maintenance of the qualities of raw materials and their derivatives, ensuring their industrial use in addition to ensuring the seasonal regulation of production.

Figure 4.6 Scanning electronic microscopy (SEM) of maize starch before kernels storage (a; b); and after storage for 12 months at 5 °C (c; d), 15 °C (e; f), 25 °C (g; h), and 35 °C (i; j). Source: Paraginski et al. (2014)

4.2.2 Granule structure

Starch granules are organized in concentric alternating semicrystalline and amorphous layers (Figure 4.7a) with a typical thickness of 120–400 nm (Le Corre et al., 2010; S. Pérez & Bertoft, 2010). This semicrystalline structure is constituted as growing rings composed of amylopectin double helixes as a unit that forms small blocklets (20–50 nm) (Le Corre et al., 2010). The effects on α, (1–6) bonds are aggregated in the amorphous lamella within the semicrystalline rings. All A-chains are external chains (12 glucose units long), being wholly situated radially outside the outermost branches. These external chains are arranged as double helixes in the crystalline lamella, with 4–6 nm (Bertoft et al., 2010). The X-ray diffraction reveals the periodical 9–10 nm amorphous and crystalline lamellas, resulting in starch allomorph patterns (A, B e C) (S. Pérez & Bertoft, 2010).

A-type allomorph has only eight water molecules per monoclinic crystal unit with tightly packed helixes (Figure 4.7b). In contrast, the B-type allomorph has a more open packing of helixes and contains 36 water molecules in each hexagonal space group of six double helix crystal units (Imberty et al., 1987; Imberty & Perez, 1988). The C-type allomorph is considered a mixed type of crystals, with B-type in the centre and A-type at the periphery of the granule (Cai et al., 2014).

The polymorphic patterns revealed by the X-ray diffractogram pattern for A-type starches include sharp diffraction peaks at 15° and 23° and a double peak at 17° and 18° (2θ), whereas B-type starches include single

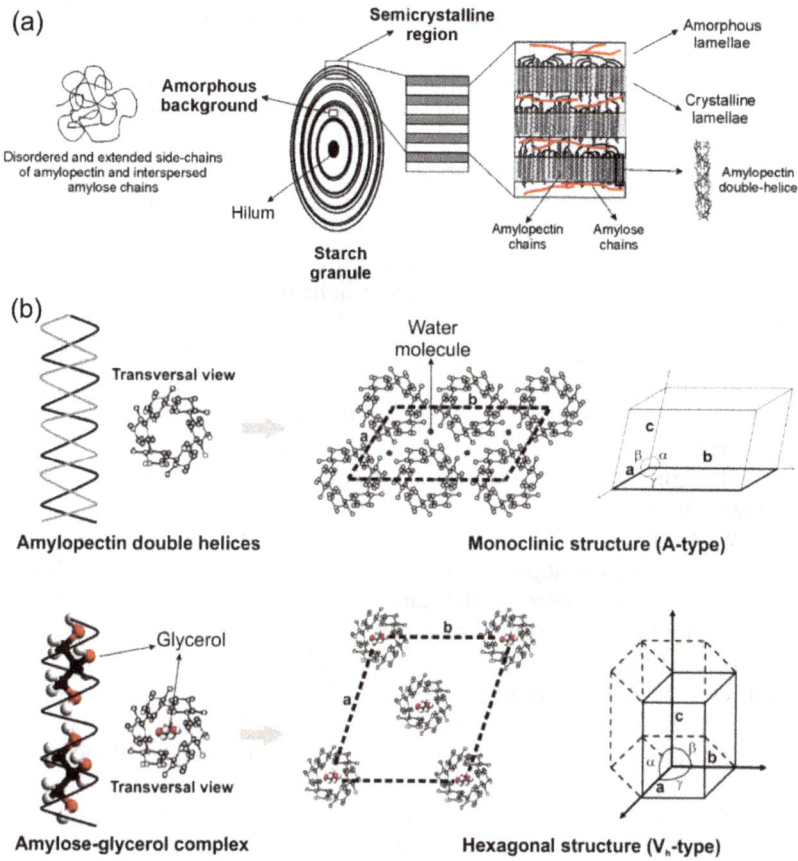

Figure 4.7 a) Architecture of corn starch granules and b) A- and V_h-type crystalline structures. Source: Castillo et al. (2019)

peaks around 5.6°, 17°, 22°, and 26°. Regular and waxy maize starches exhibit A-type crystalline patterns (Vamadevan et al., 2014). Such structure is typical of corn starch due to its cropping in a relatively dry environment (Qiao et al., 2017). High-amylose starches exhibit B-type crystalline patterns (Vamadevan et al., 2014). In Hylon® starches, an intense peak at 20° 2θ is attributed to the highly ordered crystalline structure of amylose–lipid complexes (Morrison, 1988), reflecting its V-type crystallinity.

4.2.3 Pasting and rheological properties

Intact starch granules are not soluble in cold water; however, they may retain small amounts of water, causing a slight swelling reversibly by drying (Vamadevan & Bertoft, 2015). The granules' insolubility is due to the strong hydrogen bonds which keep the starch chains closed (Vamadevan & Bertoft, 2020). At high concentrations, starch becomes a non-Newtonian dilatant fluid, where the flow is not proportional to the pressure exerted (Steffe, 1996). The shear thickening effect of maize starch solutions is expressed by a fast thickening, and therefore stabilization, of the liquid when exposed to pressure (Schneider & Gärtner, 2013).

When the starch is heated to above a gelatinization temperature, there is a thermal disordering of the crystal structures and the birefringence lost (Cai et al., 2014; Castanha et al., 2021; Rincón-Londoño et al., 2016). The starch gelatinization exhibits two-stage behaviour. First is limited swelling and low solubilization, located around the gelatinization temperature of 60–75 °C. The second step occurs above 90 °C, where the granules swell and disrupt, leading to more or less complete solubilization (30–60%) (Bertolini, 2010). The granules' rupture occurs due to the relaxing of the hydrogen bonds, and the water molecules interact with amylose and amylopectin hydroxyls, accompanied by a partially swelling and increasing granule size (Hoover, 2001).

When the temperature is cooled down after the starch gelatinization, the amylose and amylopectin start reassociating through new hydrogen bonding. Amylose reassociates rapidly during cooling, while amylopectin slowly recrystallizes during storage, favouring an ordered structure formation (Mali et al., 2010; Rincón-Londoño et al., 2016; V. Singh et al., 1995; Soest et al., 1996; Tester et al., 2004; S. Wang et al., 2015). The pasting properties are often studied by the pasting profile of a starch-water suspension as a function of the time and temperature to grasp the concept and behaviour of gelatinization and retrogradation.

Rapid Visco Analyser (RVA) is one of the most popular rheological analyzers. Starch's slurry concentration for RVA usually is fixed at ~10% (w/w), corrected to 14% moisture on a wet basis (Balet et al., 2019). Two standard temperature profiles are used: standard 1: holding the temperature at 50 °C for 1 min, heating to 95 °C in 3.5 min, and then holding at 95 °C for 2.5 min, followed by cooling to 50 °C in 4 min and then holding at 50 °C for 1 min; standard 2: 50 °C holding for 1 min, heating to 95 °C for 7.3 min, holding this temperature for 5 min, cooling to 50 °C in 7.4 min, and holding it for 2 min. The rotating speed was set to 960 rpm for 10 s and then maintained at 160 rpm in both profiles. The main drawback of the empirical viscometer gelatinization profile study is that it cannot be used with limited or insufficient water content (Schirmer et al., 2015).

The RVA starch has a typical viscosity profile, as described in Figure 4.8. The viscosity peak, holding viscosity (minimum viscosity), breakdown, final viscosity, and setback and pasting temperature are extracted from RVA curves. Starches show a gradual increase in viscosity with temperature increases (Schirmer et al., 2015). The increase in viscosity with temperature may be attributed to removing water from the

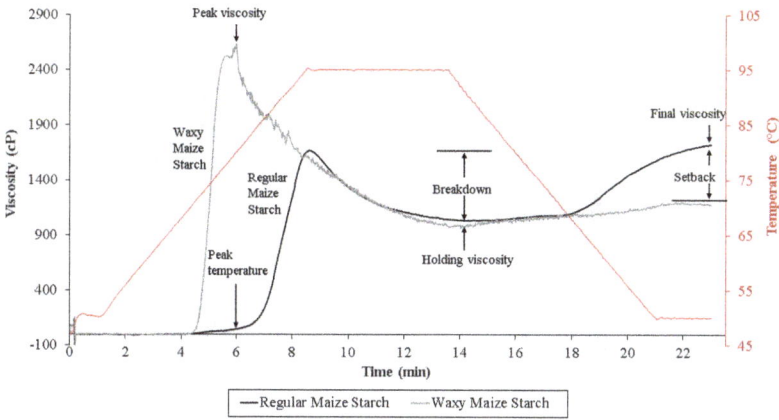

Figure 4.8 Typical pasting profile of regular maize and waxy maize starch conventional definitions used in the rapid viscosity analysis (RVA). Starch suspension of 10% (correct to 14% of moisture), temperature profile at Standard 2 (authors' unpublished data)

exuded amylose by the granules as they swell (Ghiasi et al., 1982). The parameters measured by the RVA are similar to the Brabender Amylograph (Limpisut & Jindal, 2002; Suh & Jane, 2003). Table 4.1 provides a comparison of regular and waxy maize starch from the two amylograph machines.

The starch concentration and the amylose and amylopectin ratios are what is mainly responsible for pasting and retrogradation properties. The rigidity of the starch granules, which in turn affects the granule swelling and the amount of amylose leaching out in the solution, also affects these properties (Castanha et al., 2021; Hossen et al., 2011; Mua & Jackson, 1998; Sandhu & Singh, 2007; Takeiti et al., 2007). Maize starch is widely studied, and the amylose to amylopectin ratio shows range pasting behaviour (Balet et al., 2019; Castanha et al., 2021; Rincón-Londoño et al., 2016; Sandhu & Singh, 2007). Increased solids cause higher peak viscosities as more starch granules are available for swelling to increase viscosity (Balet et al., 2019).

The typical viscosity profile of regular maize starch is described in Figure 4.8. It is characterized as a consistent gel with high retrogradation. In general, stronger starch gels are associated with a higher amylose content (Ishiguro et al., 2000). Amylose-based networks provide starch gels with elasticity and strength against deformation (M. C. Tang & Copeland, 2007). In contrast, soft gels containing aggregates in the absence of networks display easy penetrability and good stickiness and adhesiveness (S. Wang et al., 2015). The chain length and amylopectin organization of internal unit chains also influence the gelatinization and retrogradation

111

Table 4.1 Pasting properties of the regular/normal and waxy corn starches from Brabender Amylograph and rapid viscosity analyzer (RVA) at temperature range 50 °C–95 °C

Brabender Amylograph	MV 95 °C (mPa s)	FV 50 °C (mPa s)	V_m 95 °C (mPa s)	Breakdown (mPa s)	Retrogradation (mPa s)	Tp (°C)
Regular starch*	2,937	3,153	1,934	1,002	1,218	75.02
Waxy starch*	3,950	2,056	1,667	2,282	389	71.18
RVA	Peak viscosity (mPa s)	Holding viscosity (mPa s)	Breakdown (mPa s)	Final viscosity (mPa s)	Setback (mPa s)	Tp (°C)
Regular starch**	1,674	1,034	640	1,727	693	84.05
Waxy starch**	2,633	963	1,670	1,178	215	70.65

Source: *Weber et al. (2009); ** authors' unpublished data.
MV 95 °C = maximum viscosity at 95 °C; V_m 95 °C = minimal viscosity at 95 °C; FV 50 °C = final viscosity at 50 °C; V_m 95 °C = minimum viscosity at 95 °C; Tp = pasting temperature.

of the starches, which mainly contribute to the viscosifying properties (Vamadevan & Bertoft, 2020).

The maize waxy starch pastes form a soft gel on retrogradation, containing aggregates but no network (M. C. Tang & Copeland, 2007). Improving waxy maize quality is one of the primary aims in breeding programmes, and pasting viscosity traits are used as quality criteria (Balet et al., 2019). The high pasting temperature indicates the high resistance towards swelling starch (Sandhu & Singh, 2007).

High-amylose starch does not gelatinize under RVA standard temperature profiles (50 °C–95 °C). Hylon VII under atmospheric cooking temperatures required heating above 154 °C for gelatinization (Sjoo & Nilsson, 2004). It is associated with the long chains of amylopectin association, resulting in many crystallite formations. Crystallite formation increases granular stability, reducing the granule swelling and increasing the gelatinization temperature (Miao et al., 2010). The large amylopectin chains are not easy to leach. They may participate in double helixes with adjacent amylopectin or may be entangled within the intricate architecture of the starch granule (Vamadevan & Bertoft, 2015).

The molecular interactions (hydrogen bonding between starch chains) that occur after cooling the gelatinized starch pasting are known as retrogradation (Hoover, 2001). The degree of starch retrogradation and starch crystallites formed are also influenced by refrigeration, freeze-thaw cycling (Takeiti et al., 2007), the storage time and temperature, starch concentration, botanical origin of the starch, the molecular ratio of amylose to amylopectin, and structures of amylose and amylopectin molecules (Zhou et al., 2010). Retrogradation (%) of maize starches from different varieties ranged from 40% to 60% (Sandhu & Singh, 2007). The retrogradation transition temperatures and ΔH are lower than the gelatinization temperatures of their native counterparts (Sandhu & Singh, 2007). Retrogradation is often enhanced when starch gels are subjected to freezing and thawing treatments. The formation of ice crystals and the freeze-thaw cycles lead to the formation of larger ice crystals embedded in a sponge-like network (Yuan & Thompson, 1998). Upon thawing, the water can be easily expressed from the network, giving rise to syneresis (Jacobson et al., 1997; Tian et al., 2011). Starches that exhibit hard gels tend to have higher amylose content and longer amylopectin chains (Mua & Jackson, 1997). Maize waxy starches showed detectable syneresis after the second (Takeiti et al., 2007) and third freeze-thaw cycles (Yuan & Thompson, 1998), while regular maize starch showed better results in the first cycle (Takeiti et al., 2007).

Several environmental and storage factors may affect starch quality. Some defects from maize kernels have been studied: (1) broken kernels; (2) fermented kernels; (3) rotten kernels; (4) mouldy kernels; (5) germinated kernels; (6) insect-damaged kernels; and (7) shrunken and immature kernels (Paraginski et al., 2019). Also, based on an increase in drying temperatures, the residual protein content in maize starch contributes to the pasting temperature for yellow and white floury maize starch. Temperatures below 50 °C are suggested for the drying of maize kernels to avoid starch characteristics changing (Timm et al., 2020).

The defect kernels have some effect on isolated starch. Inappropriate storage of kernels promotes endosperm compound complexation, changing the starch viscosity properties (Figure 4.9). Breakdown viscosity and setback decrease when the maize kernel is germinated (Paraginski et al., 2019) due to the activity of the amylolytic enzymes. High breakdown and setback are related to the rapid aggregation of leached amylose chains and high starch granules swelling power, respectively (Hughes et al., 2009). The starch setback also reduces when isolated from shrunken and immature maize kernels (Figure 4.9). Cropping temperatures higher than 25°C adversely affect the activity of soluble starch synthase (SSS) and amyloplastic enzyme in the endosperm of wheat (*Triticum aestivum* L. cv. Mardler). This effect results from photosynthesis, assimilate partitioning, pollination, grain set, endosperm cell division, or from the grain-filling process itself (Keeling et al., 1993) and may change the starch pasting profile from optimum cropping conditions.

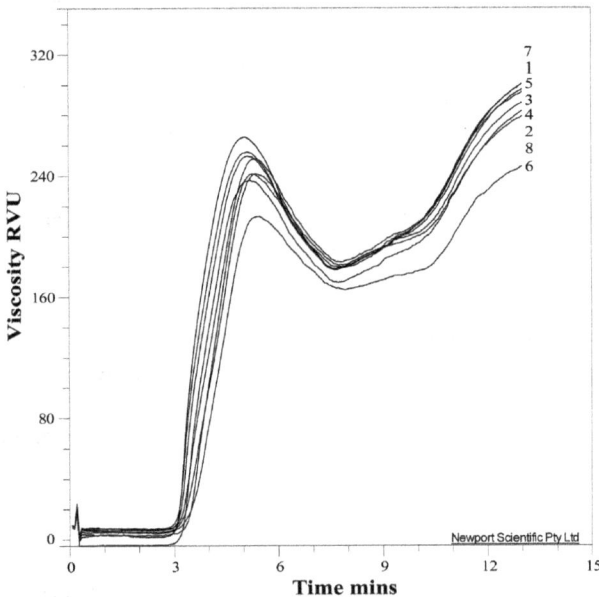

Figure 4.9 Pasting profile of maize starch isolated from defected kernels: (1) non-defect maize kernel; (2) broken kernel; (3) fermented; (3) rotten kernel; (4) mouldy kernel; (5) germinated kernel; (6) insect-damaged kernel; (7) shrunken and immature kernels. Source: Paraginski et al. (2019)

The starch and its gels also are examined by thermal analysis techniques, using differential scanning calorimetry (DSC). This is used to determine the temperatures and the enthalpy of the gelatinization process. Changes on a microscopic scale co-occur with nanoscale changes during gelatinization, and crystallinity losses can be detected as a DSC endothermic event (Bertolini, 2010). The measurements are the gelatinization transition temperatures (onset [To], melting or peak [Tm/Tp], and conclusion [Tc]), temperature range (Tc–To or ΔH), and the enthalpy (ΔH) of gelatinization. The typical DSC thermogram of native maize starch gelatinization shows a single endothermic peak (Figure 4.10).

The endothermic event primarily reflects the melting of starch helical and semicrystalline structures within starch granules (Haiteng Li et al., 2020). Tc is generally used to describe the crystalline perfection, while ΔT indicates the crystallites thermal stability extent (Adebowale et al., 2005; Klein et al., 2013; Vamadevan et al., 2013), and ΔH represents the rupture of the H-bonds between glucan strands and the loss of double-helical order (Cooke & Gidley, 1992).

The length of the external chains of amylopectin has an enormous impact on melting enthalpy; the internal unit-chain structure of amylopectin has the most influence on gelatinization transition temperatures of starch granules (Vamadevan et al., 2013). Based on it, genetic variability can be seen in the physical properties of starch, quickly detected by DSC

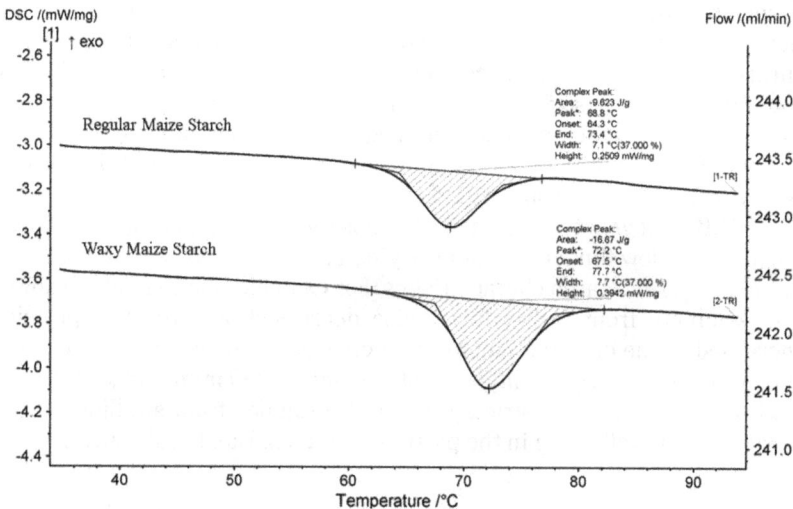

Figure 4.10 Typical DSC thermogram of regular maize starch and waxy maize starch in water (20%, by weight) with heating from 30 °C to 110 °C at 10 °C/min, using a Netzch DSC (authors' unpublished data)

analysis. This can help to create programmes to screen germplasm for desired properties of starch (Biliaderis, 2009).

4.2.4 Swelling power and solubility

The swelling power indicates the hydration ability of the granules under heating. Solubility is expressed as the percentage (by weight) of the starch leached after heating (Leach et al., 1959). The more soluble components, such as amylose, dissociate and leach out of the granule during swelling. The amylose leaching is a transitional phase of order and disorder within the starch granule and occurs when the starch is heated with water (Tester & Morrison, 1990). The swelling power is associated with the granule structure and chemical composition, the amylose and lipid content, and not much by the granule size (Lindeboom et al., 2004; Zheng et al., 1998). The leach amylose concentration is affected by starch concentration and heating temperature (Lii et al., 1995) and is restricted by the amylose-inclusion complex of fat acids (Putseys et al., 2010; Zheng et al., 1998).

Waxy starch has a high swelling power due to its open structure and low amylose content (J. Li & Yeh, 2001). At the same time, high-amylose starches have restricted swelling and solubility, even after a prolonged period of heating. The hydrogen bonding between the amylose chains results in internal associations that structurally hinder water uptake. This behaviour can be attributed to the presence of many crystallites from the association between long chains of amylopectin (J. Li & Yeh, 2001; Miao et al., 2010). The swelling pattern of the granules is also related to the structural type of the amylopectin component (Vamadevan & Bertoft, 2020). The swelling power is associated with the ratio of the relative molar distribution of amylopectin branch-chains with a degree of polymerization (DP) of 6–12 to that of chains with DP 6–24 (amylopectin unit-chain ratio), increasing with the ratio (Srichuwong et al., 2005).

Different genotypes of maize kernels (yellow floury corn, white floury corn, and yellow flint corn) under drying at 30 °C, 50 °C, 70 °C, and 90 °C creates different starch characteristics. For example, starch swelling power and solubility from yellow flint maize decreased, and residual proteins increased as the drying temperature increased; however, yellow and white floury maize starch solubility was not affected by it (Timm et al., 2020). The high protein residual restricts the starch granules from swelling during gelatinization, reflecting in the pasting profile (El Halal et al., 2019).

4.2.5 Resistant starch

According to the starch *in vitro* digestion behaviour, there are three different classifications: rapidly digestible starch (RDS), slowly digestible starch

(SDS), and resistant starch (RS). The RDS is the fraction hydrolyzed within 20 min of incubation, SDS is digested during 20 and 120 min, and RS is not hydrolyzed within 120 min (Englyst et al., 1992). RS is not digestible by amylase enzymes and escapes digestion in the small intestine, serving as a fermentation substrate for beneficial colonic bacteria (Lopez-Rubio et al., 2008). It modulates the absorption of carbohydrates, determining a lowering of the glycemic index (GI) and, indirectly, a decrease in blood lipid levels (Cione et al., 2021). RS was first classified into four categories by Englyst et al. (1992), with a later new RS describing what became the fifth kind of RS (Hasjim et al., 2010), leading to a classification of five different RS.

RS1 is a physically inaccessible starch, common in whole grains; RS2 is from B- or C-type starch, mainly uncooked starch, and has a high-amylose maize starch; RS3 is retrograded starch from cooked and cooled starch; RS4 is chemically modified starch, very often produced by cross-linked starch in thickeners; RS5 is from amylose–lipid complex, especially at palmitic acid–amylose complexes (Cione et al., 2021; McCleary & Monaghan, 2002).

Resistant starch in corn is highly correlated with amylose levels (Plumier et al., 2015). High-amylose maize starch is a commercial resistant starch patent by McNaught et al. (1998), produced by a single cross F1 hybrid, capable of producing grain having a very high-amylose content and to this grain. This RS is a native maize starch having an amylose content of more than 80% w/v. Native starch is preferred as a food ingredient because it can be considered safe and therefore does not have any regulatory issues.

The RS of maize grains depends not only on genetic breeding but also on the growing environment, handling, storage, and processing practices that may encourage its development (Plumier et al., 2015). RS3 and RS4 are often produced, such as making corn into tortillas, and RS continues to increase with storage time (García-Rosas et al., 2009; Santiago-Ramos et al., 2015). High-temperature processing techniques such as baking, boiling, and roasting also increased RS in corn (Rendon-Villalobos et al., 2002). The enzyme-resistant starch does not refer to the rearrangement of amylose chains into enzyme-resistant structures of higher crystallinity from modified or retrograded starch. A specifically processed starch resistance to enzymes' digestion results from a competition between the kinetics of enzyme hydrolysis and the kinetics of amylose retrogradation (Lopez-Rubio et al., 2008; Peng et al., 2022).

The enterprise Ingredion® has a comprehensive series of resistant starch ingredients (Ingredion.com, 2021). Hi-maize® 260 is a high-amylose resistant starch, comprising approximately 53% resistant starch. Hylon® VII is unmodified and derived from high-amylose corn (70%). Novelose® 330 is a nongranular, retrograded maize starch (RS3) from corn starch with an RS content of up to 30%.

The effects of Hylon® VII, Hi-maize® 1043, Hi-maize® 240, Hi-maize® 260, and Novelose® 330 maize starches were studied on colonic fermentation and apoptotic response to DNA damage in the colon of rats. All RS diets prevented mucosal atrophy as seen in the rats fed the control diet, short-chain fatty acids increased, and pH in caecal content and faeces decreased (Le Leu et al., 2009). Their low water-holding capacity, neutral taste, smooth mouthfeel, and relative thermal stability allow these RS ingredients to be used in a wide variety of food products (Sjoo & Nilsson, 2004).

Flour or starch processing and storing can improve the RS content in maize products, e.g. increasing through all the steps of tortilla production (Santiago-Ramos et al., 2015). Maize tortilla flours produced by traditional and nixtamalized processes have higher RS content than commercial nixtamalized maize flour and commercial corn tortilla flour from Mexican industrial processing (Rojas-Molina et al., 2020). Nixtamal, masa, and tortilla samples stored from 24 to 96 hours showed retrograded resistant starch (RS3). Masa and nixtamal RS ranged from 2.1–2.6%, and only minor increases were described after 24 h of storage. This suggests that the retrogradation phenomenon in these samples takes place very rapidly, being more pronounced in the tortilla samples (3.1–3.9%) (Rendon-Villalobos et al., 2002).

4.3 Extraction, processing, and use of starch

The maize starch is isolated using a wet milling process that can separate starch granules and corn protein (gluten) particles by particle density differences. Wet milling produces high purity starches (99.95%), and more than 85% of corn starch is produced by wet milling (R. Zhang et al., 2021). The process is based on six main steps: kernel steeping, first milling, germ separation, second milling, fibre separation, and starch-gluten separation (El Halal et al., 2019; Ji et al., 2004; N. Singh & Eckhoff, 1996).

The wet milling process starts with clean maize kernels (Figure 4.11). First, the kernels are steeped in 0.1–0.2% sulphur dioxide (SO_2) or bisulphite solution (aqueous SO_2) for 30–40 h at 48–52 °C. This process promotes water absorption by kernels up to 45% of moisture. Steeping is the primary step because it promotes kernels' hydration. The SO_2 breaks the inter- and intra- molecular disulphide bonds of the protein matrix surrounding the starch granules, making physical and protein separation easier (Ji et al., 2004; R. Zhang et al., 2021). Also, some endogenous proteases activate in the endosperm, which helps solubilize the protein matrix and release the starch granules (El Halal et al., 2019; Lopes-Filho et al., 1997; N. Singh & Eckhoff, 1996). Lactic acid added or produced by microorganisms improves the proteins solubilizing (Jackson & Shandera Jr., 1995; O. E. Pérez et al., 2001).

About 9% of the endosperm protein can be solubilized during steeping. Half of the starch and oil-rich germ is released from the kernel during the

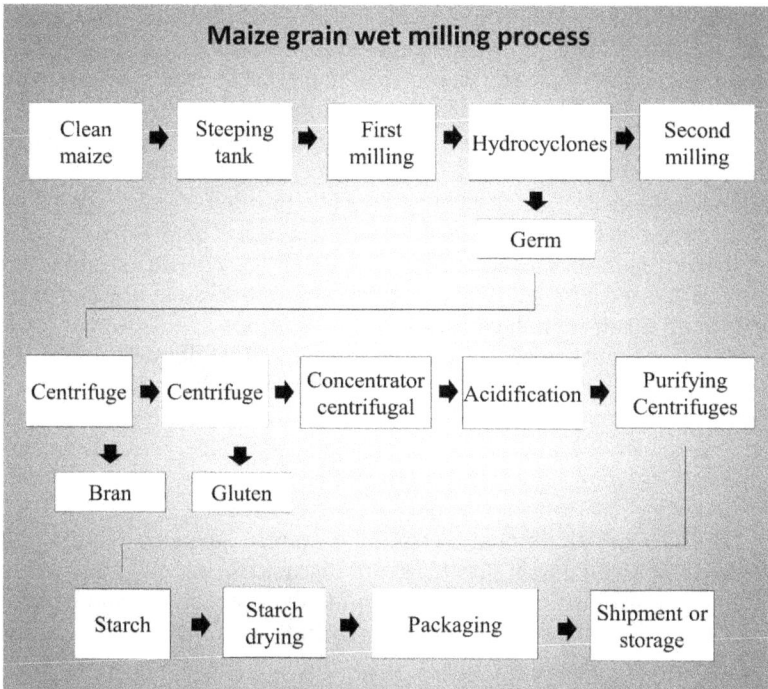

Figure 4.11 Simplified process of maize wet milling starch extraction

first milling, where the separation is followed by continuous flow-through liquid cyclones or hydro cyclone banks (R. Zhang et al., 2021). The germ-rich phase is light and overflows out of the top of the hydro cyclone. After separation of the germ, the degermed corn slurry is screened, using pressure-fed screens to separate pericarp and cell walls (or bran) from the starch and gluten (Rausch et al., 2018).

The germ-lean material is milled into a fine slurry during the second milling step to release the starch granules from the protein matrix and pericarp bran. The starch-protein slurry and soluble impurities are separated in high-speed nozzle centrifuges (Rausch et al., 2018). Separation results from density differences between gluten (1.06 specific gravity) and starch (1.6 specific gravity) (Jackson & Shandera Jr., 1995). The germ leads to pressing and drying (Gwirtz & Garcia-Casal, 2014). The slurry is washed off the remaining gluten (heavy gluten) with freshwater, vacuum-filtered to 40% solids, and finally dried to 88%–90% solids (Rausch et al., 2018).

Starch from the primary centrifuges contains 3%–5% protein and small amounts of other soluble and insoluble impurities. This starch slurry (16–20 °Be) is acidified and processed continuously by centrifugal methods, including concentrators and hydrocyclone washers, before being recovered as a starch cake on basket-type centrifuges (Eliasson, 2004). The removed protein consists primarily of starch-protein complexes, termed middling, which are recycled back into the primary separation step (Jackson & Shandera Jr., 1995). The washed starch should contain <0.30% total protein and 0.01% soluble protein to carry out downstream processes efficiently. These protein levels are attainable with normal maize. The purified starch from the wet milling operation contains about 40% solids (Rausch et al., 2018).

Starch slurry or cake at 40%–50% solids may be dried directly or modified. It is fed into belt-type dryers with 65–150 °C steam-heated air rising through the dryer belt perforations. Flash dryers are also generally fed from large basket-type centrifuges or belt filters that dewater to 55%–67% solids. The dryer heating medium is steam from coils or direct-fired gas furnaces furnishing 220 °C air. The products are generally ground in air-swept mills, blended, and shipped in bags, hopper trucks, and railroad hopper cars (Jackson & Shandera Jr., 1995; Serna-Saldivar, 2018) for food or non-food uses.

Besides the traditional wet milling starch isolation, some studies have been described using proteases to disrupt the starch-proteins inter- and intra- molecular bounding. The proteolytic enzymes could replace or reduce the use of sulphur dioxide (El Halal et al., 2019). Wet milling using L-cysteine is an environmentally friendly procedure; however, its cost is about 30 times higher than traditional ones (Xiaona Li et al., 2015). Therefore, choosing environmentally friendly and more cost-effective methods is quite challenging.

Lab-scale starch extraction yield from maize kernel storage at 35°C for 12 months was lower than at 5 °C, 15 °C, or 25 °C (Table 4.2). Proteins and oxidized lipids can interact during storage time, decreasing the extraction yield (Singh Sodhi & Singh, 2003). Also, amylose α-helixes and fat form helicoidal-complexes, changing starch properties, including extraction yielding (Salman & Copeland, 2007).

Colour parameters as L* (black-white) and b* (blue-yellow) values of starch isolated from maize kernels stored at 35 °C also increased after 12 months of storage, being yellow in colour and increasing with the content of protein residual and the presence of some lipids (Paraginski et al., 2019). The residual content of protein and the presence of lipids in the starch granules can cause restriction of the swelling power during starch gelatinization (Debet & Gidley, 2006). During storage, interactions between proteins, oxidized lipids, and starch can occur (Sodhi & Singh, 2003). Amylose chains can form helical complexes, changing starch properties and reducing extraction yield (Salman & Copeland, 2007).

Short exposure at low temperatures during the early stage of starch synthesis in wheat grains indicates that the grain starch accumulation rate and starch content decreased with decreasing temperatures. Also,

Table 4.2 Extraction yield (%), value L*, value b*, residual protein (%), and residual fat (%) content of starch isolated from maize kernels stored for 12 months at 5 °C, 15 °C, 25 °C and 35 °C with 14% of moisture

Storage	Yield (%)	L*	b*	Residual protein (%)	Residual fat (%)
Before	59.07 ± 0.31a	96.26 ± 0.49b	6.27 ± 1.46b	0.23 ± 0.03b	0.61 ± 0.08a
5 °C	62.88 ± 1.25a	97.47 ± 0.52a	6.55 ± 0.40b	0.27 ± 0.08b	0.63 ± 0.03a
15 °C	66.94 ± 0.71a	97.34 ± 0.90a	5.98 ± 0.69b	0.32 ± 0.00b	0.62 ± 0.04a
25 °C	63.36 ± 2.32a	96.82 ± 0.25ab	6.00 ± 0.44b	0.29 ± 0.06b	0.60 ± 0.04a
35 °C	45.99 ± 6.58b	92.44 ± 0.27c	10.67 ± 0.87a	0.74 ± 0.01a	0.40 ± 0.04b

Source: Paraginski et al. (2019).
* Simple arithmetic means ± standard deviation, followed by equal lowercase letters in the same column, do not differ by Tukey test at 5% significance ($p \leq 0.05$).

the duration of grain filling increased, and as maximum filling rates were reduced, this confirmed that temperature changes reduced starch accumulation, decreased filling rate, and decreased grain dry matter accumulation, and consequently grain yield (W. Zhang et al., 2021).

4.3.1 Food maize starch uses

Starch is the main ingredient responsible for the technological properties that characterize most food products, as it contributes to several texture properties in foods as a thickener, colloid stabilizer, and gelling and volume agent, affecting stickiness and water retention, among other technological properties (Cornejo-Ramírez et al., 2018; Eliasson, 2004; Sajilata & Singhal, 2005; Schirmer et al., 2015; J. Singh & Dartois, 2010). The main starch uses are summarized in Figure 4.12. Only 15% of the total maize starch is used for food. However, around 70% of the total starch in corn is used for food (Ranum et al., 2014). Regular maize starch has consistent gel, widely used in dehydrated soups and gravies, which need hot viscosity. Waxy maize starch is more stable under cooling and freezing than regular starch due to its low amylose and syneresis. This starch gel is weak, highly viscous, and clear and cohesive (Bemiller, 2019).

Corn syrup is produced using starch slurry in water (Figure 4.13) by thermally stable α-amylase hydrolysis at a constant pH of 5.6–6.0. The

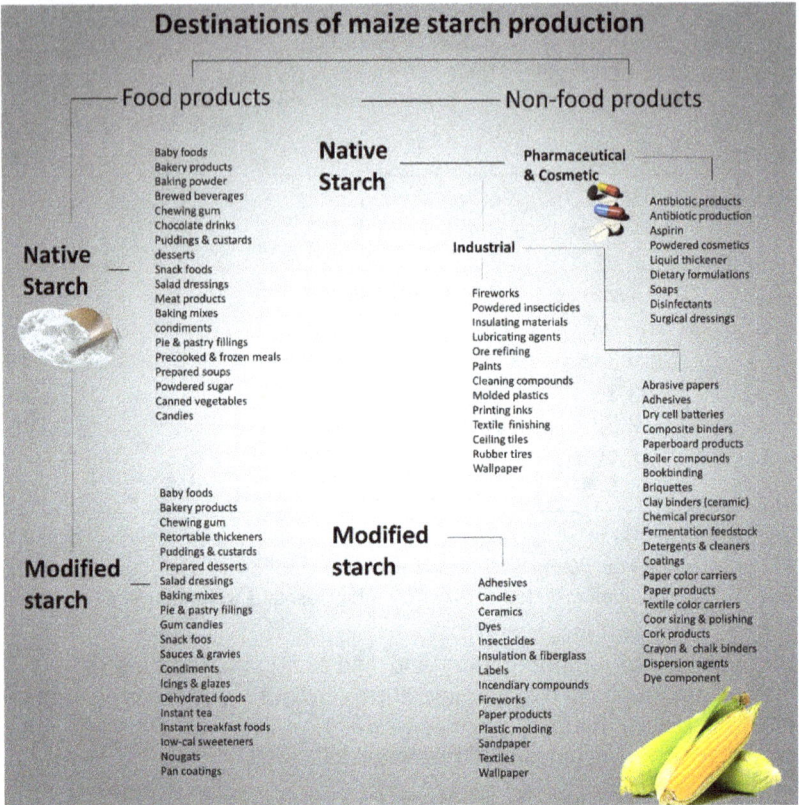

Destinations of maize starch production

Figure 4.12 Maize starch uses in food and non-food products

starch is gelatinized using a jet cooker for enzyme-catalyzed pasting liquefaction (Bemiller, 2019). Glucose syrup is composed of D-glucose and D-fructose purified, with a dextrin equivalent (DE) greater than 20 (Bemiller, 2019). The syrup is a purified and concentrated aqueous solution of nutritive saccharides from food-grade starch, with a dextrin equivalent (DE) greater than 20 (Bemiller, 2019; Helstad, 2018).

About 4% of total maize makes high-glucose corn syrup for industrial applications, while some make whisky and other alcoholic beverages. The range of sweetness from corn syrup and dried corn syrup products depends on the degree of starch hydrolysis and dextrose isomerization to fructose (Maldonado-Guzmán et al., 1995; Spencer et al., 1996). Corn syrup with sweetness around 130 is based on high fructose levels. It is the complete starch hydrolysis to dextrose and maximal dextrose enzyme conversion into fructose. The low sweetness of corn syrup products is the partial

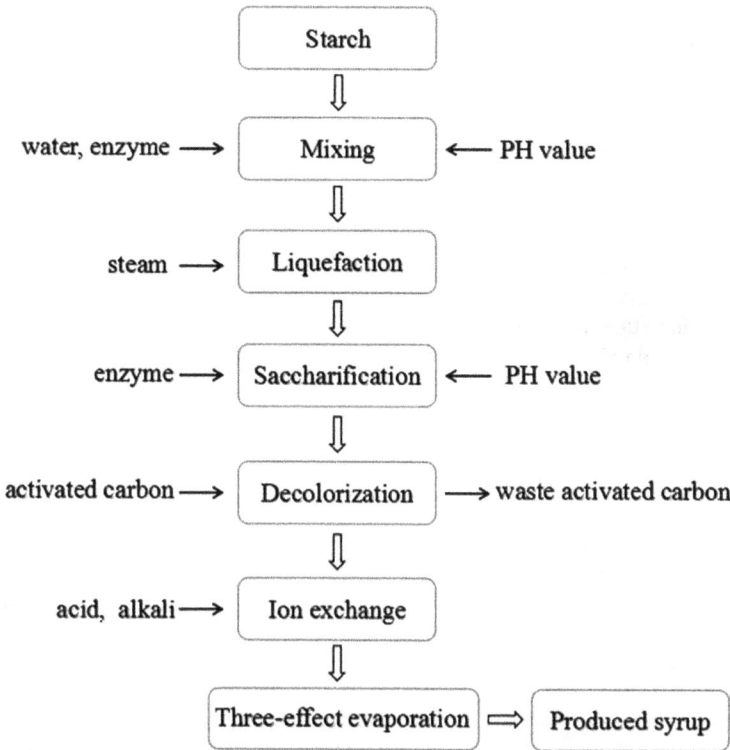

Figure 4.13 Simplified maize starch syrup-making process

corn starch hydrolysis to dextrose and no conversion of dextrose to fructose. This syrup is more used for water-binding abilities than sweetness (Gahlawat et al., 2017; Spencer et al., 1996).

4.3.2 Non-food maize starch uses

4.3.2.1 General non-food maize starch uses

It is estimated that nearly 40% of total maize production is used to make ethanol for fuel, 27% becoming ethanol and 12% distillers' dry grain residue (Ranum et al., 2014). Moreover, 90% of corn bioethanol is processed from dry-milled corn directly from maize flour (R. Zhang et al., 2021). The primary substrate for ethanol production is sugar from corn starch, sugarcane, and beets. The ethanol is produced by yeast alcoholic fermentation. This process converts one mole of glucose into two moles of ethanol and two moles of carbon dioxide, producing two moles of ATP (Mohd Azhar et al., 2017).

The use of starch and its derivatives by the paper industry is presently high. They are widely used for paper and paper coatings due to their completely biodegradable nature, wide availability, and low cost (Haiming Li et al., 2019). The main uses are furnishing preparations prior to sheet formation, surface sizing, coating, adhesive for corrugation, and laminating (Maurer & Kearney, 1998). The paper surface sizing uses native and modified starches to improve printability, physical strength, oil and grease resistance, and optical properties (Khwaldia et al., 2010). However, dry conditions for starch binder coating must be controlled to avoid non-uniform porosity and print mottle (Maurer & Kearney, 1998).

Maize starch is frequently used in conventional granulation and tablet production, especially as an excipient due to its low cost, versatility, and non-toxic and non-irritant properties. The uses are also in diluents, glidants, disintegrants, binders, lubricants, advanced drug delivery, and targeting specific sites in the body (Builders & Arhewoh, 2016). Its modification can improve the native starch characteristics and properties in the drug's use (Zámostný et al., 2012). The starch granule size has some critical influences on the functional application of some native starches (Sujka et al., 2018). In this way, starch modification can reduce granules' size and improve their use (Thanyapanich et al., 2021).

On the other hand, consumer demand for natural ingredients has increased to avoid possible adverse effects on health. The industry is developing cosmetics based on natural ingredients to meet this consumer behavioural change (Amberg & Fogarassy, 2019). Starch is an attractive natural alternative to substitute for talcum and is used in cosmetic preparation (Amberg & Fogarassy, 2019; Barbulova et al., 2015).

4.3.2.2 Trends in maize starch uses

4.3.2.2.1 Biodegradable packaging Starch derivate materials are a promising solution to petroleum plastic (Altayan, Al Darouich, and Karabet, 2020) due to starch abundance, low cost, biodegradability, and easy handling (Bemiller, 2019; Giuri et al., 2018; Molina-boisseau et al., 2006; Nafchi et al., 2013). There are some limitations to starch technology approaches related to starch's hydrophilic behaviour and high water-binding ability (Khan et al., 2017; Nafchi et al., 2013). Maize starch is often used to produce packing peanuts, balls, or loose-fill packaging, banking foams (Figure 4.14), thermoplastic films, and edible films and coatings, among others (Table 4.1) (Jiang et al., 2020; Khan et al., 2017; Lawton et al., 2004; Y. Lin et al., 1995; Sadeghizadeh-Yazdi et al., 2019).

The development of maize starch films is based on the amylose gel- and film-forming capacity (Mali et al., 2010). Amylose content is mandatory to promote structural stability to the hydrogen bonding network (WANG

Figure 4.14 Photos of starch foams: from loose-fill packaging to form sheet and post-processed products. Source: Jiang et al. (2020)

et al., 2015) and has highlighted the influence on films' tensile strain and elongation at break points (Koch et al., 2010; Xiaojing Li et al., 2015). On the other hand, amylopectin favours inter-chain hydrogen bonding (H. Wang et al., 2016). So, the amylose/amylopectin ratio plays an important role in biodegradable films (Altayan, Al Darouich, and Karabet 2020; Altayan et al., 2020; Basiak et al., 2017; García et al., 2000; M. Li et al., 2011). The high-amylose proportion results in strong films with low water vapour permeation (V. D. Alves et al., 2007). Table 4.3 summarizes some maize starch coating and films for food application.

Casting films and edible coatings is wet plasticizing processing. Edible or not, films are used as wrap packaging, while edible coatings are directly applied to the surface of food and fruits. Both processes need a dry step to create the packaging (Aguirre et al., 2018; Cazón et al., 2017; Suhag et al., 2020). The starch-based casting and coating solutions use water, polyols, fat acids, essential oils, and vegetal extracts as plasticizers (Thakur et al., 2019). These films' components are edible and GRAS (generally recognized as safe). The starch thermoplastic is flexible due to the inter- and intramolecular hydrogen bonds dissociation and plasticizer inclusion

Table 4.3 Summary of maize starch films

Starch films composition	Processing method	References
Waxy maize starch (4.3% of amylose content), regular starch (29% of amylose content), Gelose 50 (61.5% of amylose content) and Gelose 80 (77.4% of amylose content)	Lip die twin-screw extruder (starch-water slurry or separated at feeding rate of 1.2–2.4 kg/h and the extruder screw speed range 30–120 rpm, study zones temperature 115 °C–180 °C)	(M. Li et al., 2011)
Regular starch plasticized by glycerol	Lip die twin-screw extruder (zones temperatures 150 °C–160 °C, followed by calendar stretching)	(Giuri et al., 2018)
Regular maize starch (31% of amylose content) and wheat starch (22% of amylose content); 50% starch–30% glycerol–20% water (w/w)	Lip die twin-screw extruder (twin counter-rotating internal mixer turning at 3:2 differential speed rotation speed of (60 rpm) with simultaneous heating up to (160 °C)	(Altayan et al., 2020)
Maize starch (27% of amylose content), potato starch (20% of amylose content), wheat starch (25% of amylose content), glycerol	Casting of film-forming solutions using 5 g of starch and 30% of glycerol (starch weight), heating and stirring at 85 °C, oven-drying at 25 °C, 30% UR for 48 h	(Basiak et al., 2017)
Regular maize starch (25% of amylose content) and Amylomaize VII (65% of amylose content), glycerol, and sorbitol	Casting of film-forming solutions using 20 g of NaOH (1%) for cold gelatinization and 20 g/L of glycerol or sorbitol, oven-drying at 60 °C for 8 h	(García et al., 2000)
Maize starch (33% of amylose content) glycerol and thymol	Casting of film-forming solutions using 4 g of starch 25% and glycerol and/or 5% of thymol (starch weight) heating and stirring at 85 °C, room temperature drying (21 °C) for 48 h	(Nordin et al., 2020)
Regular maize starch, glycerol, and sorbitol	Casting of film-forming solutions using 10 g of starch and 0, 30, 45, or 60% (w/w) of glycerol and/or sorbitol, heating and stirring at 85 °C, oven-drying at 65 °C for 15 h	(Hazrol et al., 2021)

between the polymer chains, increasing the intermolecular spacing and mobility of the molecular starch chains.

The starch gelatinization step is required to disrupt the granules and plasticize the solution before casting it by drying it on a smooth surface or mould to produce the films (Thakur et al., 2019). Under light microscopy,

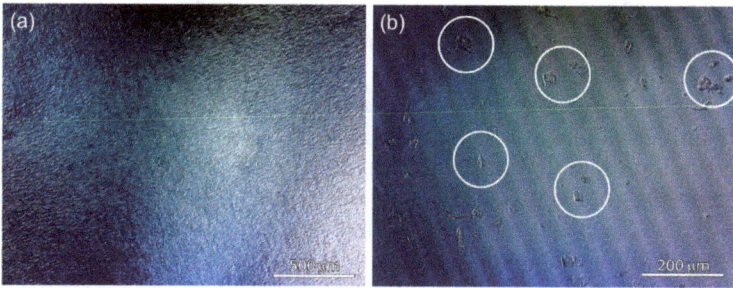

Figure 4.15 Starch biodegradable casting films (a) 40× magnification; (b) 100× magnification by incomplete granules' gelatinization. Source:Lenhani et al. (2021)

using cross-polarized light, it is possible to notice the incomplete gelatinization of the starch granules (Figure 4.15). This incomplete granular disruption results from the time and temperature used for gelatinization, resulting in low-resistance films due to the intact granules and thermoplastic phases interface (Corradini et al., 2005; Lenhani et al., 2021).

The application of edible coatings improves physical protection for fruits and fresh vegetables, especially to retain colour and flavour. Also, there is a partial barrier of solutes diffusion and transfer, oxygen, water, and carbon dioxide, decreasing the weight and water loss, respiration, and oxidation kinetics (Baldwin et al., 2016; Fitch-Vargas et al., 2019; Sapper & Chiralt, 2018). These coatings can decrease the development of the food-borne and photogenic microorganisms and improve food safety (Hassan et al., 2018; Sapper & Chiralt, 2018).

The starch thermoplastic extrusion (TPS) turns brittle starch into thermoplastic starch materials. It allows the production of films and foam goods and precedes the moulding processes. TPS is a thermomechanical treatment that uses temperature and high shear with an appropriated plasticizer (Altayan, Al Darouich, and Karabet, 2020; Forato et al., 2013; Khan et al., 2017; Nafchi et al., 2013; Selling, 2010; Verbeek & Van Den Berg, 2010). The processing is absent of solvents, and it is easy to handle high viscous polymers, has an extensive range of processing parameters (pressure, shear, temperature), and high homogeneous TPS material (Jiménez et al., 2012; Liu et al., 2009). Starch TPS processing promotes granules' gelatinization at 90 °C–180 °C under high shear, melting them with the plasticizer and turning the starch into an amorphous thermoplastic material (Corradini et al., 2005; Liu et al., 2009; Morin et al., 2021). The brittleness is caused by TPS's relatively high Tg and lack of a sub-Tg main chain relaxation area. During storage, this brittleness increases because of starch retrogradation (Nafchi et al., 2013).

The starch and TPS starch sensibility to UR has been the most significant disadvantage for commercializing these materials, like the packaging. This is due to the films' low mechanical properties and their UR dependence (Bemiller, 2019; Giuri et al., 2018; Khan et al., 2017; Morin et al., 2021; Nafchi et al., 2013). High-amylose TPS maize starches usually have improved mechanical properties; however, their extrusion is quite challenging, requiring higher temperatures, moisture, conditioning time, and mould pressure than regular starches due to the high viscous nature and instability of the processing flow (M. Li et al., 2011).

Green composites based on thermoplastic starch are a way to improve mechanical properties and water resistance (Rivadeneira-Velasco et al., 2021). Cellulose fibres as fillers in starch-based films increase the tensile strength and the elasticity module and decrease the elongation at break points (Ali et al., 2018; J. S. Alves et al., 2015; Debiagi et al., 2010; Lomelí-Ramírez, 2014; Müller et al., 2009; Teacă et al., 2013). Besides this, the water vapour barrier is improved in almost all cellulose-starch composite films (Campos et al., 2017; Fazeli et al., 2018; Müller et al., 2009; Pelissari et al., 2017; Sun et al., 2016), and water sorption decreases with a low cellulose filling concentration (Curvelo et al., 2001; Carvalho et al., 2018). Cellulose and starch are plant-based materials that are low cost, easy to handle, and have a closed chemical composition, increasing their affinity and enabling improved green materials (Debiagi et al., 2010; Lenhani et al., 2021; Lomelí-Ramírez et al., 2014).

4.3.2.2.2 Starch nanomaterials

Starch nanocrystals (SNC) are insoluble crystalline platelets isolated from the granule thin crystalline lamella (Figure 4.7). The most common method for extracting SNC is by mild acid hydrolysis of the amorphous parts of native granular starch (Molina-boisseau & Dufresne, 2006). The first report on starch nanocrystals was published by Dufresne et al. (1996). The hierarchical organization of starch granules and their semicrystalline structure has been a critical factor in preparing nanocrystals through controlled acid hydrolysis (Bel Haaj et al., 2016).

The insoluble residue from acid hydrolysis is segregated by high-speed centrifugation and is water washed to complete removing the acid solution and soluble fractions. The standard processing yield range from 5% to 15%, and the particle size (thickness or diameter) ranges from 20 to 400 nm (Angellier et al., 2004; Gonçalves et al., 2014; Kristo & Biliaderis, 2007; Le Corre & Angellier-Coussy, 2014; Le Corre et al., 2010; N. Lin et al., 2011; Pinto et al., 2021). Several studies have shown that they could be used as fillers to improve mechanical and barrier properties of biocomposites (Campelo et al., 2020) and as black carbon replacement for tires (Le Corre & Angellier-Coussy, 2014; Le Corre et al., 2010).

The SNC production by acid hydrolysis depends on many factors, being the amylose content and starch botanical origin the main interferents (Angellier et al., 2004). Waxy maize starch is the most used for SNC production due to its amylose-free characteristic (Wei et al., 2014). Higher crystallinity was expected for hydrolyzed amylopectin-rich (waxy) starch since there is less amorphous starch to hydrolyze. Indeed, amylopectin-rich starches are more crystalline than amylose-rich ones (Le Corre et al., 2011). As suggested in Figure 4.16, the initial starch crystallinity results in SNC with high crystallinity, which is highly desired for this nanomaterial. The insoluble hydrolyzed residue obtained from waxy maize was composed of crystalline nanoplatelets around 5–7 nm thick with a length of 20–40 nm and a width of 15–30 nm (Molina-Boisseau & Dufresne, 2006).

Nano-biocomposites are combined biomaterials based on two phases: a continuous matrix and a filler or reinforcement discontinuous one. Maize starch is frequently studied for nanocomposite development. It is generally used as the continuous phase due to its homogeneous film-forming ability (Palanisamy et al., 2020). The materials combined may result in synergic properties from each material, especially its processability and mechanical properties (Almeida et al., 2015; Mathew et al., 2005). The nanocomposites are also promising materials for petroleum-based polymers and plastics, at least in part, replacing food and non-food applications (Vieira et al., 2011; Li et al. 2015; L. Yu et al., 2006).

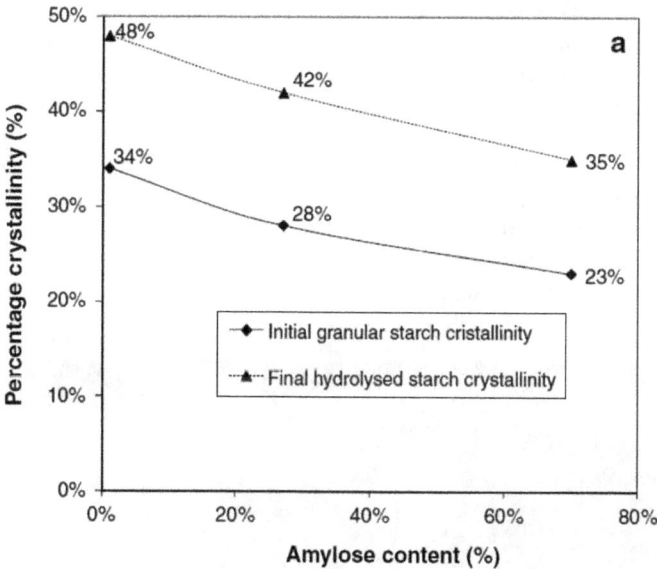

Figure 4.16 Original crystallinity (diamond) and final crystallinity (triangle) for maize starches with different amylose content. Source: Le Corre et al. (2011)

There are several scientific reports about maize starch nanocomposites. The main interest in these materials is to use lower filler concentration (0.1–10%) rather than traditional material (5–50%) and improve properties for each specific application. Briefly, these can be described as nanofillers from nanofibres (cellulose, hemicellulose, chitin, and chitosan, electrospun fibres), inorganic nanoparticles (clays, silicates, carbonates), nanometals (silver, copper, gold), even SNC as a crystalline structure (Palanisamy et al., 2020; Anukiruthika et al., 2020; Ibrahim et al., 2019; Lagaron, 2019; Le Corre & Angellier-Coussy, 2014; X. Tang et al., 2008).

Electrospinning is an electrohydrodynamic process that exploits electrostatic forces used to produce long and ultrafine fibres with micro- to nanometric diameters (Formhals, 1934; Lim et al., 2019). The setup is based on a polymer infusion system, AC/DC power supply (5–30 kv) using a charged spinneret or a free-surface polymer release and a grounded target (Noruzi, 2016; C. Zhang et al., 2020).

Due to the starch amylose linearity, it can be electrospun into fibres readily. Pure high-amylose starches have been successfully electrospun into fibres (Figure 4.17) using dimethylsulphoxide (DMSO) or DMSO/water mixtures as solvents (Kong & Ziegler, 2012a, 2012b, 2013, 2014b, 2014a) or formic acid (Lancuški et al., 2017; Vasilyev et al., 2015). Commercial starches have been electrospun and include Gelose 80, Hylon V, and Hylon VII, with amylose contents of ~80%, 70%, and 55%, respectively (Kong & Ziegler, 2012a, 2012b, 2013, 2014b, 2014a; Lancuški et al., 2017; Vasilyev et al., 2015).

DMSO or DMSO/water solvents are relatively non-volatile compared to other solvents commonly used for electrospinning, being unable to deposit solid fibres on a grounded metal mesh collector, as the DMSO solution cannot be evaporated sufficiently under standard ambient conditions (Kong & Ziegler, 2014a). Turning the setup into "electro-wet-spinning", the starch solution jet reaches the coagulation bath, DMSO is extracted, and a fibrous form can be collected (Kong & Ziegler, 2012a, 2012b, 2013, 2014b, 2014a).

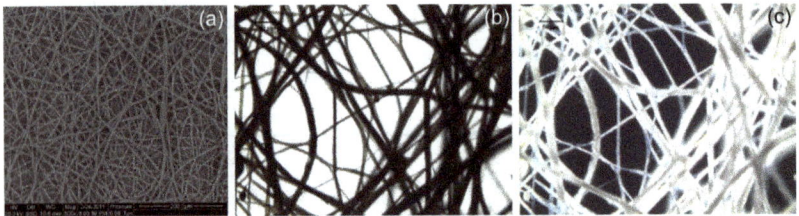

Figure 4.17 Scanning electron micrograph (a) and optical micrographs under normal light (b) and between crossed polarizers (c) of pure high-amylose maize starch fibres produced by electrospin. Source: Kong & Ziegler (2014a)

A typical setup or a spinneret with two concentric capillaries can be used to electrospin starch fibres instead of a coagulation bath, using formic acid as solvent (Lancuški et al., 2017; Vasilyev et al., 2015). Formic acid disrupts the starch granule structure, solubilizes and esterifies starch into starch-formate, and disperses for electrospinning (Vasilyev et al., 2015).

These fibres can archive diameters ranging from 80 to 300 nm using formic acid and in the order of micros using DMSO. Electrospun fibres have a high superficial area due to their diameters, which favours their applications in tissue engineering, and in automotive, energy, environmental, electronics, biotechnology, chemical, pharmaceutical, cosmetic, and food industries (Lim et al., 2019; Persano et al., 2013).

4.4 Conclusion

Maize is the most important starch source and used as a model for many kinds of research over the last years. Regular and mutant starches are commercially available and aid food and non-food industries in developing several products. Starch granule morphology, pasting and rheological properties, swelling power, and solubility depend entirely on the amylose and amylopectin ratios, an extensive list of deep structural characteristics, and intrinsic and extrinsic factors. RDS, SDS, and RS development are the consequence of the breeding of regular and mutant starches; enzyme hydrolysis and the kinetics of amylose retrogradation; natural RS occurrence; and starch cropping and processing. Resistant starch is a competition between the kinetics of enzyme hydrolysis and the kinetics of amylose retrogradation. These starch fractions have important health-promoting properties, and RS can modulate carbohydrates absorption, lower GI, and decrease lipid levels in the blood. Maize starch versatility is translated to its food and non-food applications. Traditional products are produced daily using starch as raw material, and about 70% of it is from maize. Novel products are popping up every day, and the most promising uses are for packaging and nanomaterials. There is some biodegradable starch packaging already commercially available as transport loose-fill packaging and baking foams. Also, there have been several efforts to transform starch thermoplastic into packaging and use this as a replacement for petroleum-based plastics. Thousands of studies are being undertaken to produce starch nanomaterials for packaging, advanced drug delivery targeting, cosmetics, and other applications.

References

Adebowale, K. O., Afolabi, T. A., & Olu-Owolabi, B. I. (2005). Hydrothermal treatments of finger millet (*Eleusine coracana*) starch. *Food Hydrocolloids, 19*(6), 974–983. https://doi.org/10.1016/j.foodhyd.2004.12.007

Aguirre, J., Leonzapata, M., Alvarezperez, O., Torres, C., Nieto-Oropeza, D., Ventura-Sobrevilla, J., ... Elena, M. (2018). Basic and applied concepts of edible packaging for foods. In A. M. Grumezescu & A. M. Holban (Eds.), *Food packaging and preservation* (Issue October, pp. 1–61). Elsevier Inc. https://doi.org/10.1016/B978-0-12-811516-9/00001-4

Ali, A., Xie, F., Yu, L., Liu, H., Meng, L., Khalid, S., & Chen, L. (2018). Preparation and characterization of starch-based composite films reinfoced by polysaccharide-based crystals. *Composites Part B: Engineering, 133*, 122–128. https://doi.org/10.1016/j.compositesb.2017.09.017

Almeida, A. C. S., Franco, E. A. N., Peixoto, F. M., Pessanha, K. L. F., & Melo, N. R. (2015). Aplicação de nanitecnologia em embalagens de alimentos. *Polimeros, 25*, 89–97.

Altayan, M. M., Al Darouich, T., & Karabet, F. (2020). Thermoplastic starch from corn and wheat: A comparative study based on amylose content. *Polymer Bulletin, 89*, 0123456789. https://doi.org/10.1007/s00289-020-03262-9

Altayan, M. M., Ayaso, M., Al Darouich, T., & Karabet, F. (2020). The effect of increasing soaking time on the properties of premixing starch–glycerol–water suspension before melt-blending process: Comparative study on the behavior of wheat and corn starches. *Polymer Bulletin, 77*(4), 1695–1706. https://doi.org/10.1007/s00289-019-02826-8

Alves, J. S., Dos Reis, K. C., Menezes, E. G. T., Pereira, F. V., & Pereira, J. (2015). Effect of cellulose nanocrystals and gelatin in corn starch plasticized films. In *Carbohydrate polymers* (Vol. 115, pp. 215–222). https://doi.org/10.1016/j.carbpol.2014.08.057

Alves, V. D., Mali, S., Beléia, A., & Grossmann, M. V. E. (2007). Effect of glycerol and amylose enrichment on cassava starch film properties. *Journal of Food Engineering, 78*(3), 941–946. https://doi.org/10.1016/j.jfoodeng.2005.12.007

Amberg, N., & Fogarassy, C. (2019). Green consumer behaviour in cosmetic market. *Resources, 8*(137), 1–19.

Angellier, H., Choisnard, L., Molina-Boisseau, S., Ozil, P., & Dufresne, A. (2004). Optimization of the preparation of aqueous suspensions of waxy maize starch nanocrystals using a response surface methodology. *Biomacromolecules, 5*(4), 1545–1551. https://doi.org/10.1021/bm049914u

Anukiruthika, T., Sethupathy, P., Wilson, A., Kashampur, K., Moses, J. A., & Anandharamakrishnan, C. (2020). Multilayer packaging: Advances in preparation techniques and emerging food applications. *Comprehensive Reviews in Food Science and Food Safety*, October 2019, 1156–1186. https://doi.org/10.1111/1541-4337.12556

Bahaji, A., Li, J., Sánchez-López, Á. M., Baroja-Fernández, E., José, F., Ovecka, M., ... Pozueta-Romero, J. (2014). Starch biosynthesis, its regulation and biotechnological approaches to improve crop yields. *Biotechnology Advances, 32*(1), 87–106. https://doi.org/10.1016/j.biotechadv.2013.06.006

Baldwin, E. A., Hagenmaier, R., & Bai, J. (2016). *Edible coatings and films to improve food quality* (2nd ed.) (E. A. Baldwin, R. Hagenmaier, & J. Bai, Eds.). Boca Raton, FL: CRC Press.

Balet, S., Guelpa, A., Fox, G., & Manley, M. (2019). Rapid visco analyser (RVA) as a tool for measuring starch-related physiochemical properties in cereals: A review. *Food Analytical Methods, 12*(10), 2344–2360. https://doi.org/10.1007/s12161-019-01581-w

Barbulova, A., Colucci, G., & Apone, F. (2015). New trends in cosmetics: By-products of plant origin and their potential use as cosmetic active ingredients. *Cosmetics, 2*(2), 82–92. https://doi.org/10.3390/cosmetics2020082

Basiak, E., Lenart, A., & Debeaufort, F. (2017). Effect of starch type on the physico-chemical properties of edible films. *International Journal of Biological Macromolecules, 98,* 348–356. https://doi.org/10.1016/j.ijbiomac.2017.01.122

Bel Haaj, S., Thielemans, W., Magnin, A., & Boufi, S. (2016). Starch nanocrystals and starch nanoparticles from waxy maize as nanoreinforcement: A comparative study. *Carbohydrate Polymers, 143,* 310–317. https://doi.org/10.1016/j.carbpol.2016.01.061

Bemiller, J. N. (2019). Carbohydrate chemistry for food scientists. In J. N. BeMiller (Ed.), *Carbohydrate chemistry for food scientists* (3rd ed.). Woodhead Publishing. https://doi.org/10.1002/star.200890023

Bertoft, E., Laohaphatanalert, K., Piyachomkwan, K., & Sriroth, K. (2010). The fine structure of cassava starch amylopectin. Part 2: Building block structure of clusters. *International Journal of Biological Macromolecules, 47*(3), 325–335. https://doi.org/10.1016/j.ijbiomac.2010.05.018

Bertolini, A. C. (2010). *Starches: Characterization, properties and applications* (1st ed.). Boca Raton, FL: CRC Press.

Biliaderis, C. G. (2009). Structural transitions and related physical properties of starch. In *Starch* (3rd ed.). Elsevier Inc. https://doi.org/10.1016/B978-0-12-746275-2.00008-2

Boyer, C. D., Channon, J. C., Garwood, D. L., & Creech, R. G. (1976). Changes in starch granule size and amylose percentage during kernel development in several Zea mays L. genotypes. *Cereal Chemistry, 53,* 327–337.

Builders, P. F., & Arhewoh, M. I. (2016). Pharmaceutical applications of native starch in conventional drug delivery. *Starch/Staerke, 68*(9–10), 864–873. https://doi.org/10.1002/star.201500337

Cai, J., Cai, C., Man, J., Zhou, W., & Wei, C. (2014). Structural and functional properties of C-type starches. *Carbohydrate Polymers, 101,* 289–300. https://doi.org/10.1016/j.carbpol.2013.09.058

Campelo, P. H., Sant'Ana, A. S., & Pedrosa Silva Clerici, M. T. (2020). Starch nanoparticles: Production methods, structure, and properties for food applications. *Current Opinion in Food Science, 33*(i), 136–140. https://doi.org/10.1016/j.cofs.2020.04.007

Campos, A. de, Sena Neto, A. R. D., Rodrigues, V. B., Luchesi, B. R., Moreira, F. K. V., Correa, A. C., ... Marconcini, J. M. (2017). Bionanocomposites produced from cassava starch and oil palm mesocarp cellulose nanowhiskers. *Carbohydrate Polymers*, *175*, 330–336. https://doi.org /10.1016/j.carbpol.2017.07.080

Carvalho, R. A., Santos, T. A., de Azevedo, V. M., Felix, P. H. C., Dias, M. V., & Borges, S. V. (2018). Bio-nanocomposites for food packaging applications: Effect of cellulose nanofibers on morphological, mechanical, optical and barrier properties. *Polymer International*, *67*(4), 386–392. https://doi.org/10.1002/pi.5518

Castanha, N., Rojas, M. L., & Augusto, P. E. D. (2021). An insight into the pasting properties and gel strength of starches from different sources: Effect of starch concentration. *Scientia Agropecuaria*, *24*(2), 203–212. https://doi.org/10.17268/SCI.AGROPECU.2021.023

Castillo, L. A., López, O. V., García, M. A., Barbosa, S. E., & Villar, M. A. (2019). Crystalline morphology of thermoplastic starch/talc nanocomposites induced by thermal processing. *Heliyon*, *5*(6). https://doi .org/10.1016/j.heliyon.2019.e01877

Cazón, P., Velazquez, G., Ramírez, J. A., & Vázquez, M. (2017). Polysaccharide-based films and coatings for food packaging: A review. *Food Hydrocolloids*, *68*, 136–148. https://doi.org/10.1016/j.foodhyd .2016.09.009

Cione, E., Fazio, A., Curcio, R., Tucci, P., Lauria, G., Cappello, A. R., & Dolce, V. (2021). Resistant starches and non-communicable disease: A focus on mediterranean diet. *Foods*, *10*(9). https://doi.org/10.3390 /foods10092062

Cooke, D., & Gidley, M. J. (1992). Loss of crystalline and molecular order during starch gelatinisation: Origin of the enthalpic transition. *Carbohydrate Research*, *227*(C), 103–112. https://doi.org/10.1016 /0008-6215(92)85063-6

Cornejo-Ramírez, Y. I., Martínez-Cruz, O., Del Toro-Sánchez, C. L., Wong-Corral, F. J., Borboa-Flores, J., & Cinco-Moroyoqui, F. J. (2018). The structural characteristics of starches and their functional properties. *CyTA - Journal of Food*, *16*(1), 1003–1017. https://doi.org/10.1080 /19476337.2018.1518343

Corradini, E., Lotti, C., Medeiros, E. S. de, Carvalho, A. J. F., Curvelo, A. A. S., & Mattoso, L. H. C. (2005). Estudo comparativo de amidos termoplásticos derivados do milho com diferentes teores de amilose. *Polímeros: Ciência e Tecnologia*, *15*(4), 268–273. https://doi.org/10 .1590/S0104-14282005000400011

Curvelo, A. A. S., Carvalho, A. J. F. de, & Agnelli, J. A. M. (2001). Thermoplastic starch–cellulose fibers composites: Preliminary results. *Carbohydrate Polymers*, *45*(2), 183–188.

Debet, M. R., & Gidley, M. J. (2006). Three classes of starch granule swelling: Influence of surface proteins and lipids. *Carbohydrate Polymers*, *64*(3), 452–465. https://doi.org/10.1016/j.carbpol.2005.12.011

Debiagi, F., Mali, S., Grossmann, M. V. E., & Yamashita, F. (2010). Efeito de fibras vegetais nas propriedades de compósitos biodegradáveis de amido de mandioca produzidos via extrusão. *Ciência Agrotec, 34*(6), 1522–1529.

Delcour, J. A., & Hoseney, R. (2010). *Principles of cereal science and technology* (3rd ed.) (J. A. Delcour & R. Hoseney, Eds.). American Association of Cereal Chemists.

Dhital, S., Shrestha, A. K., Hasjim, J., & Gidley, M. J. (2011). Physicochemical and structural properties of maize and potato starches as a function of granule size. *Journal of Agricultural and Food Chemistry, 59*(18), 10151–10161. https://doi.org/10.1021/jf202293s

Dufresne, A., Cavaille, J.-Y., & Helbert, W. (1996). New nanocomposite materials: Microcrystalline starch reinforced thermoplastic. *Macromolecules, 29*(1), 7624–7626. https://doi.org/10.1021/ja0130 4a065

El Halal, S. L. M., Kringel, D. H., Zavareze, E. da R., & Dias, A. R. G. (2019). Methods for extracting cereal starches from different sources: A review. *Starch/Staerke, 71*(11–12), 1–14. https://doi.org/10.1002/star .201900128

Eliasson, A.-C. (2004). Starch in food: Structure, function and application. In A.-C. Eliasson (Ed.), *Starch in food: Structure, function and applications*. Boca Raton, FL: CRC Press. https://doi.org/10.1016/B978-1 -85573-731-0.50013-X

Englyst, H. N., Kingman, S. M., & Cummings, J. H. (1992). Classification and measurement of nutritionally important starch fractions. *European Journal of Clinical Nutrition, 46*(Suppl. 2), 33–50.

FAO - Food and Agriculture Organization. (2018). Maize, starch. *FAOSTAT.* Retrieved from http://www.fao.org/faostat/en/#data/ QC/visualize

Fazeli, M., Keley, M., & Biazar, E. (2018). Preparation and characterization of starch-based composite films reinforced by cellulose nanofibers. *International Journal of Biological Macromolecules, 116*(2017), 272–280. https://doi.org/10.1016/j.ijbiomac.2018.04.186

Fitch-Vargas, P. R., Aguilar-Palazuelos, E., Vega-García, M. O., Zazueta-Morales, J. J., Calderón-Castro, A., Montoya-Rodríguez, A., ... Camacho-Hernández, I. L. (2019). Effect of a corn starch coating obtained by the combination of extrusion process and casting technique on the postharvest quality of tomato. *Revista Mexicana de Ingeniera Quimica, 18*(3), 789–801. https://doi.org/10.24275/uam/izt /dcbi/revmexingquim/2019v18n3/Fitch

Forato, L. A., Britto, D. de, Scramin, J. A., Colnago, L. A., & Assis, O. B. G. (2013). Propriedades Mecânicas e Molhabilidade de filmes de Zeínas Extraídas de Glúten de Milho mechanical and wetting properties of Zein films extracted from from corn gluten meal, *23*, 42–48.

Formhals, A. (1934). *Process and apparatus for preparing artificial threads* (Patent No. US Patent 1975504).

Gahlawat, S. K., Salar, R. K., Siwach, P., Duhan, J. S., Kumar, S., & Kaur, P. (2017). Plant biotechnology: Recent advancements and developments. *Plant Biotechnology: Recent Advancements and Developments*, 1–390. https://doi.org/10.1007/978-981-10-4732-9

García, M. A., Martino, M. N., & Zaritzky, N. E. (2000). Microstructural characterization of plasticized starch-based films. *Starch/ Staerke, 52*(4), 118–124. https://doi.org/10.1002/1521-379X(20000 6)52:4<118::AID-STAR118>3.0.CO;2-0

García-Lara, S., Chuck-Hernandez, C., & Serna-Saldivar, S. O. (2018). Development and structure of the corn kernel. In S. O. Serna-Saldivar (Ed.), *Corn: Chemistry and technology* (3rd ed., p. 674). Duxford: Woodhead Publishing.

García-Rosas, M., Bello-Pérez, A., Yee-Madeira, H., Ramos, G., Flores-Morales, A., & Mora-Escobedo, R. (2009). Resistant starch content and structural changes in Maize (*Zea mays*) tortillas during storage. *Starch/Staerke, 61*(7), 414–421. https://doi.org/10.1002/star .200800147

Ghiasi, K., Varriano-Marston, E., & Hoseney, R. C. (1982). Gelatinization of wheat starch. II. Starch-surfactant interaction. *Cereal chemistry, 59*(2), 86–88.

Giuri, A., Colella, S., Listorti, A., Rizzo, A., & Esposito Corcione, C. (2018). Biodegradable extruded thermoplastic maize starch for outdoor applications. *Journal of Thermal Analysis and Calorimetry, 134*(1), 549–558. https://doi.org/10.1007/s10973-018-7404-7

Gonçalves, P. M., Noreña, C. P. Z., da Silveira, N. P., & Brandelli, A. (2014). Characterization of starch nanoparticles obtained from Araucaria angustifolia seeds by acid hydrolysis and ultrasound. *LWT - Food Science and Technology, 58*(1), 21–27. https://doi.org/10.1016/j.lwt .2014.03.015

Gwirtz, J. A., & Garcia-Casal, M. N. (2014). Processing maize flour and corn meal food products. *Annals of the New York Academy of Sciences, 1312*(1), 66–75. https://doi.org/10.1111/nyas.12299

Hamaker, B. R., Tuncil, Y. E., & Shen, X. (2018). Carbohydrates of the kernel. In *Corn: Chemistry and technology* (3rd ed., pp. 305–318). Elsevier Inc. https://doi.org/10.1016/B978-0-12-811971-6.00011-5

Hasjim, J., Lee, S. O., Hendrich, S., Setiawan, S., Ai, Y., & Jane, J. L. (2010). Characterization of a novel resistant-starch and its effects on postprandial plasma-glucose and insulin responses. *Cereal Chemistry, 87*(4), 257–262. https://doi.org/10.1094/CCHEM-87-4-0257

Hassan, B., Ali, S., Chatha, S., Hussain, A. I., & Zia, K. M. (2018). Recent advances on polysaccharides, lipids and protein based edible films and coatings: A review. *International Journal of Biological Macromolecules, 109*, 1095–1107. https://doi.org/10.1016/j.ijbiomac.2017.11.097

Hazrol, M. D., Sapuan, S. M., Zainudin, E. S., & Zuhri, M. Y. M. (2021). Corn starch (Zea mays) biopolymer plastic reaction in combination with Sorbitol and Glycerol. *Polymers, 13*, 242.

Helstad, S. (2018). Corn sweeteners. In S. O. Serna-Saldivar (Ed.), *Corn: Chemistry and technology* (3rd ed., pp. 551–591). Duxford: Woodhead Publishing.

Hoover, R. (2001). Composition, molecular structure, and physicochemical properties of tuber and root starches: A review. *Carbohydrate Polymers*, *45*(3), 253–267. https://doi.org/10.1016/S0144-8617(00)00260-5

Hossen, M. S., Sotome, I., Takenaka, M., Isobe, S., Nakajima, M., & Okadome, H. (2011). Effect of particle size of different crop starches and their flours on pasting properties. *Japan Journal of Food Engineering*, *12*(1), 29–35. https://doi.org/10.11301/jsfe.12.29

Huber, K. C., & BeMiller, J. N. (1997). Visualization of channels and cavities of corn and sorghum starch granules. *Cereal Chemistry*, *74*(5), 537–541. https://doi.org/10.1094/CCHEM.1997.74.5.537

Hughes, T., Hoover, R., Liu, Q., Donner, E., Chibbar, R., & Jaiswal, S. (2009). Composition, morphology, molecular structure, and physicochemical properties of starches from newly released chickpea (*Cicer arietinum* L.) cultivars grown in Canada. *Food Research International*, *42*(5–6), 627–635. https://doi.org/10.1016/j.foodres.2009.01.008

Ibrahim, M. I. J., Sapuan, S. M., Zainudin, E. S., & Zuhri, M. Y. M. (2019). Potential of using multiscale corn husk fiber as reinforcing filler in cornstarch-based biocomposites. *International Journal of Biological Macromolecules*, *139*, 596–604. https://doi.org/10.1016/j.ijbiomac.2019.08.015

Imam, S. H., Wood, D. F., Abdelwahab, M. A., Chiou, B.-S., Williams, T. G., Glenn, G. M., & Orts, W. J. (2012). Starch chemistry, microstructure, processing, and enzymatic degradation. In J. Ahmed, B. K. Tiwari, S. H. Imam, & M. A. Rao (Eds.), *Starch-based polymeric materials and nanocomposites* (pp. 5–25). Boca Raton, FL: CRC Press.

Imberty, A., Chanzy, H., Pérez, S., Buléon, A., & Vinh, T. (1987). New three-dimensional structure for A-type starch. *Macromolecular Symposia*, *20*(10), 2634–2636.

Imberty, A., & Perez, S. (1988). A revisit to the three-dimensional structure of B-type starch. *Biopolymers*, *27*(8), 1205–1221.

Ingredion.com. (2021). Starches. Retrieved from www.ingredion.com

Ishiguro, K., Noda, T., Kitahara, K., & Yamakawa, O. (2000). Retrogradation of sweetpotato starch. *Starch/Staerke*, *52*(1), 13–17. https://doi.org/10.1002/(SICI)1521-379X(200001)52:1<13::AID-STAR13>3.0.CO;2-E

Jackson, D. S., & Shandera Jr., D. L. (1995). Corn wet milling: Separation chemistry and technology. In John E. Kinsella, Steve L. Taylor (Eds.), *Advances in food and nutrition research* (Vol. 38, pp. 1487–1488). Cambridge: Academic Press.

Jacobson, M. R., Obanni, M., & Bemiller, J. N. (1997). Retrogradation of starches from different botanical sources. *Cereal Chemistry*, *74*(5), 511–518.

Jayakody, L., & Hoover, R. (2008). Effect of annealing on the molecular structure and physicochemical properties of starches from different botanical origins - A review. *Carbohydrate Polymers*, *74*(3), 691–703. https://doi.org/10.1016/j.carbpol.2008.04.032

Ji, Y., Seetharaman, K., & White, P. J. (2004). Optimizing a small-scale cornstarch extraction method for use in the laboratory. *Cereal Chemistry*, *81*(1), 55–58. https://doi.org/10.1094/CCHEM.2004.81.1.55

Jiang, T., Duan, Q., Zhu, J., Liu, H., & Yu, L. (2020). Starch-based biodegradable materials: Challenges and opportunities. *Advanced Industrial and Engineering Polymer Research*, *3*(1), 8–18. https://doi.org/10.1016/j.aiepr.2019.11.003

Jiménez, A., Fabra, M. J., Talens, P., & Chiralt, A. (2012). Edible and biodegradable starch films: A review. *Food and Bioprocess Technology*, *5*(6), 2058–2076. https://doi.org/10.1007/s11947-012-0835-4

Keeling, P. L., Bacon, P. J., & Holt, D. C. (1993). Elevated temperature reduces starch deposition in wheat endosperm by reducing the activity of soluble starch synthase. *Planta*, *191*(3), 342–348. https://doi.org/10.1007/BF00195691

Khan, B., Bilal Khan Niazi, M., Samin, G., & Jahan, Z. (2017). Thermoplastic starch: A possible biodegradable food packaging material—A review. *Journal of Food Process Engineering*, *40*(3). https://doi.org/10.1111/jfpe.12447

Khwaldia, K., Arab-Tehrany, E., & Desobry, S. (2010). Biopolymer coatings on paper packaging materials. *Comprehensive Reviews in Food Science and Food Safety*, *9*(1), 82–91. https://doi.org/10.1111/j.1541-4337.2009.00095.x

Kim, K., & Kim, J. (2021). Understanding wheat starch metabolism in properties, environmental stress condition, and molecular approaches for value-added utilization. *Plants*, *10*(2282), 1–23.

Klein, B., Pinto, V. Z., Vanier, N. L., Zavareze, E. D. R., Colussi, R., do Evangelho, J. A., … Dias, A. R. G. (2013). Effect of single and dual heat-moisture treatments on properties of rice, cassava, and pinhao starches. *Carbohydrate Polymers*, *98*(2), 1578–1584. https://doi.org/10.1016/j.carbpol.2013.07.036

Koch, K., Gillgren, T., Stading, M., & Andersson, R. (2010). Mechanical and structural properties of solution-cast high-amylose maize starch films. *International Journal of Biological Macromolecules*, *46*(1), 13–19. https://doi.org/10.1016/j.ijbiomac.2009.10.002

Kong, L., & Ziegler, G. R. (2012a). Patents on fiber spinning from starches. *Recent Patents on Food, Nutrition & Agriculturee*, *4*(3), 210–219. https://doi.org/10.2174/2212798411204030210

Kong, L., & Ziegler, G. R. (2012b). Role of Molecular Entanglements in Starch Fiber Formation by Electrospinning. *Biomacromolecules*, *13*, 2247–2253.

Kong, L., & Ziegler, G. R. (2013). Quantitative relationship between electrospinning parameters and starch fiber diameter. *Carbohydrate Polymers*, *92*(2), 1416–1422. https://doi.org/10.1016/j.carbpol.2012.09.026

Kong, L., & Ziegler, G. R. (2014a). Fabrication of pure starch fibers by electrospinning. *Food Hydrocolloids*, *36*, 20–25. https://doi.org/10.1016/j.foodhyd.2013.08.021

Kong, L., & Ziegler, G. R. (2014b). Formation of starch-guest inclusion complexes in electrospun starch fibers. *Food Hydrocolloids*, *38*, 211–219. https://doi.org/10.1016/j.foodhyd.2013.12.018

Kristo, E., & Biliaderis, C. G. (2007). Physical properties of starch nanocrystal-reinforced pullulan films. *Carbohydrate Polymers*, *68*(1), 146–158. https://doi.org/10.1016/j.carbpol.2006.07.021

Lagaron, J.-M. (2019). *Multifunctional and nanoreinforced polymers for food packaging* (Vol. 1, Issue). Woodhead Publishing. https://doi.org/10.1017/CBO9781107415324.004

Lancuški, A., Aiman, A. A., Avrahami, R., Vilensky, R., Vasilyev, G., & Zussman, E. (2017). Design of starch-formate compound fibers as encapsulation platform for biotherapeutics. *Carbohydrate Polymers*, *158*, 68–76. https://doi.org/10.1016/j.carbpol.2016.12.003

Lawton, J. W., Shogren, R. L., & Tiefenbacher, K. F. (2004). Aspen fiber addition improves the mechanical properties of baked cornstarch foams. *Industrial Crops and Products*, *19*(1), 41–48. https://doi.org/10.1016/S0926-6690(03)00079-7

Le Corre, D., & Angellier-Coussy, H. (2014). Preparation and application of starch nanoparticles for nanocomposites: A review. *Reactive and Functional Polymers*, *85*, 97–120. https://doi.org/10.1016/j.reactfunctpolym.2014.09.020

Le Leu, R. K., Hu, Y., Brown, I. L., & Young, G. P. (2009). Effect of high amylose maize starches on colonic fermentation and apoptotic response to DNA-damage in the colon of rats. *Nutrition and Metabolism*, *6*, 1–9. https://doi.org/10.1186/1743-7075-6-11

Leach, H. W., Mcwen, L. D., & Schoch, T. J. (1959). Structure of the starch granule. I. Swelling and solubility patterns of various starches. *Cereal Chemistry*, *36*, 534–544.

Le Corre, D. S., Bras, J., & Dufresne, A. (2010). Starch nanoparticles: A review. *Biomacromolecules*, *11*(5), 1139–1153.

Le Corre, D., Bras, J., & Dufresne, A. (2011). Influence of botanic origin and amylose content on the morphology of starch nanocrystals. *Journal of Nanoparticle Research*, *13*(12), 7193–7208. https://doi.org/10.1007/s11051-011-0634-2

Lenhani, G. C., dos Santos, D. F., Koester, D. L., Biduski, B., Deon, V. G., Machado Junior, M., & Pinto, V. Z. (2021). Application of corn fibers from harvest residues in biocomposite films. *Journal of Polymers and the Environment*. https://doi.org/10.1007/s10924-021-02078-6

Li, H., Dhital, S., Flanagan, B. M., Mata, J., Gilbert, E. P., & Gidley, M. J. (2020). High-amylose wheat and maize starches have distinctly different granule organization and annealing behaviour: A key role for chain mobility. *Food Hydrocolloids*, *105*(December 2019), 105820. https://doi.org/10.1016/j.foodhyd.2020.105820

Li, H., Qi, Y., Zhao, Y., Chi, J., & Cheng, S. (2019). Starch and its derivatives for paper coatings: A review. *Progress in Organic Coatings*, *135*(April), 213–227. https://doi.org/10.1016/j.porgcoat.2019.05.015

Li, J., & Yeh, A. (2001). Relationships between thermal, rheological characteristics and swelling power for various starches. *Journal of Food Engineering, 50*(3), 141–148.

Li, L., Blanco, M., & Jane, J. L. (2007). Physicochemical properties of endosperm and pericarp starches during maize development. *Carbohydrate Polymers, 67*(4), 630–639. https://doi.org/10.1016/j.carbpol.2006.08.013

Li, M., Liu, P., Zou, W., Yu, L., Xie, F., Pu, H., … Chen, L. (2011). Extrusion processing and characterization of edible starch films with different amylose contents. *Journal of Food Engineering, 106*(1), 95–101. https://doi.org/10.1016/j.jfoodeng.2011.04.021

Li, X., Qiu, C., Ji, N., Sun, C., Xiong, L., & Sun, Q. (2015). Mechanical, barrier and morphological properties of starch nanocrystals-reinforced pea starch films. *Carbohydrate Polymers, 121*, 155–162. https://doi.org/10.1016/j.carbpol.2014.12.040

Li, X., Wang, C., Lu, F., Zhang, L., Yang, Q., & Li, X. (2015). Effect of bonding forces on corn starch isolation. *Cereal Chemistry, 92*(4), 418–425. https://doi.org/10.1094/CCHEM-05-14-0092-R

Lii, C.-Y., Shao, Y.-Y., & Tseng, K.-H. (1995). Gelation mechanism of rice starch. *Cereal Chemistry, 73*(4), 393–400.

Lim, L. T., Mendes, A. C., & Chronakis, I. S. (2019). Electrospinning and electrospraying technologies for food applications. In *Advances in food and nutrition research* (1st ed., Vol. 88). Elsevier Inc. https://doi.org/10.1016/bs.afnr.2019.02.005

Limpisut, P., & Jindal, V. K. (2002). Comparison of rice flour pasting properties using Brabender Viscoamylograph and rapid visco analyser for evaluating cooked rice texture. *Starch/Staerke, 54*(8), 350–357. https://doi.org/10.1002/1521-379X(200208)54:8<350::AID-STAR350>3.0.CO;2-R

Lin, N., Huang, J., Chang, P. R., Anderson, D. P., & Yu, J. (2011). Preparation, modification, and application of starch nanocrystals in nanomaterials: A review. *Journal of Nanomaterials, 2011*. https://doi.org/10.1155/2011/573687

Lin, Y., Huff, H. E., Parsons, M. H., Iannotti, E., & Hsieh, F. (1995). Mechanical properties of extruded high amylose starch for loose-fill packaging material. *LWT - Food Science and Technology, 28*(2), 163–168. https://doi.org/10.1016/S0023-6438(95)91294-0

Lindeboom, N., Chang, P. R., & Tyler, R. T. (2004). Analytical, biochemical and physicochemical aspects of starch granule size, with emphasis on small granule starches: A review. *Starch/Stärke, 56*(34), 89–99. https://doi.org/10.1002/star.200300218

Liu, H., Xie, F., Yu, L., Chen, L., & Li, L. (2009). Thermal processing of starch-based polymers. *Progress in Polymer Science Journal, 34*(12), 1348–1368. https://doi.org/10.1016/j.progpolymsci.2009.07.001

Lomelí-Ramírez, M. G., Kestur, S. G., Manríquez-González, R., Iwakiri, S., De Muniz, G. B., & Flores-Sahagun, T. S. (2014). Bio-composites of cassava starch-green coconut fiber: Part II - Structure and properties. *Carbohydrate Polymers, 102*(1), 576–583. https://doi.org/10.1016/j.carbpol.2013.11.020

Lopes-Filho, J. F., Buriak, P., Tumbleson, M. E., & Eckhoff, S. R. (1997). Intermittent milling and dynamic steeping process for corn starch recovery. *Cereal Chemistry*, *74*(5), 633–638. https://doi.org/10.1094/CCHEM.1997.74.5.633

Lopez-Rubio, A., Flanagan, B. M., Shrestha, A. K., Gidley, M. J., & Gilbert, E. P. (2008). Molecular rearrangement of starch during in vitro digestion: Toward a better understanding of enzyme resistant starch formation in processed starches. *Biomacromolecules*, *9*(7), 1951–1958. https://doi.org/10.1021/bm800213h

Maldonado-Guzmán, H., López-Paredes, O., & Biliaderis, G. C. (1995). Critical reviews in food science and nutrition amylolytic enzymes and products derived from starch: A review amylolytic enzymes and products derived from starch: A review. *Critical Reviews in Food Science and Nutrition*, *35*(5), 373–403.

Mali, S., Grossmann, M. V. E., & Yamashita, F. (2010). Filmes de amido: Produção, propriedades e potencial de utilização. *Semina: Ciencias Agrarias*, *31*(1), 137–156. https://doi.org/10.5433/1679-0359.2010v31n1p137

Mathew, A. P., Oksman, K., & Sain, M. (2005). Mechanical properties of biodegradable composites from poly lactic acid (PLA) and microcrystalline cellulose (MCC). *Journal of Applied Polymer Science*, *97*(5), 2014–2025. https://doi.org/10.1002/app.21779

Maurer, H. W., & Kearney, R. L. (1998). Opportunities and challenges for starch in the paper industry. *Starch/Staerke*, *50*(9), 396–402. https://doi.org/10.1002/(SICI)1521-379X(199809)50:9<396::AID-STAR396>3.0.CO;2-8

McCleary, B. V., & Monaghan, D. A. (2002). Measurement of resistant starch. *Journal of AOAC International*, *85*(3), 665–675. https://doi.org/10.1093/jaoac/85.3.665

McNaught, K. J., Moloney, E., Brown, I. L., & Knight, A. T. (1998). High amylose starch and resistant starch fractions (US Patent No. 5,714,600).

Miao, M., Zhang, T., Mu, W., & Jiang, B. (2010). Effect of controlled gelatinization in excess water on digestibility of waxy maize starch. *Food Chemistry*, *119*(1), 41–48. https://doi.org/10.1016/j.foodchem.2009.05.035

Mohd Azhar, S. H., Abdulla, R., Jambo, S. A., Marbawi, H., Gansau, J. A., Mohd Faik, A. A., & Rodrigues, K. F. (2017). Yeasts in sustainable bioethanol production: A review. *Biochemistry and Biophysics Reports*, *10*(November 2016), 52–61. https://doi.org/10.1016/j.bbrep.2017.03.003

Molina-Boisseau, S., Dole, P., Dufresne, A., & Dufresne, A. (2006). Thermoplastic starch - Waxy maize starch nanocrystals nanocomposites. *Biomacromolecules*, *7*(2), 531–539.

Molina-Boisseau, S., & Dufresne, A. (2006). Waxy maize starch nanocrystals as filler in natural rubber. *Macromolecular Symposia*, *233*, 132–136. https://doi.org/10.1002/masy.200650117

Morin, S., Dumoulin, L., Delahaye, L., Jacquet, N., & Richel, A. (2021). Green composites based on thermoplastic starches and various natural plant fibers: Impacting parameters of the mechanical properties using machine-learning. *Polymer Composites*, April, 3458–3467. https://doi.org/10.1002/pc.26071

Morrison, W. R. (1988). Lipids in cereal starches: A review. *Journal of Cereal Science, 8*(1), 1–15. https://doi.org/10.1016/S0733-5210(88)80044-4

Mua, J. P., & Jackson, D. S. (1997). Relationships between functional attributes and molecular structures of amylose and amylopectin fractions from corn starch. *Journal of Agricultural and Food Chemistry, 45*(10), 3848–3854. https://doi.org/10.1021/jf9608783

Mua, J. P., & Jackson, D. S. (1998). Retrogradation and gel textural attributes of corn starch amylose and amylopectin fractions. *Journal of Cereal Science, 27*(2), 157–166. https://doi.org/10.1006/jcrs.1997.0161

Müller, C. M. O., Laurindo, J. B., & Yamashita, F. (2009). Effect of cellulose fibers addition on the mechanical properties and water vapor barrier of starch-based films. *Food Hydrocolloids, 23*(5), 1328–1333. https://doi.org/10.1016/j.foodhyd.2008.09.002

Nafchi, A. M., Moradpour, M., Saeidi, M., & Alias, A. K. (2013). Thermoplastic starches: Properties, challenges, and prospects. *Starch/Staerke, 65*(1–2), 61–72. https://doi.org/10.1002/star.201200201

Nordin, N., Othman, S. H., Rashid, S. A., & Basha, R. K. (2020). Effects of glycerol and thymol on physical, mechanical, and thermal properties of corn starch films. *Food Hydrocolloids, 106*(March), 105884. https://doi.org/10.1016/j.foodhyd.2020.105884

Noruzi, M. (2016). Electrospun nanofibres in agriculture and the food industry: A review. *Journal of the Science of Food and Agriculture, 96*(14), 4663–4678. https://doi.org/10.1002/jsfa.7737

Paes, M. C. D. (2006). Aspectos físicos, químicos e tecnológia do grão de milho. *Circular Técnica, 75*, 1–6.

Palanisamy, C. P., Cui, B., Zhang, H., Jayaraman, S., & Muthukaliannan, G. K. (2020). A comprehensive review on corn starch-based nanomaterials: Properties, simulations, and applications. *Polymers, 12*(9). https://doi.org/10.3390/POLYM12092161

Pan, D., & Jane, J. L. (2000). Internal structure of normal maize starch granules revealed by chemical surface gelatinization. *Biomacromolecules, 1*(1), 126–132. https://doi.org/10.1021/bm990016l

Paraginski, R. T., Colussi, R., Dias, A. R. G., da Rosa Zavareze, E., Elias, M. C., & Vanier, N. L. (2019). Physicochemical, pasting, crystallinity, and morphological properties of starches isolated from maize kernels exhibiting different types of defects. *Food Chemistry, 274*(September 2018), 330–336. https://doi.org/10.1016/j.foodchem.2018.09.026

Paraginski, R. T., Vanier, N. L., Moomand, K., De Oliveira, M., Zavareze, E. D. R., e Silva, R. M., ... Elias, M. C. (2014). Characteristics of starch isolated from maize as a function of grain storage temperature. *Carbohydrate Polymers, 102*(1), 88–94. https://doi.org/10.1016/j.carbpol.2013.11.019

Pelissari, F. M., Andrade-Mahecha, M. M., Sobral, P. J. do A., & Menegalli, F. C. (2017). Nanocomposites based on banana starch reinforced with cellulose nanofibers isolated from banana peels. *Journal of Colloid and Interface Science, 505*, 154–167. https://doi.org/10.1016/j.jcis.2017.05.106

Peng, Y., Yao, T., Xu, Q., & Janaswamy, S. (2022). Preparation and characterization of corn flours with variable starch digestion. *Food Chemistry,*
366(July 2021), 130609. https://doi.org/10.1016/j.foodchem.2021
.130609

Pérez, O. E., Haros, M., & Suarez, C. (2001). Corn steeping: Influence of
time and lactic acid on isolation and thermal properties of starch.
Journal of Food Engineering, 48(3), 251–256. https://doi.org/10.1016/
S0260-8774(00)00165-5

Pérez, S., & Bertoft, E. (2010). The molecular structures of starch components and their contribution to the architecture of starch granules: A
comprehensive review. *Starch/Stärke, 62*, 389–420. https://doi.org/10
.1002/star.201000013

Persano, L., Camposeo, A., Tekmen, C., & Pisignano, D. (2013). Industrial
upscaling of electrospinning and applications of polymer nanofibers:
A review. *Macromolecular Materials and Engineering, 298*(5), 504–
520. https://doi.org/10.1002/mame.201200290

Pinto, V. Z., Moomand, K., Deon, V. G., Biduski, B., Lenhani, G. C., Zavareze,
E. D. R., ... Dias, A. R. G. (2021). Effect of physical pretreatments on
the structural and physicochemical properties of pinhão starch nanocrystals. *Starch/Stärke.* https://doi.org./10.1002/star.202000008

Plumier, B. M., Danao, M. G. C., Rausch, K. D., & Singh, V. (2015). Changes
in unreacted starch content in corn during storage. *Journal of Stored*
Products Research, 61, 85–89. https://doi.org/10.1016/j.jspr.2014.11
.006

Putseys, J. A., Lamberts, L., & Delcour, J. A. (2010). Amylose-inclusion
complexes: Formation, identity and physico-chemical properties.
Journal of Cereal Science, 51(3), 238–247. https://doi.org/10.1016/j.jcs
.2010.01.011

Qiao, D., Zhang, B., Huang, J., Xie, F., Wang, D. K., Jiang, F., ... Zhu, J.
(2017). Hydration-induced crystalline transformation of starch polymer under ambient conditions. *International Journal of Biological*
Macromolecules, 103, 152–157. https://doi.org/10.1016/j.ijbiomac
.2017.05.008

Ranum, P., Peña-Rosas, J. P., & Garcia-Casal, M. N. (2014). Global maize
production, utilization, and consumption. *Annals of the New York*
Academy of Sciences, 1312(1), 105–112. https://doi.org/10.1111/nyas
.12396

Rausch, K. D., Hummel, D., Johnson, L. A., & May, J. B. (2018). Wet milling:
The basis for corn biorefineries. In S. O. Serna-Saldivar (Ed.), *Corn:*
Chemistry and technology (3rd ed., pp. 503–535). Duxford: Woodhead
Publishing.

Rendon-Villalobos, R., Bello-Pérez, L. A., Osorio-Díaz, P., Tovar, J., &
Paredes-López, O. (2002). Effect of storage time on in vitro digestibility and resistant starch content of nixtamal, masa, and tortilla. *Cereal*
Chemistry, 79(3), 340–344. https://doi.org/10.1094/CCHEM.2002.79
.3.340

Rincón-Londoño, N., Vega-Rojas, L. J., Contreras-Padilla, M., Acosta-Osorio, A. A., & Rodríguez-García, M. E. (2016). Analysis of the pasting profile in corn starch: Structural, morphological, and thermal transformations, part I. *International Journal of Biological Macromolecules, 91,* 106–114. https://doi.org/10.1016/j.ijbiomac.2016.05.070

Rivadeneira-Velasco, K. E., Utreras-Silva, C. A., Díaz-Barrios, A., Sommer-Márquez, A. E., Tafur, J. P., & Michell, R. M. (2021). Green nanocomposites based on thermoplastic starch: A review. *Polymers, 13*(19). https://doi.org/10.3390/polym13193227

Rojas-Molina, I., Mendoza-Avila, M., Cornejo-Villegas, M. D. L. Á., Real-López, A. Del, Rivera-Muñoz, E., Rodríguez-García, M., & Gutiérrez-Cortez, E. (2020). Physicochemical properties and resistant starch content of corn tortilla flours refrigerated at different storage times. *Foods, 9*(4), 1–20. https://doi.org/10.3390/foods9040469

Sadeghizadeh-Yazdi, J., Habibi, M., Kamali, A. A., & Banaei, M. (2019). Application of edible and biodegradable starch-based films in food packaging: A systematic review and meta-analysis. *Current Research in Nutrition and Food Science, 7*(3), 624–637. https://doi.org/10.12944/CRNFSJ.7.3.03

Sajilata, M. G., & Singhal, R. S. (2005). Specialty starches for snack foods. *Carbohydrate Polymers, 59*(2), 131–151. https://doi.org/10.1016/j.carbpol.2004.08.012

Salman, H., & Copeland, L. (2007). Effect of storage on fat acidity and pasting characteristics of wheat flour. *Cereal Chemistry, 84*(6), 600–606. https://doi.org/10.1094/CCHEM-84-6-0600

Sandhu, K. S., & Singh, N. (2007). Some properties of corn starches II: Physicochemical, gelatinization, retrogradation, pasting and gel textural properties. *Food Chemistry, 101*(4), 1499–1507. https://doi.org/10.1016/j.foodchem.2006.01.060

Santiago-Ramos, D., Figueroa-Cárdenas, J. D. D., Véles-Medina, J. J., Mariscal-Moreno, R. M., Reynoso-Camacho, R., Ramos-Gómez, M., ... Morales-Sánchez, E. (2015). Resistant starch formation in tortillas from an ecological nixtamalization process. *Cereal Chemistry, 92*(2), 185–192.

Sapper, M., & Chiralt, A. (2018). Starch-based coatings for preservation of fruits and vegetables. *Coatings, 8*(5). https://doi.org/10.3390/coatings8050152

Schirmer, M., Jekle, M., & Becker, T. (2015). Starch gelatinization and its complexity for analysis. *Starch/Staerke, 67*(1–2), 30–41. https://doi.org/10.1002/star.201400071

Schneider, L., & Gärtner, H. (2013). The advantage of using a starch based non-Newtonian fluid to prepare micro sections. *Dendrochronologia, 31*(3), 175–178. https://doi.org/10.1016/j.dendro.2013.04.002

Selling, G. W. (2010). The effect of extrusion processing on Zein. *Polymer Degradation and Stability, 95*(12), 2241–2249. https://doi.org/10.1016/j.polymdegradstab.2010.09.013

Serna-Saldivar, S. O. (2018). *Corn: Chemistry and technology* (3rd ed., Vol. 148) (S. O. Serna-Saldivar, Ed.). Duxford: Woodhead Publishing.

Singh Sodhi, N., & Singh, N. (2003). Morphological, thermal and rheological properties of starches separated from rice cultivars grown in India. *Food Chemistry, 80*(1), 99–108. https://doi.org/10.1016/S0308 -8146(02)00246-7

Singh, J., Dartois, A., & Kaur, L. (2010). Starch digestibility in food matrix: A review. *Trends in Food Science and Technology, 21*(4), 168–180. https://doi.org/10.1016/j.tifs.2009.12.001

Singh, N., & Eckhoff, S. R. (1996). Wet milling of corn - A review of laboratory-scale and pilot plant- scale procedures. *Cereal Chemistry, 73*(6), 659–667.

Singh, V., Urs, R. G., Somashekarappa, H., Ali, S. Z., & Somashekar, R. (1995). X-ray analysis of different starch granules. *Bulletin of Material Science, 18*(5), 549–555.

Sjoo, M., & Nilsson, L. (2004). Starch in food. In *Starch in food* (2nd ed.). Woodhead Publishing. https://doi.org/10.1201/9781439823347

Soest, J. J. G. Van, Benes, K., & De, D. W. (1996). Influence of starch molecular mass on the properties of extruded thermoplastic starch. *Polymer, 37*(16), 3543–3552.

Spencer, I. K. C., Rojak, P. A., & Sabatini, K. S. (1996). Method of producing high fructose corn syrup from glucose using noble gases. United States Patent.

Srichuwong, S., Candra, T., Mishima, T., Isono, N., & Hisamatsu, M. (2005). Starches from different botanical sources II: Contribution of starch structure to swelling and pasting properties, *62*(1), 25–34. https://doi .org/10.1016/j.carbpol.2005.07.003

Steffe, J. F. (1996). *Rheological methods in food process engineering* (2nd ed.) (J. F. Steffe, Ed.). San Francisco, CA: Freeman Press.

Suh, D. S., & Jane, J. L. (2003). Comparison of starch pasting properties at various cooking conditions using the micro visco-amylo-graph and the rapid visco analyser. *Cereal Chemistry, 80*(6), 745–749. https://doi .org/10.1094/CCHEM.2003.80.6.745

Suhag, R., Kumar, N., Trajkovska, A., & Upadhyay, A. (2020). Film formation and deposition methods of edible coating on food products: A review. *Food Research International, 136*(March), 109582. https://doi .org/10.1016/j.foodres.2020.109582

Sujka, M., Pankiewicz, U., Kowalski, R., Nowosad, K., & Noszczyk-Nowak, A. (2018). Porous starch and its application in drug delivery systems. *Polimery w Medycynie, 48*(1). Retrieved from www.polimery.umed .wroc.pl

Sun, H., Shao, X., & Ma, Z. (2016). Effect of incorporation nanocrystalline corn straw cellulose and polyethylene glycol on properties of biodegradable films. *Journal of Food Science, 81*(10), E2529–E2537. https:// doi.org/10.1111/1750-3841.13427

Svihus, B., Uhlen, A. K., & Harstad, O. M. (2005). Effect of starch granule structure, associated components and processing on nutritive value of cereal starch: A review. *Animal Feed Science and Technology, 122*(3–4), 303–320. https://doi.org/10.1016/j.anifeedsci.2005.02.025

Taiz, L., Zeiger, E., Møller, I. M., & Murphy, A. (2017). Fisiologia e desenvolvimento vegetal. In *Biochemical education*. Retrieved from https://linkinghub.elsevier.com/retrieve/pii/0307441276901217

Takeiti, C., Fakhouri, F., Ormenese, R., Steel, C., & Collares, F. (2007). Freeze-thaw stability of gels prepared from starches of non-conventional sources. *Starch/Staerke, 59*(3–4), 156–160. https://doi.org/10.1002/star.200600544

Tang, M. C., & Copeland, L. (2007). Investigation of starch retrogradation using atomic force microscopy. *Carbohydrate Polymers, 70*(1), 1–7. https://doi.org/10.1016/j.carbpol.2007.02.025

Tang, X., Alavi, S., & Herald, T. J. (2008). Barrier and mechanical properties of starch-clay nanocomposite films. *Cereal Chemistry, 85*(3), 433–439. https://doi.org/10.1094/CCHEM-85-3-0433

Teacă, C. A., Bodîrlău, R., & Spiridon, I. (2013). Effect of cellulose reinforcement on the properties of organic acid modified starch microparticles/plasticized starch bio-composite films. *Carbohydrate Polymers, 93*(1), 307–315. https://doi.org/10.1016/j.carbpol.2012.10.020

Tester, R. F., Karkalas, J., & Qi, X. (2004). Starch — Composition, fine structure and architecture. *Journal of Cereal Science, 39*(2), 151–165. https://doi.org/10.1016/j.jcs.2003.12.001

Tester, R. F., & Morrison, W. R. (1990). Swelling and gelatinization of cereal starches. I. Effects of amylopectin, amylose, and lipids. *Cereal Chemistry, 6*(67), 551–557.

Thakur, R., Pristijono, P., Scarlett, C. J., Bowyer, M., Singh, S. P., & Vuong, Q. V. (2019). Starch-based films: Major factors affecting their properties. *International Journal of Biological Macromolecules, 132*, 1079–1089. https://doi.org/10.1016/j.ijbiomac.2019.03.190

Thanyapanich, N., Jimtaisong, A., & Rawdkuen, S. (2021). Functional properties of banana starch (Musa spp.) and its utilization in cosmetics. *Molecules, 26*(12), 1–16. https://doi.org/10.3390/molecules26123637

Tian, Y., Li, Y., Xu, X., & Jin, Z. (2011). Starch retrogradation studied by thermogravimetric analysis (TGA). *Carbohydrate Polymers, 84*(3), 1165–1168. https://doi.org/10.1016/j.carbpol.2011.01.006

Timm, N. da S., Ramos, A. H., Ferreira, C. D., Biduski, B., Eicholz, E. D., & Oliveira, M. de (2020). Effects of drying temperature and genotype on morphology and technological, thermal, and pasting properties of corn starch. *International Journal of Biological Macromolecules, 165*, 354–364. https://doi.org/10.1016/j.ijbiomac.2020.09.197

Vamadevan, V., & Bertoft, E. (2015). Structure-function relationships of starch components. *Starch/Stärke, 67*, 55–68. https://doi.org/10.1002/star.201400188

Vamadevan, V., & Bertoft, E. (2020). Observations on the impact of amylopectin and amylose structure on the swelling of starch granules. *Food Hydrocolloids*, *103*(October 2019), 105663. https://doi.org/10.1016/j.foodhyd.2020.105663

Vamadevan, V., Bertoft, E., & Seetharaman, K. (2013). On the importance of organization of glucan chains on thermal properties of starch. *Carbohydrate Polymers*, *92*(2), 1653–1659. https://doi.org/10.1016/j.carbpol.2012.11.003

Vamadevan, V., Hoover, R., Bertoft, E., & Seetharaman, K. (2014). Hydrothermal treatment and iodine binding provide insights into the organization of glucan chains within the semi-crystalline lamellae of corn starch granules. *Biopolymers*, *101*(8), 871–885. https://doi.org/10.1002/bip.22468

Vasilyev, G., Putaux, J., Zussman, E., & Zussman, E. (2015). Rheological properties and electrospinnability of high-amylose starch in formic acid. *Biomacromolecules*, *16*(8), 2529–2536. https://doi.org/10.1021/acs.biomac.5b00817

Verbeek, C. J. R., & Van Den Berg, L. E. (2010). Extrusion processing and properties of protein-based thermoplastics. *Macromolecular Materials and Engineering*, *295*(1), 10–21. https://doi.org/10.1002/mame.200900167

Vieira, M. G. A., Da Silva, M. A., Dos Santos, L. O., & Beppu, M. M. (2011). Natural-based plasticizers and biopolymer films: A review. *European Polymer Journal*, *47*(3), 254–263. https://doi.org/10.1016/j.eurpolymj.2010.12.011

Wang, H., Zhang, B., Chen, L., & Li, X. (2016). Understanding the structure and digestibility of heat-moisture treated starch. *International Journal of Biological Macromolecules*, *88*, 1–8. https://doi.org/10.1016/j.ijbiomac.2016.03.046

Wang, S., Li, C., Copeland, L., Niu, Q., & Wang, S. (2015). Starch retrogradation: A comprehensive review. *Comprehensive Reviews in Food Science and Food Safety*, *14*(5), 568–585. https://doi.org/10.1111/1541-4337.12143

Weber, F. H., Clerici, M. T. P. S., Collares-Queiroz, F. P., & Chang, Y. K. (2009). Interaction of guar and xanthan gums with starch in the gels obtained from normal, Waxy and high-amylose corn starches. *Starch/Staerke*, *61*(1), 28–34. https://doi.org/10.1002/star.200700655

Wei, B., Hu, X., Li, H., Wu, C., Xu, X., Jin, Z., & Tian, Y. (2014). Effect of pHs on dispersity of maize starch nanocrystals in aqueous medium. *Food Hydrocolloids*, *36*, 369–373. https://doi.org/10.1016/j.foodhyd.2013.08.015

Yao, Y., Thompson, D. B., & Guiltinan, M. J. (2004). Maize starch-branching enzyme isoforms and amylopectin structure. In the absence of starch-branching enzyme IIb, the further absence of starch-branching enzyme Ia leads to increased branching. *Plant Physiology*, *136*(3), 3515–3523. https://doi.org/10.1104/pp.104.043315

Yu, L., Dean, K., & Li, L. (2006). Polymer blends and composites from renewable resources. *Progress in Polymer Science, 31*(6), 576–602. https://doi.org/10.1016/j.progpolymsci.2006.03.002

Yu, X., Yu, H., Zhang, J., Shao, S., Xiong, F., & Wang, Z. (2015). Endosperm structure and physicochemical properties of starches from normal, waxy, and super-sweet maize. *International Journal of Food Properties, 18*(12), 2825–2839. https://doi.org/10.1080/10942912.2015.1015732

Yuan, R. C., & Thompson, D. B. (1998). Freeze-thaw stability of three waxy maize starch pastes measured by centrifugation and calorimetry. *Cereal Chemistry, 75*(4), 571–573. https://doi.org/10.1094/CCHEM.1998.75.4.571

Yuan, R. C., Thompson, D. B., & Boyer, C. D. (1993). Fine structure of amylopectin in relation to gelatinization and retrogradation behavior of maize starches from three wx-containing genotypes in two inbred lines. *Cereal Chemistry, 70*(1), 81–89. Retrieved from http://ukpmc.ac.uk/abstract/AGR/IND93048322

Zámostný, P., Petrů, J., & Majerová, D. (2012). Effect of maize starch excipient properties on drug release rate. *Procedia Engineering, 42*(August), 482–488. https://doi.org/10.1016/j.proeng.2012.07.439

Zeeman, S. C., Kossmann, J., & Smith, A. M. (2010). Starch: Its metabolism, evolution, and biotechnological modification in plants. *Annual Review of Plant Biology, 61*, 209–234. https://doi.org/10.1146/annurev-arplant-042809-112301

Zhang, C., Li, Y., Wang, P., & Zhang, H. (2020). Electrospinning of nanofibers: Potentials and perspectives for active food packaging. *Comprehensive Reviews in Food Science and Food Safety, 19*(2), 479–502. https://doi.org/10.1111/1541-4337.12536

Zhang, R., Ma, S., Li, L., Zhang, M., Tian, S., Wang, D., … Wang, X. (2021). Comprehensive utilization of corn starch processing by-products: A review. *Grain & Oil Science and Technology, 4*(3), 89–107. https://doi.org/10.1016/j.gaost.2021.08.003

Zhang, W., Zhao, Y., Li, L., Xu, X., Yang, L., Luo, Z., … Huang, Z. (2021). The effects of short-term exposure to low temperatures during the booting stage on starch synthesis and yields in wheat grain. *Frontiers in Plant Science, 12*(July), 1–18. https://doi.org/10.3389/fpls.2021.684784

Zheng, G. H., Sosulskib, F. W., & Tyler, R. T. (1998). Wet-milling, composition and functional properties of starch and protein isolated from buckwheat groats. *Food Research International, 30*(I), 493–502.

Zhou, X., Baik, B. K., Wang, R., & Lim, S. T. (2010). Retrogradation of waxy and normal corn starch gels by temperature cycling. *Journal of Cereal Science, 51*(1), 57–65. https://doi.org/10.1016/j.jcs.2009.09.005

Ziegler, V., Paraginski, R. T., & Ferreira, C. D. (2021). Grain storage systems and effects of moisture, temperature and time on grain quality - A review. *Journal of Stored Products Research, 91*, 101770. https://doi.org/10.1016/j.jspr.2021.101770

Žilić, S., Serpen, A., Akillioğlu, G., Gökmen, V., & Vančetović, J. (2012). Phenolic compounds, carotenoids, anthocyanins, and antioxidant capacity of colored maize (*Zea mays* L.) kernels. *Journal of Agricultural and Food Chemistry, 60*(5), 1224–1231. https://doi.org/10.1021/jf204367z

Chapter 5

Maize starch modifications and industrial uses

*Anil Kumar Siroha, Sneh Punia Bangar,
Sukriti Singh, and Sukhvinder Singh Purewal*

DOI: 10.1201/9781003245230-5

5.1 Introduction

Zea mays L. is a member of *Poaceae* family that is generally known as corn and maize. It has been developed as a staple food in several regions of the globe (Scott & Emery, 2015). Corn is broadly categorized into six varieties, namely dent corn, flint corn, pod corn, popcorn, flour corn, and sweetcorn (Palanisamy et al., 2020). The industrial food applications of native starches are limited due to their low shear resistance, thermal decomposition, and high tendency towards retrogradation (Hui et al., 2009). To improve these properties, starch is generally modified with physical, chemical, and enzymatic methods and explored for specific applications in food (Rafiq et al., 2016) (see Figure 5.1). Modification is defined as the process that brings changes in starch structure by various factors, i.e. environmental, operational, processing, etc. According to a Mordor Intelligence report, it is estimated that the global market for modified starch could reach USD 13.18 billion by 2024, with an annual growth rate of 4.4% from 2019 to 2024 (MI, 2019).

Physical treatments offer various benefits; some methods contribute to changes in starch molecules and chemical structure without any chemical reagents. Physical modifications seem to be associated with the concept of "clean label" food ingredients for consumers (Zhu et al., 2021). Due to high efficiency, fewer by-products, and easy operation, physical modifications have the potential for industrial applications. However, issues of weak repeatability and high energy consumption remain unsolved (BeMiller & Huber, 2015; Chizoba et al., 2018). In the early 1940s, studies of chemical modifications became popular in the world of modern science and research (Haq et al., 2019). Chemical modification includes introduction of functional groups (carboxylic, ester, ether, and amino groups) to starch molecules without any physical alteration in the shape and size of molecules, resulting

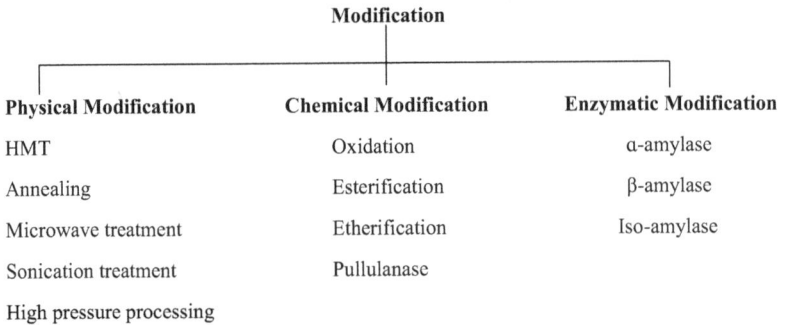

	Modification	
Physical Modification	**Chemical Modification**	**Enzymatic Modification**
HMT	Oxidation	α-amylase
Annealing	Esterification	β-amylase
Microwave treatment	Etherification	Iso-amylase
Sonication treatment	Pullulanase	
High pressure processing		

Figure 5.1 Overview of starch modification

in significantly altered physicochemical properties (Mohammed & Bin, 2020). Chemical modification is widely carried out, but physical modification has a growing relevance, particularly for its use in foods. The key benefit of modification is that starch is considered an extremely safe natural resource and ingredient (Bemiller, 1997). Enzymatic modifications are progressing as a greener alternative to chemical modification due to environmental concerns. Modifications of starch with enzymes primarily use hydrolyzing enzymes (amylases) to degrade starch to various extents for modification (Bangar et al., 2021a). In this chapter different modification techniques for maize starch will be elaborated.

5.2 Physical modification

Food industries have shown their interest in producing more natural food components, so there is an increasing need to improve the properties of native starches without using chemical modifications (Ortega-Ojeda & Eliasson, 2001). These modifications are taken into consideration as they are simple, cost-effective, eco-friendly, and safe methods, and because chemicals or biological agents, which are harmful for human consumption, are not required (Punia, 2020). Therefore, physically modified starches are a current requirement for food industries (see Table 5.1).

5.2.1 Heat moisture treatment

The first study to use heat moisture treatment (HMT) to modify corn starch occurred in 1944 (Sair & Fetzer, 1944). HMT is a physical method and is carried out under a restricted moisture content (10–30%), at higher temperatures (90–120 °C), and for a period ranging from 15 min to 16 h (Maache-Rezzoug et al., 2008) (see Figure 5.2). HMT is considered to be a green technique to alter the molecular, crystalline, and granule structure of starches, therefore regulating the swelling, amylose leaching, and viscosity of starch granules (Hoover, 2010; Jiranuntakul et al., 2012). Both annealing and HMT are performed at conditions above the glass transition temperature (Tg) and below the gelatinization temperature (To) of starch so as to promote molecular rearrangements of the starch chains without disrupting the starch granules (Zavareze and Dias, 2011). The HMT changes the properties of starch by promoting degradation of starch crystallites and/ or interactions between the starch chains in amorphous and crystalline regions (Piecyk & Domian, 2021). The extent of the changes varies with modification parameters (moisture, temperature, and duration), the plant source of the starch, the ratio of amylose to amylopectin, and the structures of these two components (da Rosa Zavareze & Dias, 2011). Significant

Table 5.1 Physical modification of maize starch

Modification	Major results and methodology	References
HMT	Pasting properties and gelatinization temperature reduced after treatment. (Moisture content 25%, temperature 105 °C, time 4 h)	Zhang et al. (2021)
HMT	Amylose content, swelling power, and peak viscosity reduced whereas peak temperature and particle size increased after HMT (moisture content 28%, temperature 110 °C, time 2.5 h)	Chandla et al. (2017)
HMT	Solubility increased while reverse is observed for swelling power and the gelatinization enthalpy of the samples (starch sample is dispersed in guar gum solution, after 30 min. Starch is dried at 45 °C)	Xie et al. (2018)
ANN	Increase in crystallinity is observed. Morphology of ANN starches also altered and the numbers of pores on the granule surface, observed by SEM. (Starch suspension (5% w/v), temperature 62 and 63 °C, time 24 h, vacuum filtered, and dried)	Rocha et al. (2012)
ANN	Swelling power and amylose leaching decreased on ANN while reverse is observed for gelatinization temperature. RS content increased after modification. (Moisture content 70%, temperature 10–15 °C, time 24 h, centrifuged and dried)	Chung et al. (2009)
ANN	Increase in the granular size of ANN starches and gelatinization enthalpy of the amylose-rich starches is observed. (Starch suspension (1:10), temperature 30 or 50 °C, time 72 h, centrifuged and dried)	Liu et al. (2009)
Microwave	PV and AP increase and AM content decrease after treatment. (HAMS, moister 30%, 1.2 kW power, time 1–4 min)	Zhong et al. (2019)

(Continued)

Table 5.1 (Continued) Physical modification of maize starch

Modification	Major results and methodology	References
Microwave	Viscosity increases while gelatinization reduced after treatment. (Starch 45.5 g dissolved in 0.5 g xanthan gum dissolved in 100 ml water, dried and treated with microwave oven at 600 W for 4–6 min)	Sun et al. (2014)
Microwave	Pasting properties reduced and transition temperature increased after treatment. (Moisture 15–40%, 1 h at microwave power of either 0.17 W/g or 0.5 W/g)	Stevenson et al. (2005)
HPP	Swelling, viscosity, and gelatinization temperature decreased after treatment. (The starch samples (20% w/w) were treated at 100, 300, and 500 MPa for 15 and 30 min at room temperature).	Rahman et al. (2020)
HPP	Starch lamellae and morphology of starches are affected after treatment (samples were held at a maximum pressure level of 100, 200, 300, 400, 500, or 600 MPa for pressurization duration of 30 min)	Yang et al. (2016)

($p<0.05$) effects are observed after the HMT on starch properties. Amylose content and swelling power reduced after HMT of corn starch (Chandla et al., 2017). Xie et al. (2018) also observed a decrease in swelling and an increase in solubility of the maize starches. This reduction in swelling power has also been attributed to increased interactions between the amylose and amylopectin molecules, strengthened intramolecular bonds (Jacobs et al., 1995), the formation of amylose-lipid complexes (Waduge et al., 2006), and changes in the arrangements of the crystalline regions of starch (Hoover & Vasanthan, 1994).

Pasting properties of starches altered after HMT. Zhang et al. (2021) evaluated that after treatment peak viscosity (PV), breakdown viscosity (BV), setback viscosity (SV) and final viscosity (FV) reduced for corn starch. Similar results are observed by Sui et al. (2015) and Sandhu et al. (2020) for heat moisture-treated corn and pearl millet starches. Malumba et al. (2010) reported the effect of pre-treatment (60–130 °C) on pasting temperature and observed an increase in pasting temperature after HMT.

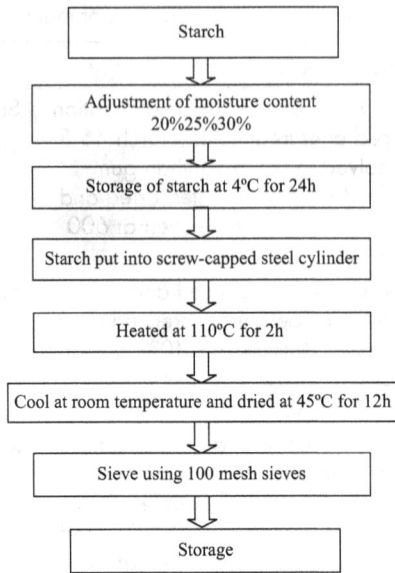

Figure 5.2 Method for heat moisture treatment (Liu et al., 2019)

Xing et al. (2017) evaluated the effect of treatment time on pasting properties of corn starch and observed a significant effect on pasting parameters.

Sui et al. (2015) and Chen et al. (2017) confirmed that gelatinization temperatures (To, Tc, Tp) of HMT starches were higher than native starches. The authors also reported that these effects were the result of the more orderly crystalline structure formed after physical modification. This increase could be attributed to the strengthened interactions of amylose–amylose, amylose–amylopectin, and amylose–lipid (Ambigaipalan et al., 2014; Xing et al., 2017). Yan et al. (2019) evaluated the effect of extrusion on HMT corn starch and observed an increase in gelatinization temperature. In the DSC studies by Takaya et al. (2000), the HMT of maize starch increased the level and rate of retrogradation during storage at 5 °C, while X-ray diffraction studies also showed that retrogradation of maize and potato starches significantly increased during HMT.

In HMT-modified starches, morphological changes are closely related to the moisture, time, and temperature used in the treatment and source of the starch (Schafranski et al., 2021). Chen et al. (2017) reported that HMT caused the destruction and agglomeration of starch granules. Moreover, such a phenomenon became more serious with the increase in treating time. HMT modified starches showed the Maltese cross centred at the hilum. Compared to the native starches, the heat moisture-treated samples exhibited a weaker and fuzzier Maltese cross. Yang et al. (2019) observed that after treatment starch granules became more irregular and appeared less granular as a consequence of extrusion treatment. Bangar et al. (2021b) evaluated cavities and indentation after HMT for pearl millet starch.

5.2.2 Annealing

Annealing (ANN) is a physical modification that involves the treatment of starch granules in the presence of excess (>60% w/w) or intermediate ($\approx$40% w/w) moisture levels for a relatively long period (24–48 h) (Jayakody & Hoover, 2008). Kweon et al. (2008) have shown that treatment of corn starch for 15 min at elevated pressures (600 MPa) resulted in annealing even at 25 °C. The above authors suggested that elevated pressure can induce annealing at temperatures significantly lower than the characteristic Tg at atmospheric pressure. ANN also leads to elevation and sharpening of the gelatinization range and causes little solubilization of the α -glucan (Tester & Debon, 2000). The ANN process leads to a reorganization of starch molecules and amylopectin double helices such that the structure acquires a more organized configuration (Gomes et al., 2005). As a crucial physical modification technology, annealing changes molecular weight, microscopic crystalline structure, physicochemical properties, and *in vitro* digestibility of starch, while maintaining the relative integrity of grain structure (da Rosa Zavareze & Dias, 2011). Rocha et al. (2012) reported that annealing caused an increase in crystallinity in the waxy corn starch, whereas the numbers of pores on the granule surface are observed by SEM. Amylose and amylopectin chains of the annealed normal corn starch are degraded to a greater extent during enzymatic hydrolysis than those of native starch. It has been observed that the granular size of corn starches increased after the annealing process, but the size variation rates are different, with higher amylopectin content resulting in a higher diameter growth rate and final accretion ratio. Increases in the gelatinization enthalpy of the amylose-rich starches have also been observed after the ANN process (Liu et al., 2009). Chung et al. (2009) observed a decrease in swelling power and amylose leaching capacity after ANN treatment of corn starch. Gelatinization temperature, slow digestive starch (SDS), and resistance starch (RS) increased while rapidly digestible starch (RDS) decreased after treatment. Hydrothermally treated starches could be utilized in canned frozen food and in noodles (Hormdok & Noomhorm, 2007; Jayakody & Hoover, 2008).

5.2.3 Microwave treatment

Recently, microwave treatment, a physical modification, has been widely used in modifying starch due to its advantages of high heating speed, high efficiency, and environmental protection. Microwave heating of starch involves the application of electro-magnetic waves within the frequency range of 300 GHz and 300 MHz (Braşoveanu & Nemţanu, 2014). The electric field of microwave radiation vibrates at around 4.9×10^9 times per second, causing polar molecules and ions to rearrange continuously, rub together, and collide with surrounding molecules via electromagnetic induction, thus generating heat energy (El Khaled et al., 2018). Polar materials, like starch, are capable of absorbing microwave energy, consequently aligning them with their electric field (Fan et al., 2012). Microwave irradiation causes polar molecules to vibrate and causes the accumulation of heat; the presence of water plays the main role in this treatment. The higher availability of water molecules will cause more vibrations in the starch matrices and generate more heat (Zailani et al., 2021). Microwave treatment can alter the physicochemical properties of starch, including WAC and swelling power, granular properties, ratio of amylose to amylopectin, thermodynamics, dielectric properties, gelatinization, and pasting properties (Bilbao-Sainz et al., 2007; Man et al., 2012; Nawaz et al., 2018; Wang et al., 2019; Zhong et al., 2019; Bangar et al., 2021b). The method of microwave treatment is described in Figure 5.3.

In 2014, Sun et al. evaluated the effect of microwave treatment on gum addition in corn starch. It was observed that the viscosity of both the normal corn starch (CS) and waxy corn starch (WCS) with xanthan increased compared with untreated samples after microwave-assisted dry heating (MADH), and the effect on WCS was more obvious. Stevenson et al. (2005) observed that the pasting properties of treated CS, PV, BV, and SV reduced after treatment while the reverse was observed for PT. The extent of the decrease in these pasting properties was found to depend on the microwave heating time. A higher microwave heating time was found to result in a greater decrease in the pasting properties of starch (Colman et al., 2014; Yang et al., 2017; Oyeyinka et al., 2019). Zhong et al. (2019) observed that an increase in PV occurred after the increase in treatment time (three and four minutes).

Starch gelatinization is the most important functional characteristic with regard to food processing. Sun et al. (2014) determined the effect of gum addition on microwave treatment of starch. It is observed that MADH with xanthan reduced the To, Tc, Tp, and ΔH values of both the CS and WCS, while Yang et al. (2017) also reported reverse results and observed an increase in gelatinization temperatures and a decrease in ΔH after microwave treatment. This increase could be explained by the enhancement of the interactions among the starch chains in amorphous regions (Li et al., 2019). The higher the relative crystallinity of starch granules, the more ordered the crystal structure and the higher the ΔH value of starch (Ek et al., 2014). Yang et al. (2017) also evaluated that microwave irradiation decreased starch molecular weight and the relative crystallinity of

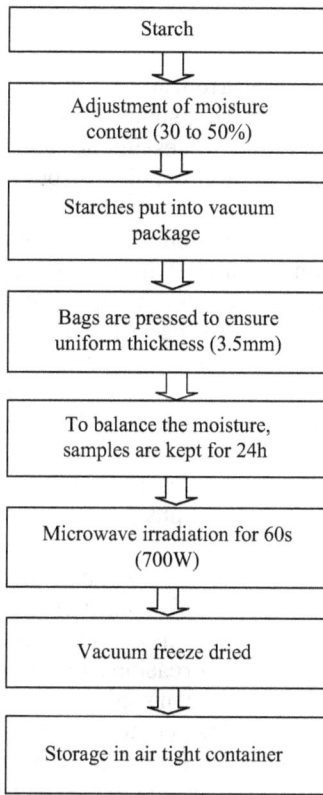

Figure 5.3 Method for preparation of microwave treatment of starch (Li et al., 2019)

waxy maize starch. HNMR data showed α-(1,6) glycosidic linkages were destroyed more easily than α-(1,4) glycosidic linkages during microwave treatment.

Luo et al. (2006) concluded that microwave treatment did not alter the size and shape of these starch granules, which was consistent with the results of Zhong et al. (2019) (high-amylose maize starch, 30% moisture, wb). Ouyang et al. (2021) also reported on the morphological properties of corn starch and observed that ultrasonic treatment resulted in a rough surface, with the characteristic pores of the native corn surface having almost disappeared, which could be because sonication temperature trimmed off the external chains of the amylopectin and amylose molecules, thereby removing some amorphous regions (Ye et al., 2018). Zhang et al. (2021) evaluated the digestive behaviour of microwave-treated corn starch and observed an increase in the resistant starch content.

5.2.4 Sonication treatment

Power ultrasound is considered a tool for the physical modification of starch. Serving as an eco-friendly method, some of the remarkable advantages of ultrasound applications in food processing are an increase in the reaction rate, reduction in the processing time, high throughput, and low consumption of energy (Zhu, 2015). Sonication is a process in which sound waves are used to agitate particles in a solution. When the ultrasonic energy propagates in the liquid, cavitation bubbles are formed due to pressure changes. These bubbles collapse violently in subsequent cycles of compression as the sound waves propagate, resulting in regions of high temperature and pressure (Dias et al., 2015). Ultrasound techniques use mechanical waves with frequencies above 16 kHz. Ultrasound techniques have beneficial effects on food processing and preservation by modifying the composition and structure of products (Patist & Bates, 2008). Physical depolymerization is the main driving force that alters starch structure during sonication (Han et al., 2021). The changes in physicochemical properties and structures of starch can be due to the mechanical, thermal, and cavitation effects and shear forces, micro jets, shock waves, and turbulence resulting from the collapse of cavitation bubbles (Li et al., 2019; Xu et al., 2020).

A gradual decrease in apparent amylose content of sonicated corn starch is observed following the increasing trend of sonication temperature, suggesting that a stepwise increase in sonication temperature led to a stepwise decrease in apparent amylose content of corn starch (Ouyang et al., 2021). Zhang et al. (2021) also observed a decrease in the amylose content of sonicated corn starch. According to previous studies, the decreased apparent amylose content could be attributed to the rearrangement of the dispersed starch chains (Bao et al., 2018).

These results are the opposite of what was reported by Falsafi et al. (2019), who found that ultrasonic treatment increased the amylose content of oat starch after ultrasonication, indicating the molecular scission of chains and depolymerization of amylopectin by the ultrasound treatment contributed to the strengthened linear fractions. Wang et al. (2020) reported a similar decrease in amylose content for sweet potato starch.

Gelatinization temperatures are very significant for selecting specific properties of starches according to requirements in various food applications (Morales-Martínez et al., 2014). Gelatinization temperature also reduced (Tp & Tc) after sonication treatment while To value increased after sonication treatment for corn starches (Ouyang et al., 2021; Zhang et al., 2021). The increase in To values was related to the dissolution of the weak crystalline structures within starch granules influenced by ultrasonic treatment, which could make the structure of starch granule more stable (Yang et al., 2019). These changes might be because ultrasonic treatment disrupted the ordered double-helical structures in crystalline regions of corn starch by the energy released from the collapsing bubbles (Zhang et al., 2021)

Morphological characteristics are affected by sonication treatment. Ultrasonic treatment resulted in a rough surface, with the characteristic pores of the native corn starch surface having almost disappeared (Ouyang et al., 2021). Flores Silva et al. (2017) observed the effect of different times of sonication treatment on corn starch. After sonication treatment, fissures, cracks, and disruption of the granules surfaces were observed. It has been also suggested that the appearance of the disrupted granules may partially reflect gelatinization (Montalbo-Lomboy et al., 2010) due to temperature increase by ultrasound cavitations.

Digestibility of starch is an essential metabolic process, and the rate and extent of starch digestion in the small bowel/intestine determines the eventual blood glucose level (Jenkins et al., 1982). After treatment, resistant starch is increased as compared to native starch (Flores-Silva et al., 2017; Zhang et al., 2021). Zhang et al. (2021) reported the effect of sonication power on digestibility of corn starch and observed that sonication power decreased the SDS content of sonicated corn starch.

5.2.5 High-pressure processing

High pressure (HP) processing is a technology that subjects a product to high pressures (up to 1000 MPa) for a controlled time and temperature (Leite et al., 2017). High hydrostatic pressure (HHP), which is a non-thermal technology, can be used to create new food products with unique textures or tastes with minimum effects on flavour, colour, and nutritional value (Pei-Ling et al., 2010). HPP is a green and environmental-friendly technology and can alter non-covalent chemical linkages with minimal effects on covalent linkages (Castro et al., 2020). To achieve the desired product functionality and texture, an understanding of pressure-induced gelatinization of starch is vital for applications of high-pressure treatment in starch-containing products (Vallons & Arendt, 2009). HPP affects only non-covalent bonds and can cause serious structural damage to biopolymers including protein denaturation and starch gelatinization (Hu et al., 2011).

HHP affects the physicochemical properties of starches. Rahman et al. (2020) determined an increase in the amylose content of maize starch after HHP. The increase of amylose content after HHP treatment might be associated with the degradation of amylopectin induced by HHP, and HHP could also limit the amylose leaching by affecting the amylose-amylopectin interaction (Liu et al., 2016). In 2018, Li and Zhu reported reduced swelling and water solubility index of HP-treated maize starches. Similar results were observed by Rahman et al. (2020) for HHP-treated maize starch.

HHP treatment reduced the amylose leaching and promoted the formation of amylose–lipid complexes, which stabilized the granule structure and reduced the SP and WSI above gelatinization temperatures (Katopo et al., 2002; Oh et al., 2008; Li et al., 2012).

According to Liu et al. (2016), the structural changes of starch by the application of pressure restrict the leaching of amylose and amylopectin, increasing pasting temperature and reducing viscosity. It has been observed that an increase in the pressure decreases the swelling power of maize starches, while the reverse has been observed for setback ratio (Li & Zhu, 2018). Rahman et al. (2020) reported that all viscosity parameters of maize starch increased except PT_S and PT_K at a pressure up to 300 MPa and further decreased at 500 MPa compared to native starch. The increase of PT_S and PT_K and decrease of other viscosity parameters at 500 MPa could be attributed to granular structure changes during the transformation of the crystalline structure. Results showed that the G', G'', and apparent viscosity values of starch/gum (RS/LBG) mixtures were enhanced with an increased pressure level and demonstrated a bi-phasic behaviour. HP-treated RS/LBG samples were predominantly either solid-like ($G' > G''$) or viscous ($G'' > G'$), depending on the pressure level and LBG concentrations (Hussain et al., 2016).

HPP, an emerging technology, can be used to promote the gelatinization of starch granules. The different starch sources, starch types, amylose and amylopectin content, and the process condition (time, temperature, and pressure), explain these differences observed in the degree of gelatinization induced by HPP (Leite et al., 2017). Additionally to changes in the degree of gelatinization, some studies noted that HPP was able to modify the temperature of gelatinization-modified waxy maize starch (Ahmed et al., 2014). Rahman et al. (2020) reported a decrease in gelatinization temperature at a pressure up to 300 MPa, and no peaks were observed after starch-treated at 500 MPa, which showed complete gelatinization and loss of the crystalline pattern of starch granules.

Significant effects are observed after HHP treatment. Rahman et al. (2020) observed that starch granules disintegrated after HHP treatment. Shen et al. (2018) observed minor cracks with rough surfaces for maize starch granules after pressure treatment of 200 and 400 MPa. Liu et al. (2016) stated that high pressure contributed to a strong interaction between the amylose and amylopectin chains, which led to the formation of cavities and fractures on the surface of the starch (Table 5.1).

5.3 Chemical modification of starch

Chemical modification of starches improves the functionality of starches in their granular or cooked forms, which cannot be achieved by the native form of starch. The non-destructive nature of this modification and increased

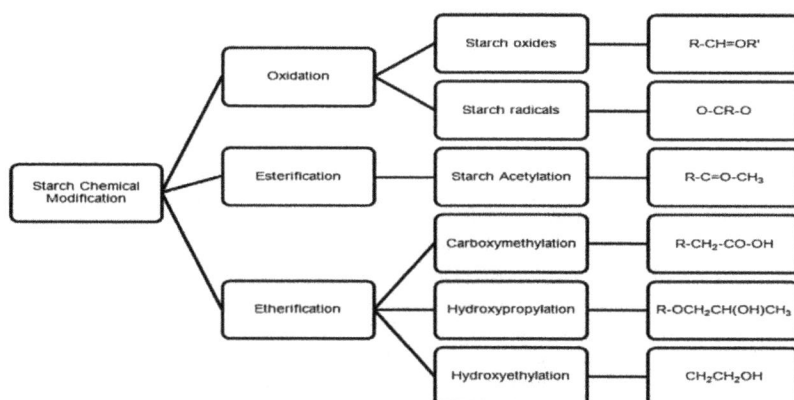

Figure 5.4 Schematic summarizing the classical chemical methods for starch modification (Masina et al., 2017). Reprinted with permission

functionality of starch have made chemical modification the commonly used method (Masina et al., 2017). In the food industry, the main objective of chemical modification is to enhance freeze–thaw stability, gel clarity, and cold storage stability, to provide heat stability and desirable texture, and to increase or decrease peak viscosity, gel formation, and viscosity (Mason, 2009). Changes brought by chemical modification in the functional properties of starches depend upon the source of starch, adequate conditions for reactions to occur, degree of substitution, and substituent's distribution in starch molecules. Chemical modification of starches can be accomplished by dervitization of starches by various chemical reactions such as acetylation, cationization, cross linking, acid or enzymatic hydrolysis, oxidation, and etherification (Jayakody & Hoover, 2002) (see Figure 5.4). Chemical modification of starch has been reported to improve physicochemical properties such as thermal stability and enzyme resistance and depends on the degree of substitution (DS) and the type of functional groups introduced (Li et al., 2011) (see Table 5.2).

Mono-functional or bi-functional reagents are used for chemical modifications (Wolf et al., 1999), where mono-functional reagents change the pasting and gelatinization properties of starch, and bi-functional reagents can react with more than one hydroxyl group and stabilize starch by altering its solubility and swelling ability (Jeon et al., 1999; Tharanathan, 2005; Xiao, 2012).

5.3.1 Oxidation

Oxidation is a process in which oxidizing agents such as hypochlorite, hydrogen peroxide, chromic acid, and bromine react with starch at a specific

Table 5.2 Chemical modification of maize starch

Modification	Major results and methodology	References
Oxidation	Improved emulsification, freeze-thaw stability, and cavities or pits appeared on the surface of modified starches (corn starch+$CuCl_2$+H_2O_2 added to the kneading reactor finally washed with ethanol and dried)	Li et al. (2021)
Oxidation	Improved paste clarity and adhesion property, low viscosity and temperature stability (30 °C, pH 8.5, 50 g (2.5 g Cl/100 g starch, 2.5% w/w) sodium hypochlorite (NaOCl)	Xiao et al. (2011)
Oxidation	Oxidized waxy corn starch showed decreased swelling power, solubility, and gelatinization enthalpy but showed increased pasting viscosity (Starch 35%; pH 9.5, Sodium hypochlorite (1 g/100g), 30 min)	Sandhu et al. (2008)
Esterification	Improved paste viscosity, thermo stability, and emulsion stabilizing property (catalyst mixed with anhydride or acid reagent, heated at 120 for 5 min, then temp set 90–150 °C, addition of freeze dried corn starch, reaction for 1–16 h)	Imre and Vilaplana (2020)
Esterification	OSA treatment induced the formation of more complex starch chains, offering more resistance for amylolytic reaction (corn starch, OSA (3%), pH 8.75, 6 h at 25 °C)	Lopez-Silva et al. (2020)
Esterification	DS increased with increase in temperature, improved structure, increased crystallinity and viscosity (Starch+glacial acetic acid+acetic anhydride+methanesulphonic acid reaction for 3 h by changing the reaction temperature at 50 °C, 65 °C, 75 °C)	Chi et al. (2008)
Carboxymethylation	Improved clarity of starch paste and reduced syneresis and freeze-thaw stability (Alkylaization 10–30% NaOH, Chloro acetic acid (2.92–14.58 g), Microwave 55 °C/15 min)	Zhang et al. (2015)
Hydroxypropylation	Improved water-holding capacity, solubility, paste clarity (propylene oxide was added with concentrations of 8%, 10%, and 12% by weight of starch, 24 h (40 °C; 200 rpm).	Maulani et al. (2015)

(Continued)

Table 5.2 (Continued) Chemical modification of maize starch

Modification	Major results and methodology	References
Cross-linking	Cross-linking decreased the solubility, swelling factor, and paste clarity of corn starch. In SEM measurement, a black zone was observed on the surface granule (Corn starch (5, 10, and 12% (w/w) of STMP/STPP (99/1% w/w), pH 11, 3 h at 45 °C)	Koo et al. (2010)

pH and temperature (Chan et al., 2011; Vanier et al., 2012; Gracia-Tejeda et al., 2013). In oxidation, the hydroxyl groups (C-2, C-3, and C-6 positions) convert into carboxyl/carbonyl groups. Starches from various botanical sources show changes in their molecular structure. The starches after oxidation have been found to have better clarity and water solubility, retrograde less, and have high viscosity at a lower concentration than native starches. These unique characteristics improve the utilization of oxidized starches in the food, textile, and paper industries. Changes in the characteristics of maize starch during oxidation have been investigated by various researchers. Hung et al. (2017) studied the effect of sodium hypochlorite at various concentrations on the structure and physicochemical properties of maize starch. The study showed that the higher active chlorine concentration increases the intensity of oxidation but decreases the viscosity, although no significant difference was found in gelatinization temperature. Xiao et al. (2011) reported that oxidation of maize starch significantly decreases peak viscosity and its ability to show resistance to shear. Changes in amylose content also affect oxidation, as it varies with the amylopectin and amylose ratio of corn starches. Amylose content plays an important role in controlling oxidation efficiency. Various modifications in oxidization methods are also used to prepare oxidized corn starch. Dry methods are also used to prepare oxidized starches, as these methods are simple with low energy consumption and water liquid discharge (Hou et al., 2007; Zhou et al., 2018; Li et al., 2019). Li et al. (2021) prepared oxidized starch using the dry method in which hydrogen peroxide (oxidant) and copper chloride (catalyst) are assisted by a kneader as a horizontal stirring reactor. The studies showed an increase in light transmittance and solubility and had low viscosity and high viscosity stability, making them suitable for low viscosity applications. The semi-dry method of preparing oxidized starch using hydrogen peroxide as a catalyst showed increases in the degree of crystallinity, with increases in the carboxyl content. Semi-dried oxidized starches have high onset, peak, and conclusion temperatures than that of commercially oxidized starches. Electrochemical oxidation in an aqueous solution of sodium chloride was used to prepare oxidized starch (Dang et al., 2018). Hydroxyl groups at C-2, C-3, and C-6 positions got oxidized by adjusting currents.

The content of carboxyl and carbonyl content increased, which promoted hydration and swelling of starch. Structural changes showed a loose surface and porous structure. Oxidation done by acids hydrogen peroxide and hypochlorite reduced the molecular weight and also produced high solid pastes (BeMiller, 2019), but the oxidized starches obtained from various oxidizing agents are not similar in characteristics.

5.3.2 Esterification and etherification

These methods are substitution methods in which the hydrophobic function groups substitute the hydrophilic hydroxyl groups present in starch and result in acetylation, hydroxypropylation, hydroxyethylation, and carboxymethylation. Monofunctional agents react with starches to give ethers and esters.

5.3.2.1 Esterification

Esterification of starch is one of the effective ways of denaturation. In esterification, ether chain is formed when hydroxypropyl, carboxymethyl, and other derivative groups substitute the hydroxyl groups attached to glucose unit in starch (Masina et al., 2017). In the food industry, commonly used starch esters are starch sodium octenylsuccinate, phosphated distarch phosphate, hydroxypropyl distarch phosphate, distarch phosphate, and acetylated distarch adipate (Singh et al., 2007). Esterification is done by adding anhydrides, acids, and acid chlorides to starch at low temperatures, which leads to the disruption of inter- and intra-molecular hydrogen bonds. Commercially, the acetylation method is used to produce esterified starches by using acetic anhydrides and acetic acid as catalysts in an aqueous medium (pH 7–9). Acetylated starches have found their utilization in the food processing industry as binders, thickeners, and stabilizers due to their good emulsion stabilizing and filming properties. Acetylation of corn starch by various methods has been studied by researchers for many years (Wolf et al., 1951; Singh et al., 2007; Chi et al., 2008; Zhu et al., 2021; Han et al., 2012; Ayucitra, 2012).

Investigations done on acetylation on corn starch showed that acetylated corn starch has a higher amylose content, light transmittance, solubility, swelling power, and better paste clarity than their native counterparts. A characterization study of acetylated corn starches using techniques like XRD, FTIR, SEM, and DSC showed that with the increase in the degree of substitution, substitution of acetyl group increases. Chi et al. (2008) studied the effect of acetylation on corn starch with variable degrees of substitution and found that in Fourier transform infrared spectroscopy, with an increase in the degree of substitution, modified starch shows an increase in the characteristic absorption intensity. X-Ray diffraction showed strong peaks after esterification and showed formation of new crystalline regions. Solvent-free methods of esterification are also used by various researchers, which overcome the problem associated with variation in moisture content and effects due to the physical state of the chemical agent on the starch modification. Zhu et al. (2021) studied the esterification of corn starch using the solvent-free method and used two agents for esterification in solid form (maleic anhydride and maleic acid) and in liquid form (acetic anhydride and acetic acid). Investigations noted low degrees of esterification at low moisture content due to the low activity of the hydroxyl group and strong interaction between hydrogen bonds. The degree of substitution increased with an increase in moisture content. The dry method of esterification is used to produce esterified corn starch by using maleic anhydride as the esterifying agent (Zuo et al., 2013). In this method maleic acid is added to starch in liquid form in an anhydrous environment. Esterified starch showed weak hydrogen bonds, responsible for improved thermos-plasticity and decreases in the degree of crystallinity. Esterified starches are found to lose their highly crystalline granular structure, and the surface of starch particles becomes rough (Imre & Vilaplana, 2020). Siroha et al. (2019b) observed that higher levels of acetylation affect the external morphology of starch granules.

Crosslinking modification is achieved by using different chemicals such as epichlorohydrin (ECH), phosphorous oxychloride (POCl3), sodium trimetaphosphate (STMP), sodium tripolyphosphate (STPP), and adipic-acetic mixed anhydride. Koo et al. (2010) evaluated the properties of cross-linked corn starches. It was observed that after modification, solubility, swelling factor, and paste clarity of corn starch decreased. SEM measurement showed that a black zone was observed on the surface granule. Similar results, reduced physicochemical properties, have been reported by Siroha and Sandhu (2018), Sandhu et al. (2021), and Siroha et al. (2021). Chung et al. (2010) modified the corn starch using octenyl succinic anhydride under dry heated conditions. It is observed that after modification at high pH conditions, PV is increased, whereas it is reduced at low pH conditions. The gelatinization characteristic was not significantly altered by OSA substitution and dry heating. Results are also supported by the findings of Siroha et al. (2019a) and Punia et al. (2019).

5.3.2.2 Etherification

5.3.2.2.1 Carboxymethylation Carboxymethylation is a widely studied modification as it is an easily achieved modification that involves the reaction of starch with chloroacetic acid under alkaline conditions (Li et al., 2010). The hydroxyl groups present in linear amylose and branch amylopectin get derivatized and form carboxymethyl ethers. In the food industry, carboxymethylated starches are used as emulsion stabilizers, binders, and thickeners due to their improved aqueous dispersibility and storage stability in cold water. Carboxymethylated starches (CMS) have found their use in the food industry as they show a high range of viscosity and stability. CMS is used in the bakery industry to decrease retrogradation, in pie fillings for their stability at high temperature, as a stabilizer in ice creams, and in noodles to reduce their softening and improve their consistency (Shinde, 2005). Khalil et al. (1990) studied the carboxymethylation of corn starch under various conditions and its effect on the degree of substitution. Their results indicated that for carboxymethylation, the optimal ratio of starch to water is 1:2.5, averagely affecting the extent of reactions; the reason for this is due to miscibility and ability of etherifying agents to solubilize. Shi and Hu (2013) prepared ultrasonicated carboxymethylated corn starch and found that the degree of substitution increased by 38.7% more than the carboxymethylated corn starch without ultrasonification. Microwave-assisted carboxymethylated corn starch was produced by Zhang et al. (2015) and showed improved swelling power, solubility, and anti-retrogradation.

5.3.2.2.2 Hydroxypropylation   Hydroxypropylation of starches imparts useful physicochemical properties that have found their utility both in food as well as non-food applications. Hydropropylation is a form of etherification in which the hydroxyl group in starch gets substituted by the hydroxypropyl group when starch reacts with propylene oxide under alkaline conditions (Maulani et al., 2015). This substitution is capable of weakening granular structure, which improves the swelling power of starches. Significant changes in physicochemical properties of starches occur after hydroxypropylation, such as reduced gelatinization, improved freeze-thaw stability, clarity, and durability, and increased swelling point. Han and BeMiller (2005) studied the rate of hydroxypropylation on corn starch and found that hydroxypropylation occurred at a high rate for 0–12 h, and then the reaction slowed down due to the amount of reagent depleting after 24 h. Hongbo et al. (2017) studied the effect of hydroxypropylation on high amylose corn starch and found that hydroxypropylation mainly occurred in the amorphous region. Pal et al. (2002) investigated hydroxypropylation on corn starch and found that hydroxypropylated corn starch showed improved paste clarity, light transmittance, and freeze–thaw stability and showed the perfect utilization of hydroxypropylated starch in the frozen food industry.

5.3.2.2.3 Hydroxyethylation   Hydroxyethylation is a process of adding epoxyethane or ethylene oxide to starch, and this introduction of the hydroxyethyl group reduces the ability of starch to gelatinize (Egharevba, 2019). Hydroxyethylated starches showed improved solubility in water, clear and stable paste, and are widely used in the paper industry as a sizing and coating agent. Some health concerns are associated with hydroxyethylated starches that pose some challenges in their utilization in the food industry. However these starches have found their utilization in medicine as a plasma volume expander and in the pharmaceutical industry as an extracorporeal perfusion fluid (Trieb et al., 1999). Commercially, hydroxyethylated starches are produced by substituting hydroxyethyl for amylopectin, a d-glucose polymer obtained from sorghum or maize.

5.3.3 Dual chemical modification

Dual modification includes methods in which starches undergo chemical changes in the presence of a physical environment or undergo enzymatic changes which enhance the degree of substitution and rate of derivitization (Ashogbon, 2021). There are two types of dual modification: heterogeneous dual modification, which involves two physical (annealing/extrusion), chemical (crosslinking/succinylation), or enzymatic modifications (pullulanase/α-amylase); and heterogeneous dual modification, which involves a combination of two of each of the differently stated modifications (acetylation/heat moisture treatment, succinylation/extrusion. Basillo-Cortes et al. (2019) studied the effect of dual modification (acid hydrolysis and succinylation) on corn starch and found that there have been alterations in starch granules, pasting properties, rearrangement of amorphous and crystalline regions, and peaks are observed in FTIR confirming the substitution of the succinyl group. Ariyantoro et al. (2018) applied succinylation and annealing on corn starch and found the resultant modified starch had improved paste clarity, gelatinization temperature, swelling power, and water binding capacity but no significant effect on the starch granule. Lin et al. (2011) partially degraded the native corn starch (Acid-ethanol) and applied heat moisture treatment on the starch, the results showing that resistant starch content increased with increased duration of acid ethanol treatment. Chung et al. (2009) studied the impact of single (annealing, hydrothermal treatment) and dual modification (annealing/hydrothermal treatment), and results showed that hydrothermal treatment, annealing/hydrothermal, and hydrothermal/annealing treatment are more effective, as these treatments increased the thermal stability of starch and rapidly digestible starch content. Starch films obtained from dual modification (hydroxypropylation and debranching) showed improved tensile strength and barrier properties (Hu et al., 2019). Guo et al. (2015) prepared modified starches by applying enzyme hydrolysis and cross linking, and the results showed that there was a significant increase in pore size distribution, specific surface area, and adsorption ability of modified starches. Enzymatic hydrolyzed and acetylation of corn starch showed significant improvement in water/oil capacity and moisture content. The dual modified starch displayed a higher substitution degree than the acetylated starch and lower reducing sugar content than the hydrolyzed starch.

5.4 Enzymatic modification

In recent decades, enzymatic modifications have been adopted, partly replacing the chemical and physical methods for the preparation of modified starch because enzymes are safer and healthier than chemical methods for both the environment and food consumers (Park et al., 2018). Enzyme molecules affect the granules in two ways. First, enzymes erode the outer surface of the granule and cause the occurrence of characteristic fissures and pits (exocorrosion). Second, enzymes digest channels leading to the granule centre, weakening granule integrity and leading consequently to its breakdown (endo-corrosion) (Sujka & Jamroz, 2009). Enzymatic hydrolysis is usually combined with other treatments such as sonication, high hydrostatic pressure, and HMTs to improve the efficiency of hydrolysis and accelerate the production of porous starch (Huang et al., 2016; Deladino et al., 2017).

Xiao et al. (2022) modified the waxy corn starch using pullulanase enzyme. It is observed that the main component of modified starch is short linear glucans. PV, BV, TV, and FV of starch decreased with the incorporation of debranched waxy corn starch (DWCS). Zhao et al. (2018) also observed that after enzymatic treatment (α-amylase & amyloglucosidase) PV decreased while the reverse was observed for BV. Rheological properties (G′ and G′′) of native corn starch (NCS)-DWCS gels decreased with an increasing concentration of DWCS (Xiao et al., 2022). Dura et al. (2014) proved that α-amylase and amyloglucosidase hydrolyzed starch granules with a lower breakdown value compared with the control group.

Enzymatic hydrolysis delays the gelatinization process, which starts at higher onset, peak temperature, and gelatinization enthalpy (Zhao et al., 2018). Enzymatic hydrolysis starches exhibited increased peak gelatinization temperatures (Tp) and decreased gelatinization enthalpy (ΔHgel) upon hydrolysis. The increase in Tp indicates hydrolysis of the amorphous structure by enzymes (Jiang et al., 2011). The reduction in the ΔH gel, on the other hand, supports the hydrolysis of the crystalline and helical structures (Sandhu et al., 2007).

The enzymatic hydrolysis of starch created significant surface changes, producing pores on most corn starch granules without affecting the granules' shape (Zhao et al., 2018). Jiang et al. (2011) observed that extensive enzymatic corrosion occurs after treatment. SEM shows fragments and large pores when compared to native starch. The starch granules started to fracture, and some cracks came to appear on their surfaces due to heavy hydrolysis. The above results indicated that the core part of the starch granule was more easily attacked by glucoamylase (Wang et al., 2008). Starches were rapidly digested with enzymes which penetrated into starch granules through natural pores and disrupted the interior of the starch granules (Lin et al., 2005).

Significant effects of enzymatic modifications on digestibility are observed by many researchers. Kittisuban et al. (2014) modified different starches (corn, potato, rice, etc.) using 1,4-α-glucan branching enzyme (GBE), either alone or in combination with β-amylase, and reported that all starches treated with the enzyme combination exhibited reduced glucose release rates. Ren et al. (2020) reported a novel two-stage modification method comprising a 10-h GBE treatment, gelatinization, and a second 10-h GBE treatment and produced samples with the lowest *in vitro* digestibility. The rapidly digestible starch content was 34.2% lower than that of the control and 18.0% lower than that of the product of one-stage modification with the same duration.

5.5 Applications

Maize starch is widely used in food applications. Starch is modified to increase the applications of starches. Chung et al. (2010) modified corn starch using octenyl succinic anhydride (OSA). OSA starch was used as a fat replacement to prepare muffins. The unique physiochemical and functional characteristics of natural starches such as their good biocompatibility, biodegradability, non-toxicity, and degradation make them useful for a wide range of biomedical applications (Palanisamy et al., 2020). Many researchers prepare starch-based biodegradable films to replace traditional plastics, which helps to protect the environment by decreasing non-degradable plastic waste production (Alqahtani et al., 2021; Wang et al., 2021). Xiao et al. (2022) prepared debranched starch (DBS) from waxy corn starch. DBS is an important novel material and has widespread applications such as fillers, functional foods, packaging materials, and drug formulation in the food and pharmaceutical industries due to its great biocompatibility, biodegradability, and low toxicity (Yotsawimonwat et al., 2008; Chang et al., 2018; Qin et al., 2019). Reddy et al. (2021) evaluated the effect of de-branched maize starch for the preparation of green tea enriched with catechin, as commercial formulations of the catechin-rich ethanolic extract are limited by low aqueous solubility and high oxidation sensitivity.

Morán et al. (2021) stated that starch nano particles (SNPs) from maize starches may be used for various applications. SNPs have been used to encapsulate bioactive compounds for controlled release in different biomedical applications such as drug administration, enzyme inhibition process, and even DNA precipitation.

References

Ahmed, J., Singh, A., Ramaswamy, H. S., Pandey, P. K., & Raghavan, G. S. V. (2014). Effect of high-pressure on calorimetric, rheological and dielectric properties of selected starch dispersions. *Carbohydrate Polymers, 103*, 12–21.

Alqahtani, N., Alnemr, T., & Ali, S. (2021). Development of low-cost biodegradable films from corn starch and date palm pits (*Phoenix dactylifera*). *Food Bioscience, 42*, 101199.

Ambigaipalan, P., Hoover, R., Donner, E., & Liu, Q. (2014). Starch chain interactions within the amorphous and crystalline domains of pulse starches during heat-moisture treatment at different temperatures and their impact on physicochemical properties. *Food Chemistry, 143*, 175–184.

Ariyantoro, A. R., Katsuno, N., & Nishizu, T. (2018). Effects of dual modification with succinylation and annealing on physicochemical, thermal and morphological properties of corn starch. *Foods, 7*(9), 133.

Ashogbon, A. O. (2021). Dual modification of various starches: Synthesis, properties and applications. *Food Chemistry, 342*, 128325.

Ayucitra, A. (2012). Preparation and characterisation of acetylated corn starches. *International Journal of Chemical Engineering and Applications, 3*, 156–159.

Bangar, S. P., Kumar, M., & Whiteside, W. S. (2021a). Mango seed starch: A sustainable and eco-friendly alternative to increasing industrial requirements. *International Journal of Biological Macromolecules, 183*, 1807–1817.

Bangar, S. P., Nehra, M., Siroha, A. K., Petrů, M., Ilyas, R. A., Devi, U., & Devi, P. (2021b). Development and characterization of physical modified pearl millet starch-based films. *Foods, 10*(7), 1609.

Bao, W., Li, Q., Wu, Y., & Ouyang, J. (2018). Insights into the crystallinity and in vitro digestibility of chestnut starch during thermal processing. *Food Chemistry, 269*, 244–251.

Basilio-Cortés, U. A., González-Cruz, L., Velazquez, G., Teniente-Martínez, G., Gómez-Aldapa, C. A., Castro-Rosas, J., & Bernardino-Nicanor, A. (2019). Effect of dual modification on the spectroscopic, calorimetric, viscosimetric and morphological characteristics of corn starch. *Polymers, 11*(2), 333.

Bemiller, J. N. (1997). Starch modification: Challenges and prospects. *Starch/Stärke, 49*, 127–131.

BeMiller, J. N. (2019). Starches: Molecular and granular structures and properties. In *Carbohydrate chemistry for food scientists*. Duxford: Elsevier. ISBN: 978-0-12-812069-9.

BeMiller, J. N., & Huber, K. C. (2015). Physical modification of food starch functionalities. *Annual Reviews of Food Science and Technology, 6*, 19–69.

Bilbao-Sáinz, C., Butler, M., Weaver, T., & Bent, J. (2007). Wheat starch gelatinization under microwave irradiation and conduction heating. *Carbohydrate Polymers, 69*(2), 224–232.

Braşoveanu, M., & Nemţanu, M. R. (2014). Behaviour of starch exposed to microwave radiation treatment. *Starch/Stärke, 66*, 3–14.

Castro, L. M., Alexandre, E. M., Saraiva, J. A., & Pintado, M. (2020). Impact of high pressure on starch properties: A review. *Food Hydrocolloids, 106*, 105877.

Chan, H. T., Leh, C., Bhat, R., Senan, C., Williams, P., & Karim, A. (2011). Molecular structure, rheological and thermal characteristics of ozone-oxidized starch. *Food Chemistry, 126*(3), 1019–1024.

Chandla, N. K., Saxena, D. C., & Singh, S. (2017). Processing and evaluation of heat moisture treated (HMT) amaranth starch noodles; An inclusive comparison with corn starch noodles. *Journal of Cereal Science, 75*, 306–313.

Chang, R., Xiong, L., Li, M., Liu, J., Wang, Y., Chen, H., & Sun, Q. (2018). Fractionation of debranched starch with different molecular weights via edible alcohol precipitation. *Food Hydrocolloids, 83*, 430–437.

Chen, Y., Yang, Q., Xu, X., Qi, L., Dong, Z., Luo, Z., ... Peng, X. (2017). Structural changes of waxy and normal maize starches modified by heat moisture treatment and their relationship with starch digestibility. *Carbohydrate Polymers, 177*, 232–240.

Chi, H., Xu, K., Wu, X., Chen, Q., Xue, D., Song, C., ... Wang, P. (2008). Effect of acetylation on the properties of corn starch. *Food Chemistry, 106*(3), 923–928.

Chizoba Ekezie, F. G., Cheng, J. H., & Sun, D. W. (2018). Effects of mild oxidative and structural modifications induced by argon-plasma on physicochemical properties of actomyosin from king prawn (*Litopenaeus vannamei*). *Journal of Agricultural and Food Chemistry, 66*(50), 13285–13294.

Chung, H. J., Hoover, R., & Liu, Q. (2009). The impact of single and dual hydrothermal modifications on the molecular structure and physicochemical properties of normal corn starch. *International Journal of Biological Macromolecules, 44*(2), 203–210.

Chung, H. J., Lee, S. E., Han, J. A., & Lim, S. T. (2010). Physical properties of dry-heated octenyl succinylated waxy corn starches and its application in fat-reduced muffin. *Journal of Cereal Science, 52*(3), 496–501.

Colman, T. A. D., Demiate, I. M., & Schnitzler, E. (2014). The effect of microwave radiation on some thermal, rheological and structural properties of cassava starch. *Journal of Thermal Analysis and Calorimetry, 115*(3), 2245–2252.

da Rosa Zavareze, E., & Dias, A. R. G. (2011). Impact of heat-moisture treatment and annealing in starches: A review. *Carbohydrate Polymers, 83*(2), 317–328.

Dang, X., Chen, H., Wang, Y., & Shan, Z. (2018). Freeze-drying of oxidized corn starch: Electrochemical synthesis and characterization. *Cellulose, 25*(4), 2235–2247.

Deladino, L., Schneider Teixeira, A., Plou, F. J., Navarro, A. S., & Molina-García, A. D. (2017). Effect of high hydrostatic pressure, alkaline and combined treatments on corn starch granules metal binding: Structure, swelling behavior and thermal properties assessment. *Food and Bioproducts Processing, 102*, 241–249.

Dias, D. D. R. C., Barros, Z. M. P., de Carvalho, C. B. O., Honorato, F. A., Guerra, N. B., & Azoubel, P. M. (2015). Effect of sonication on soursop juice quality. *LWT - Food Science and Technology, 62*(1), 883–889.

Dura, A., Blaszczak, W., & Rosell, C. M. (2014). Functionality of porous starch obtained by amylase or amyloglucosidase treatments. *Carbohydrate Polymers, 101*, 837–845.

Egharevba, H. (2019). Chemical properties of starch and its application in the food industry. *Chemical Properties of Starch.* https://doi.org/10.5772/.intechopen.87777

Ek, K. L., Wang, S., Brand-Miller, J., & Copeland, L. (2014). Properties of starch from potatoes differing in glycemic index. *Food and Function, 5*(10), 2509–2515.

El Khaled, D., Novas, N., Gazquez, J. A., & Manzano-Agugliaro, F. (2018). Microwave dielectric heating: Applications on metals processing. *Renewable and Sustainableenergy Reviews, 82*, 2880–2892.

Falsafi, S. R., Maghsoudlou, Y., Rostamabadi, H., Rostamabadi, M. M., Hamedi, H., & Hosseini, S. M. H. (2019). Preparation of physically modified oat starch with different sonication treatments. *Food Hydrocolloids, 89*, 311–320.

Fan, D., Ma, S., Wang, L., Zhao, J., Zhang, H., & Chen, W. (2012). Effect of microwave heating on optical and thermal properties of rice starch. *Starch/Stärke, 64*, 740–744.

Flores-Silva, P. C., Roldan-Cruz, C. A., Chavez-Esquivel, G., Vernon-Carter, E. J., Bello-Perez, L. A., & Alvarez-Ramirez, J. (2017). *In vitro* digestibility of ultrasound-treated corn starch. *Starch/Stärke, 69*, 1700040.

García-Tejeda, Y., López-González, C., Pérez-Orozco, J. P., Rendón-Villalobos, R., Jiménez-Pérez, R., Flores-Huicochea, E., ... Andrea Bastida, E. (2013). Physicochemical and mechanical properties of extruded laminates from native and oxidized banana starch during storage. *LWT - Food Science and Technology, 54*(2), 447–455.

Gomes, A. M. M., Silva, C. E. M., & Ricardo, N. M. P. S. (2005). Effects of annealing on the physicochemical properties of fermented cassava starch (*polvilho azedo*). *Carbohydrate Polymers, 60*(1), 1–6.

Guo, L., Liu, R., Li, X., Sun, Y., & Du, X. (2015). The physical and adsorption properties of different modified corn starches. *Starch/Stärke, 67*, 237–246.

Han, F., Liu, M., Gong, H., Lu, S., & Zhang, B. (2012). Synthesis, characterisation and functional properties low substituted acetylated corn starch. *International Journal of Biological Macromolecules, 50*, 1026–1034.

Han, J., & Bemiller, J. (2005). Rate of hydroxypropylation of starches as a function of reaction time. *Starch/Stärke, 57*, 395–404.

Han, K. T., Kim, H. R., Moon, T. W., & Choi, S. J. (2021). Isothermal and temperature-cycling retrogradation of high-amylose corn starch: Impact of sonication on its structural and retrogradation properties. *Ultrasonics Sonochemistry, 76*, 105650.

Haq, F., Yu, H., Wang, L., Teng, L., Haroon, M., Khan, R. U., … Nazir, A. (2019). Advances in chemical modifications of starches and their applications. *Carbohydrate Research, 476*, 12–35.

Hongbo, T., Kun, P., Yanping, L., & Siqing, D. (2017). Effect of oxidation and hydroxypropylation on structure and properties of high-amylose corn starch, and preparation of hydroxypropyl oxidized high-amylose corn starch. *Cellulose Chemistry and Technology, 52*, 769–787.

Hoover, R. (2010). The impact of heat-moisture treatment on molecular structures and properties of starches isolated from different botanical sources. *Critical Reviews in Food Science and Nutrition, 50*(9), 835–847.

Hoover, R., & Vasanthan, T. (1994). The effect of annealing on the physicochemical properties of wheat, oat, potato and lentil starches. *Journal of Food Biochemistry, 17*, 303–325.

Hormdok, R., & Noomhorm, A. (2007). Hydrothermal treatments of rice starch for improvement of rice noodle quality. *LWT - Food Science and Technology, 40*(10), 1723–1731.

Hou, H., Dong, H., Liu, G., & Zhang, H. (2007). Preparation and properties of oxidized corn starches by semi-dry process. *Cereal Chemistry, 84*(3), 225–230.

Hu, X., Jia, X., Zhi, C., Jin, Z., & Miao, M. (2019). Improving properties of normal maize starch films using dual-modification: Combination treatment of debranching and hydroxypropylation. *International Journal of Biological Macromolecules, 130*, 197–202.

Hu, X., Xu, X., Jin, Z., Tian, Y., Bai, Y., & Xie, Z. (2011). Retrogradation properties of rice starch gelatinized by heat and high hydrostatic pressure (HHP). *Journal of Food Engineering, 106*(3), 262–266.

Huang, T. T., Zhou, D. N., Jin, Z. Y., Xu, X. M., & Chen, H. Q. (2016). Effect of repeated heat-moisture treatments on digestibility, physicochemical and structural properties of sweet potato starch. *Food Hydrocolloids, 54*, 202–210.

Hui, R., Qi-He, C., Ming-liang, F., Qiong, X., & Guo-qing, H. (2009). Preparation and properties of octenyl succinic anhydride modified potato starch. *Food Chemistry, 114*(1), 81–86.

Hung, D., Trường, N., Phuc, M. V., Nguyen, H., Trinh, T., Nguyen, V., & Do, V. (2017). Oxidized maize starch: Characterization and effect of it on the biodegradable films. ii. Infrared spectroscopy, solubility of oxidized starch and starch film solubility. *Vietnam Journal of Science and Technology, 55*(4), 395.

Hussain, R., Vatankhah, H., Singh, A., & Ramaswamy, H. S. (2016). Effect of high-pressure treatment on the structural and rheological properties of resistant corn starch/locust bean gum mixtures. *Carbohydrate Polymers, 150*, 299–307.

Imre, B., & Vilaplana, F. (2020). Organocatalytic esterification of corn starches towards enhanced thermal stability and moisture resistance. *Green Chemistry, 22*(15), 5017–5031.

Jacobs, H., Eerlingen, R. C., Clauwaert, W., & Delcour, J. A. (1995). Influence of annealing on the pasting properties of starches from varying botanical sources. *Cereal Chemistry, 72*, 480–487.

Jayakody, L., & Hoover, R. (2002). The effect of linterization on cereal starch granules. *Food Research International, 35*(7), 665–680.

Jayakody, L., & Hoover, R. (2008). Effect of annealing on the molecular structure and physicochemical properties of starches from different botanical origins - A review. *Carbohydrate Polymers, 74*(3), 691–703.

Jenkins, D. J., Thorne, M. J., Camelon, K., Jenkins, A., Rao, A. V., Taylor, R. H., ... & Francis, T. (1982). Effect of processing on digestibility and the blood glucose response: A study of lentils. *American Journal of Clinical Nutrition, 36*(6), 1093–1101.

Jeon, Y., Viswanathan, A., & Gross, R. (1999). Studies of starch esterification: Reactions with alkenyl-succinates in aqueous slurry systems. *Starch, 51*, 90–93.

Jiang, Q., Gao, W., Li, X., & Zhang, J. (2011). Characteristics of native and enzymatically hydrolyzed Zea mays L., Fritillaria ussuriensis Maxim. and Dioscorea opposita Thunb. starches. *Food Hydrocolloids, 25*(3), 521–528.

Jiranuntakul, W., Puttanlek, C., Rungsardthong, V., Puncha-arnon, S., & Uttapap, D. (2012). Amylopectin structure of heat–moisture treated starches. *Starch/Stärke, 64*, 470–480.

Katopo, H., Song, Y., & Jane, J. (2002). Effect and mechanism of ultrahigh-hydrostatic pressure on the structure and properties of starches. *Carbohydratepolymers, 47*, 233–244.

Khalil, M. I., Hashem, A., & Hebeish, A. (1990). Carboxymethylation of maize starch. *Starch/Stärke, 42*, 60–63.

Kittisuban, P., Lee, B. H., Suphantharika, M., & Hamaker, B. R. (2014). Slow glucoserelease property of enzyme-synthesized highly branched maltodextrins differs amongstarch sources. *Carbohydrate Polymers, 107*, 182–191.

Koo, S. H., Lee, K. Y., & Lee, H. G. (2010). Effect of cross-linking on the physicochemical and physiological properties of corn starch. *Food Hydrocolloids, 24*(6–7), 619–625.

Kweon, M., Slade, L., & Levine, H. (2008). Role of glassy and crystalline transitions in the responses of corn starches to heat and high pressure treatments: Prediction of solute-induced barostability from solute-induced thermostability. *Carbohydrate Polymers, 73*(2), 293–299.

Leite, T. S., de Jesus, A. L. T., Schmiele, M., Tribst, A. A., & Cristianini, M. (2017). High pressure processing (HPP) of pea starch: Effect on the gelatinization properties. *LWT - Food Science and Technology, 76*, 361–369.

Li, G., & Zhu, F. (2018). Effect of high pressure on rheological and thermal properties of quinoa and maize starches. *Food Chemistry, 241,* 380–386.

Li, J., Zhou, M., Cheng, F., Lin, Y., Hu, J., & Zhu, P. X. (2019). Characterization and properties of long-chain fatty acid starch esters prepared with regenerated starch by dry method. *Starch/Stärke, 71*, 11–12.

Li, X., Gao, W. Y., Liu, C. X., Wang, Y., Huang, L., & Liu, C. (2010). Preparation and physicochemical properties of carboxymethyl Fritillaria ussuriensis Maxim. Starches. *Carbohydrate Polymers, 80*(3), 768–769.

Li, Y., Hu, A., Zheng, J., & Wang, X. (2019). Comparative studies on structure and physiochemical changes of millet starch under microwave and ultrasound at the same power. *International Journal of Biological Macromolecules, 141*, 76–84.

Li, Y., Zhang, K., Song, Y., Cheng, F., Zhou, M., Lin, Y., & Zhu, P. (2021). Preparation of oxidized corn starch in dry method assisted by kneader. *Materials Express, 11*(1), 100–106.

Li, Y., Zhang, Z., Leeuwen, H. P., Stuart, M. C., Norde, W., & Kleijn, J. (2011). Uptake and release kinetics of lysozyme in and from an oxidized starch polymer microgel. *Soft Matter, 7*(21), 10377–10385.

Li, W., Bai, Y., Mousaa, S. A. S., Zhang, Q., & Shen, Q. (2012). Effect of high hydrostaticpressure on physicochemical and structural properties of rice starch. *Food and Bioprocess Technology, 5*(6), 2233–2241.

Lin, J. H., Lii, C. Y., & Chang, Y. H. (2005). Change of granular and molecular structures of waxy maize and potato starches after treated in alcohols with or without hydrochloric acid. *Carbohydrate Polymers, 59*(4), 507–515.

Lin, J. H., Singh, H., Wen, C. Y., & Chang, Y. H. (2011). Partial-degradation and heat-moisture dual modification on the enzymatic resistance and boiling-stable resistant starch content of corn starches. *Journal of Cereal Science, 54*(1), 83–89.

Liu, H., Guo, X., Li, Y., Li, H., Fan, H., & Wang, M. (2016). *In vitro* digestibility and changes in physicochemical and textural properties of Tartary buckwheat starch under high hydrostatic pressure. *Journal of Food Engineering, 189*, 64–71.

Liu, H., Yu, L., Simon, G., Dean, K., & Chen, L. (2009). Effects of annealing on gelatinization and microstructures of corn starches with different amylose/amylopectin ratios. *Carbohydrate Polymers, 77*(3), 662–669.

Liu, K., Zhang, B., Chen, L., Li, X., & Zheng, B. (2019). Hierarchical structure and physicochemical properties of highland barley starch following heat moisture treatment. *Food Chemistry, 271*, 102–108.

Lopez-Silva, M., Bello-Perez, L. A., Castillo-Rodriguez, V. M., Agama-Acevedo, E., & Alvarez-Ramirez, J. (2020). In vitro digestibility characteristics of octenyl succinic acid (OSA) modified starch with different amylose content. *Food Chemistry, 304*, 125434.

Luo, Z., He, X., Fu, X., Luo, F., & Gao, Q. (2006). Effect of microwave radiation on the physicochemical properties of normal maize, waxy maize and amylomaize V starches. *Starch/Stärke, 58*, 468–474.

Maache-Rezzoug, Z., Zarguili, I., Loisel, C., Queveau, D., & Buleon, A. (2008). Structural modifications and thermal transitions of standard maize starch after DIC hydrothermal treatment. *Carbohydrate Polymers, 74*(4), 802–812.

Malumba, P., Janas, S., Roiseux, O., Sinnaeve, G., Masimango, T., Sindic, M., … Béra, F. (2010). Comparative study of the effect of drying temperatures and heat-moisture treatment on the physicochemical and functional properties of corn starch. *Carbohydrate Polymers, 79*(3), 633–641.

Man, J., Qin, F., Zhu, L., Shi, Y. C., Gu, M., Liu, Q., & Wei, C. (2012). Ordered structure and thermal property of acid-modified high-amylose rice starch. *Food Chemistry, 134*(4), 2242–2248.

Masina, N., Choonara, Y. E., Kumar, P., du Toit, L. C., Govender, M., Indermun, S., & Pillay, V. (2017). A review of the chemical modification techniques of starch. *Carbohydrate Polymer, 157*, 1226–1236.

Mason, W. R. (2009). Starch use in foods. In J. BeMiller & R. Whistler (Eds.), *Starch: Chemistry and technology* (3rd ed.), 746–795. New York: Academic Press.

Maulani, R. R., Fardiaz, D., Kusnandar, F., & Sunarti, T. (2015). Characterization of chemical and physical properties of hydroxypropylated and cross-linked arrowroot (*Marantha Arundinacea*) starch. *Journal of Engineering and Technological Sciences, 45*(3), 207–221.

Mohammed, O., & Bin, X. (2020). Review on the physicochemical properties, modifications, and applications of starches and its common modified forms used in noodle products. *Food Hydrocolloids, 112*, 106286.

Montalbo-Lomboy, M., Johnson, L., Khanal, S. K., van Leeuwen, J. H., & Grewell, D. (2010). Sonication of sugary-2 corn: A potential pretreatment to enhance sugar release. *Bioresource Technology, 101*(1), 351–358.

Morales-Martínez, L. E., Bello-Pérez, L. A., Sánchez-Rivera, M. M., Ventura-Zapata, E., & Jiménez-Aparicio, A. R. (2014). Morphometric, physicochemical, thermal, and rheological properties of rice (Oryza sativa L.) Cultivars Indica × Japonica. *Food and Nutrition Sciences, 5*, 271–279.

Morán, D., Gutiérrez, G., Blanco-López, M. C., Marefati, A., Rayner, M., & Matos, M. (2021). Synthesis of starch nanoparticles and their applications for bioactive compound encapsulation. *Applied Sciences, 11*(10), 4547.

MI (Mordor Intelligence). (2019). Retrieved October 3, 2019, from http://www.mordorintelligence.com/industry-reports/industrial-starches-market

Nawaz, H., Shad, M. A., Saleem, S., Khan, M., Nishan, U., Rasheed, T., … Iqbal, H. M. N. (2018). Characteristics of starch isolated from microwave heat treated lotus (*Nelumbo nucifera*) seed flour. *International Journal of Biological Macromolecules, 113*, 219–226.

Oh, H. E., Pinder, D. N., Hemar, Y., Anema, S. G., & Wong, M. (2008). Effect ofhigh-pressure treatment on various starch-in-water suspensions. *Food Hydrocolloids, 22*, 150–155.

Ortega-Ojeda, F. E., & Eliasson, A. C. (2001). Gelatinisation and retrogradation behaviour of some starch mixtures. *Starch/Stärke, 53*, 520–529.

Ouyang, Q., Wang, X., Xiao, Y., Luo, F., Lin, Q., & Ding, Y. (2021). Structural changes of A-, B-and C-type starches of corn, potato and pea as influenced by sonication temperature and their relationships with digestibility. *Food Chemistry, 358*, 129858.

Oyeyinka, S. A., Umaru, E., Olatunde, S. J., & Joseph, J. K. (2019). Effect of short microwave heating time on physicochemical and functional properties of Bambara groundnut starch. *Food Bioscience, 28*, 36–41.

Pal, J., Singhal, R., & Kulkarni, P. (2002). Physicochemical properties of hydroxypropyl derivative from corn and amaranth starch. *Carbohydrate Polymers, 48*(1), 49–53.

Palanisamy, C. P., Cui, B., Zhang, H., Jayaraman, S., & Kodiveri Muthukaliannan, G. (2020). A comprehensive review on corn starch-based nanomaterials: Properties, simulations, and applications. *Polymers, 12*(9), 2161.

Park, S. H., Na, Y., Kim, J., Dal Kang, S., & Park, K. H. (2018). Properties and applications of starch modifying enzymes for use in the baking industry. *Food Science and Biotechnology, 27*(2), 299–312.

Patist, A., & Bates, D. (2008). Ultrasonic innovations in the food industry: From the laboratory to commercial production. *Innovative Food Science and Emerging Technologies, 9*(2), 147–154.

Pei-Ling, L., Xiao-Song, H., & Qun, S. (2010). Effect of high hydrostatic pressure on starches: A review. *Starch/Stärke, 62*, 615–628.

Piecyk, M., & Domian, K. (2021). Effects of heat–moisture treatment conditions on the physicochemical properties and digestibility of field bean starch (Vicia faba var. minor). *International Journal of Biological Macromolecules, 182*, 425–433.

Punia, S. (2020). Barley starch modifications: Physical, chemical and enzymatic–A review. *International Journal of Biological Macromolecules, 144*, 578–585.

Punia, S., Siroha, A. K., Sandhu, K. S., & Kaur, M. (2019). Rheological and pasting behavior of OSA modified mungbean starches and its utilization in cake formulation as fat replacer. *International Journal of Biological Macromolecules, 128*, 230–236.

Qin, Y., Wang, J., Qiu, C., Hu, Y., Xu, X., & Jin, Z. (2019). Effects of degree of polymerization on size, crystal structure, and digestibility of debranched starch nanoparticles and their enhanced antioxidant and antibacterial activities of curcumin. *ACS Sustainable Chemistry and Engineering, 7*(9), 8499–8511.

Rafiq, S. I., Singh, S., & Saxena, D. C. (2016). Effect of heat-moisture and acid treatment on physicochemical, pasting, thermal and morphological properties of Horse Chestnut (*Aesculus indica*) starch. *Food Hydrocolloids, 57*, 103–113.

Rahman, M. H., Mu, T. H., Zhang, M., Ma, M. M., & Sun, H. N. (2020). Comparative study of the effects of high hydrostatic pressure on physicochemical, thermal, and structural properties of maize, potato, and sweet potato starches. *Journal of Food Processing and Preservation, 44*(11), e14852.

Reddy, C. K., Son, S. Y., & Lee, C. H. (2021). Effects of pullulanase debranching and octenylsuccinic anhydride modification on the structural properties of maize starch-green tea extract complexes. *Food Hydrocolloids, 115,* 106630.

Ren, J., Chen, S., Li, C., Gu, Z., Cheng, L., Hong, Y., & Li, Z. (2020). A two-stage modification method using 1, 4-α-glucan branching enzyme lowers the *in vitro* digestibility of corn starch. *Food Chemistry, 305,* 125441.

Rocha, T. S., Felizardo, S. G., Jane, J. L., & Franco, C. M. (2012). Effect of annealing on the semicrystalline structure of normal and waxy corn starches. *Food Hydrocolloids, 29*(1), 93–99.

Sair, L., & Fetzer, W. R. (1944). Water sorption by cornstarch and commercial. Modifications of starches. *Industrial and Engineering Chemistry, 36*(4), 316–319.

Sandhu, K. S., Kaur, M., Singh, N., & Lim, S. T. (2008). A comparison of native and oxidized normal and waxy corn starches: Physicochemical, thermal, morphological and pasting properties. *LWT - Food Science and Technology, 41*(6), 1000–1010.

Sandhu, K. S., Singh, N., & Lim, S. T. (2007). A comparison of native and acid thinnednormal and waxy corn starches: Physicochemical, thermal, morphological andpasting properties. *LWT - Food Science and Technology, 40,* 1527–1536.

Sandhu, K. S., Siroha, A. K., Punia, S., & Nehra, M. (2020). Effect of heat moisture treatment on rheological and *in vitro* digestibility properties of pearl millet starches. *Carbohydrate Polymer Technologies and Applications, 1,* 100002.

Sandhu, K. S., Siroha, A. K., Punia, S., Sangwan, L., Nehra, M., & Purewal, S. S. (2021). Effect of degree of cross linking on physicochemical, rheological and morphological properties of Sorghum starch. *Carbohydrate Polymer Technologies and Applications, 2,* 100073.

Schafranski, K., Ito, V. C., & Lacerda, L. G. (2021). Impacts and potential applications: A review of the modification of starches by heat-moisture treatment (HMT). *Food Hydrocolloids, 117,* 106690.

Scott, M. P., & Emery, M. (2015). *Maize: Overview* (Vol. 1–4, 2nd ed.). Amsterdam: Elsevier Ltd.

Shen, X., Shang, W., Strappe, P., Chen, L., Li, X., Zhou, Z., & Blanchard, C. (2018). Manipulation of the internal structure of high amylose maize starch by high pressure treatment and its diverse influence on digestion. *Food Hydrocolloids, 77,* 40–48.

Shi, H., & Hu, X. (2013). Preparation and structure characterization of carboxymethyl corn starch under ultrasonic irradiation. *Cereal Chemistry, 90*(1), 24–28.

Shinde, V. V. (2005). Production kinetics and functional properties of carboxymethyl sorghum starch. *Natural Product Radiance*, 466–470.

Singh, J., Kaur, L., & McCarthy, O. J. (2007). Factors influencing the physico-chemical, morphological, thermal and rheological properties of some chemically modified starches for food applications - A review. *Food Hydrocolloids*, *21*(1), 1–22.

Siroha, A. K., Bangar, S. P., Sandhu, K. S., Trif, M., Kumar, M., & Guleria, P. (2021). Effect of cross-linking modification on structural and film-forming characteristics of pearl millet (*Pennisetum glaucum* L.) starch. *Coatings*, *11*(10), 1163.

Siroha, A. K., & Sandhu, K. S. (2018). Physicochemical, rheological, morphological, and in vitro digestibility properties of cross-linked starch from pearl millet cultivars. *International Journal of Food Properties*, *21*(1), 1371–1385.

Siroha, A. K., Sandhu, K. S., Kaur, M., & Kaur, V. (2019b). Physicochemical, rheological, morphological and in vitro digestibility properties of pearl millet starch modified at varying levels of acetylation. *International Journal of Biological Macromolecules*, *131*, 1077–1083.

Siroha, A. K., Sandhu, K. S., & Punia, S. (2019a). Impact of octenyl succinic anhydride on rheological properties of sorghum starch. *Quality Assurance and Safety of Crops and Foods*, *11*(3), 221–229.

Stevenson, D. G., Biswas, A., & Inglett, G. E. (2005). Thermal and pasting properties of microwaved corn starch. *Starch/Stärke*, *57*, 347–353.

Sui, Z., Yao, T., Zhao, Y., Ye, X., Kong, X., & Ai, L. (2015). Effects of heat–moisture treatment reaction conditions on the physicochemical and structural properties of maize starch: Moisture and length of heating. *Food Chemistry*, *173*, 1125–1132.

Sujka, M., & Jamroz, J. (2009). α-Amylolysis of native potato and corn starches-SEM, AFM, nitrogen and iodine sorption investigations. *LWT - Food Science and Technology*, *42*(7), 1219–1224.

Sun, Q., Xu, Y., & Xiong, L. (2014). Effect of microwave-assisted dry heating with xanthan on normal and waxy corn starches. *International Journal of Biological Macromolecules*, *68*, 86–91.

Takaya, T., Sano, C., & Nishinari, K. (2000). Thermal studies on the gelatinisation and retrogradation of heat–moisture treated starch. *Carbohydrate Polymers*, *41*(1), 97–100.

Tester, R. F., & Debon, S. J. (2000). Annealing of starch–a review. *International Journal of Biological Macromolecules*, *27*(1), 1–12.

Tharanathan, R. N. (2005). Starch-value addition by modification. *Critical Reviews in Food Science and Nutrition*, *45*(5), 371–384.

Treib, J., Baron, J. F., Grauer, M. T., & Strauss, R. G. (1999). An international view of hydroxyethyl starches. *Intensive Care Medicine*, *25*(3), 258–268.

Vallons, K. J., & Arendt, E. K. (2009). Effects of high pressure and temperature on the structural and rheological properties of sorghum starch. *Innovative Food Science and Emerging Technologies*, *10*(4), 449–456.

Vanier, N. L., Zavareze, E. R., Pinto, V. Z., Klein, B., Botelho, F. T., Dias, A. R. G., & Elias, M. C. (2012). Physicochemical, crystallinity, pasting and morphological properties of bean starch oxidised by different concentrations of sodium hypochlorite. *Food Chemistry, 131*(4), 1255–1262.

Waduge, R. N., Hoover, R., Vasanthan, T., Gao, J., & Li, J. (2006). Effect of annealing on the structure and physicochemical properties of barley starches of varying amylose content. *Food Research International, 39*(1), 59–77.

Wang, B., Yan, S., Gao, W., Kang, X., Yu, B., Liu, P., ... Abd El-Aty, A. M. (2021). Antibacterial activity, optical, and functional properties of corn starch-based films impregnated with bamboo leaf volatile oil. *Food Chemistry, 357*, 129743.

Wang, H., Xu, K., Ma, Y., Liang, Y., Zhang, H., & Chen, L. (2020). Impact of ultrasonication on the aggregation structure and physicochemical characteristics of sweet potato starch. *Ultrasonics Sonochemistry, 63*, 104868.

Wang, M., Sun, M., Zhang, Y., Chen, Y., Wu, Y., & Ouyang, J. (2019). Effect of microwave irradiation-retrogradation treatment on the digestive and physicochemical properties of starches with different crystallinity. *Food Chemistry, 298*, 125015.

Wang, S. J., Yu, J. L., Liu, H. Y., & Chen, W. P. (2008). Characterisation and preliminarylipid-lowering evaluation of starch from Chinese yam. *Food Chemistry, 108*(1), 176–181.

Wolf, B. W., Bauer, L. L., & Fahey, G. C. (1999). Effects of chemical modification on in vitro rate and extent of food starch digestion: An attempt to discover a slowly digested starch. *Journal of Agricultural and Food Chemistry, 47*(10), 4178–4183.

Wolf, I. A., Olds, D. W., & Hilbert, G. E. (1951). The acylation of corn starch, amylose and amylopectin. *Journal of the American Chemical Society, 73*(1), 346–349.

Xiao, C. (2012). Current advances of chemical and physical starch-based hydrogels. *Starch-Stärke, 1*, 1–7.

Xiao, H., Lin, Q., Liu, G. Q., Wu, Y., Tian, W., Wu, W., & Fu, X. (2011). Physicochemical properties of chemically modified starches from different botanical origin. *Scientific Research and Essays, 6*(21), 4517–4525.

Xiao, W., Shen, M., Ren, Y., Wen, H., Li, J., Rong, L., ... Xie, J. (2022). Controlling the pasting, rheological, gel, and structural properties of corn starch by incorporation of debranched waxy corn starch. *Food Hydrocolloids, 123*, 107136.

Xie, H., Gao, J., Xiong, X., & Gao, Q. (2018). Effect of heat-moisture treatment on the physicochemical properties and in vitro digestibility of the starch-guar complex of maize starch with varying amylose content. *Food Hydrocolloids, 83*, 213–221.

Xing, J. J., Liu, Y., Li, D., Wang, L. J., & Adhikari, B. (2017). Heat-moisture treatment and acid hydrolysis of corn starch in different sequences. *LWT - Food Science and Technology, 79*, 11–20.

Xu, B., Yuan, J., Wang, L., Lu, F., Wei, B., Azam, R. S., ... Bhandari, B. (2020). Effect of multifrequency power ultrasound (MFPU) treatment on enzyme hydrolysis of casein. *Ultrasonics Sonochemistry, 63*, 104930.

Yan, X., Wu, Z. Z., Li, M. Y., Yin, F., Ren, K. X., & Tao, H. (2019). The combined effects of extrusion and heat-moisture treatment on the physicochemical properties and digestibility of corn starch. *International Journal of Biological Macromolecules, 134*, 1108–1112.

Yang, Q., Qi, L., Luo, Z., Kong, X., Xiao, Z., Wang, P., & Peng, X. (2017). Effect of microwave irradiation on internal molecular structure and physical properties of waxy maize starch. *Food Hydrocolloids, 69*, 473–482.

Yang, Q. Y., Lu, X. X., Chen, Y. Z., Luo, Z. G., & Xiao, Z. G. (2019). Fine structure, crystalline and physicochemical properties of waxy corn starch treated by ultrasound irradiation. *Ultrasonics Sonochemistry, 51*, 350–358.

Ye, J., Liu, C., Luo, S., Hu, X., & McClements, D. J. (2018). Modification of the digestibility of extruded rice starch by enzyme treatment (β-amylolysis): An *in vitro* study. *Food Research International, 111*, 590–596.

Yotsawimonwat, S., Sriroth, K., Kaewvichit, S., Piyachomkwan, K., Jane, J. L., & Sirithunyalug, J. (2008). Effect of pH on complex formation between debranched waxy rice starch and fatty acids. *International Journal of Biological Macromolecules, 43*(2), 94–99.

Zailani, M. A., Kamilah, H., Husaini, A., Seruji, A. Z. R. A., & Sarbini, S. R. (2021). Functional and digestibility properties of sago (Metroxylon sagu) starch modified by microwave heat treatment. *Food Hydrocolloids, 122*, 107042.

Zavareze, E. d. R., & Dias, A. R. G. (2011). Impact of heat-moisture treatment and annealing in starches: A review. *Carbohydrate Polymers, 83*(2), 317–328.

Zhang, B., Xiao, Y., Wu, X., Luo, F., Lin, Q., & Ding, Y. (2021). Changes in structural, digestive, and rheological properties of corn, potato, and pea starches as influenced by different ultrasonic treatments. *International Journal of Biological Macromolecules, 185*, 206–218.

Zhang, H., Wang, J. K., Liu, W. J., & Li, F. Y. (2015). Microwave-assisted synthesis, characterization, and textile sizing property of carboxymethyl corn starch. *Fibers and Polymers, 16*(11), 2308–2317.

Zhang, Y., Zhao, X., Bao, X., Xiao, J., & Liu, H. (2021). Effects of pectin and heat-moisture treatment on structural characteristics and physicochemical properties of corn starch. *Food Hydrocolloids, 117*, 106664.

Zhao, A. Q., Yu, L., Yang, M., Wang, C. J., Wang, M. M., & Bai, X. (2018). Effects of the combination of freeze-thawing and enzymatic hydrolysis on the microstructure and physicochemical properties of porous corn starch. *Food Hydrocolloids, 83*, 465–472.

Zhong, Y., Liang, W., Pu, H., Blennow, A., Liu, X., & Guo, D. (2019). Short-time microwave treatment affects the multi-scale structure and digestive properties of high-amylose maize starch. *International Journal of Biological Macromolecules, 137,* 870–877.

Zhou, M., Shi, L., Cheng, F., Lin, Y., & Zhu, P. (2018). High efficient preparation of carboxymethyl starch via ball milling with limited solvent content. *Starch/Stärke, 70,* 1700250.

Zhu, F. (2015). Impact of ultrasound on structure, physicochemical properties, modifications, and applications of starch. *Trends in Food Science and Technology, 43*(1), 1–17.

Zhu, J., Lu, K., Liu, H., Bao, X., Yang, M., Chen, L., & Yu, L. (2021). Influence of moisture content on starch esterification by solvent-free method. *Starch/Stärke, 73,* 2100009.

Zuo, Y., Gu, J., Yang, L., Qiao, Z., Tan, H., & Zhang, Y. (2013). Synthesis and characterization of maleic anhydride esterified corn starch by the dry method. *International Journal of Biological Macromolecules, 62,* 241–247.

Chapter 6

Maize protein

Extraction, quality, and current scenario

E. Dilipan, Jaya Sharma, Manas K. Jha,
Dilip Kumar Markandey,
Sukhvinder Singh Purewal, and Arihant Yuvraaj

DOI: 10.1201/9781003245230-6

6.1 Introduction

Biofortification is a strategy that aims to enhance micronutrient concentrations in edible plant portions via crop breeding or biotechnology; we adopt a broader term here to include biofortification through agronomic methods as well (Cakmak, 2008). Biofortification is regarded as more cost-effective than other methods of providing micronutrients to rural people in developing nations (Nestel et al., 2006; Mayer et al., 2008). Biofortification is the process of allowing plants to absorb minerals (Fe and Zn) from the soil and storing them in grains to meet human nutritional needs. This method is effective, low cost, and widespread (Poletti et al., 2004). Micronutrient deficiency is common in arid and semi-arid areas where calcareous soils limit supply. Soils lacking in microelements yield cereals lacking in microelements, causing human malnutrition. Globally, approximately three billion people suffer from vitamin deficiencies. Micronutrient malnutrition is a significant issue in developing nations, causing widespread sickness and illnesses such as zinc deficiency that impairs brain development, wound healing, and increases susceptibility to infectious illnesses. Iron deficiency causes anaemia and lowers adult productivity. Soil micronutrient deficit impacts the food chain. Recently, fodder crops grown in poor soils and fed to cattle led to the cattle producing offspring with low birth weight, and people who drank the poor milk also produced offspring with low birth weight. Even when plants get sufficient micronutrients, phytic acid, an anti-nutritional agent, reduces micronutrient bioavailability in grains. A deficiency of phytase enzyme in mono-gastric animals and humans reduces dietary Zn and Fe absorption.

Micronutrients build in the aleurone layer of grains, inhibiting Zn and Fe absorption. The molar ratio of phytic acid to Zn or Fe is frequently used to assess micronutrient bioavailability. There is a significant connection between phytic acid and total P content in seeds. Reducing micronutrient efficiencies in cereal grains is a relatively new approach for improving micronutrient status. Biofortification is a novel approach to dietary micronutrient shortages. This method has shown to be sustainable, low-cost, extremely effective, and widespread, particularly in the world's poorest areas.

6.2 Maize farming in India

Maize is a cereal crop used globally for human nutrition, animal feed, industrial goods, and biofuel. The "Queen of Cereals" is a versatile crop with great production potential among cereals grown throughout a wide variety of agro-climatic zones. 70% of worldwide output comes from the US, China, Brazil, and Mexico. India is the world's fifth-biggest producer of maize, accounting for 3% of worldwide output. Almost 28% of maize produced is used for food purposes, 11% for livestock feed, 48% for poultry feed, 12% for wet milling industry for starch and oil production, and 1% for seed (Agricoop, 2007–2008). In Tamil Nadu, maize is a non-traditional crop grown on 0.20 million hectares with a 0.24 million tonne average yield (Crop Report, 2006–2007). The US and Spain produce more than 9.0 metric tons of maize per acre, whereas Brazil produces 3.75 metric tons and India just 2.5.

To satisfy rising demand, India must raise the maize growth rate by 9.51%. Increasing productivity per unit area is one approach to bridge the demand-supply imbalance. In peri-urban regions, maize is grown all year for grain, fodder, green cobs, sweet corn, baby corn, and popcorn. Andhra Pradesh (20.9%), Karnataka (16.5%), Rajasthan (9.9%), Maharashtra (9.1%), Bihar (8.9%), Uttar Pradesh (6.1%), Madhya Pradesh (5.7%), Himachal Pradesh (5.7%) are the major maize producing states that subsidize over 80% of total maize output (4.4%). Jammu and Kashmir and the north-eastern states also produce maize. Thus, maize has become an important crop in non-traditional areas, i.e. peninsular India, where Andhra Pradesh, which ranks fifth in terms of area (0.79 m ha), has the most significant output (4.14 mt) and productivity (5.26 t ha-1). However, certain Andhra Pradesh districts produce as much as the US. Hybrid maize plants are more significant, more robust, and disease resistant. They respond well to fertilizers and have a high need for plant nutrients. Due to their more extensive leaf area index (LAI) during silking, modern hybrids capture more active photosynthetic sunlight and utilize it more efficiently during grain filling (Azam et al., 2007). The National Food Security Mission should include maize alongside wheat, rice, and legumes. This is because of maize's ability to fulfil food and nutritional security needs and impact the development of related industries such as poultry and animal production. Today, maize is mostly used for livestock and poultry feed, and only 25% is for human consumption. Thus, expanding the area under quality maize grain may enhance human nutrition and low-cost, high-quality feed. This would need a variety of maize products and a shift in eating habits. Children may get quality maize grain cheaply.

6.3 Nutritional value of maize

The importance of cereal grains to the nourishment of billions of people across the world is extensively accepted. Being made up of such a large part of diets in developing countries, cereal grains cannot be considered only a source of energy, as they provide significant amounts of protein. The protein trait is limited by the shortage of some essential amino acids, mainly lysine. However, much less appreciated is the fact that some cereal grains consist of excess amounts of certain necessary amino acids that influence the ability of protein utilization. The classic example is maize. Other cereal grains have the same restraint but less obviously.

Numerous researchers have conducted extensive research into the reasons for the low quality of maize proteins. According to some studies, tryptophan, not lysine, is the first limiting amino acid in maize, which may be true for particular kinds with a high lysine concentration or maize products that have been processed.

It has been demonstrated in animal experiments that the simultaneous addition of both lysine and tryptophan significantly improves the protein quality of maize, and this has been agreed upon by all researchers. When other amino acids were added to lysine and tryptophan, the improvement in quality was minimal in some experiments and higher in others. According to animal feeding experiments, isoleucine is the limiting amino acid after lysine and tryptophan. The majority of researchers who reported such findings claimed that isoleucine supplementation was due to an excess of leucine interfering with isoleucine absorption and utilization. Niacin needs are increased by excessive consumption of leucine and the protein in maize, and this amino acid could be part of the cause of pellagra.

6.4 Maize proteins and their classification

Corn (Zea mays Linnaeus) is an essential cereal crop for human nutrition globally because of its high carbohydrates, lipids, and proteins. Maize provides protein and calories to millions of people in underdeveloped nations (FAO, 2005).

Quality protein maize has about double the amount of lysine and tryptophan as normal maize and a more nutritionally balanced amino acid profile. QPM may help avoid protein deficiency, particularly in newborn babies whose meals are mostly maize (Sintayehu, 2008). The leucine:isoleucine ratio was improved in QPM versions, which researchers thought was beneficial for releasing more tryptophan for niacin synthesis, which helps prevent pellagra. Tryptophan levels were 0.94–1.06% in 13 QPM genotypes from the CIMMYT QPM gene pool.

Traditional maize protein lacks lysine and tryptophan. Consequently, regular maize intake, especially when given to newborns, may predispose them to childhood illnesses like "kwashiorkor", a fatal illness characterized by early development failure. QPM has double the lysine and tryptophan levels of regular maize. Normal maize and QPM contain approximately 2.0% and 4% lysine in endosperm protein, respectively (Teklewold et al., 2015). Amounts may vary from 1.6% to 2.6% in normal maize and 2.7% to 4.5% in QPM endosperm protein (Vivek et al., 2008). Similarly, tryptophan levels in endosperm protein vary from 0.2 to 0.6% in traditional maize and 0.5 to 1.1% in QPM maize (Vivek et al., 2008).

The permissible percentage levels for lysine and tryptophan for QPM may be represented in whole grain or endosperm sample units. Tryptophan levels should be >0.075 and >0.070% in whole grain and endosperm, respectively (Vivek et al., 2008; Teklewold et al., 2015). Because tryptophan analysis is less costly than lysine analysis, breeders frequently utilize tryptophan levels to indicate a QPM variety's nutritional value (Villegas et al., 1992; Nurit et al., 2009). Zein is a prolamine protein mainly found in grains (the equivalent of hordein in barley and gliadin in wheat). A large portion of zein is found in the endosperm, whereas glutelin is found in the germ. The germ layer contains albumins and globulins (Shukla & Cheryan, 2001).

6.4.1 Zein as an important protein of maize

Maize, often known as Indian corn, produces zein, a valuable by-product. In aqueous alcohol, such as ethanol, prolamins are recovered (Anderson & Lamsal, 2011a). According to its solubility, amino acid sequence, and molecular weight, zein is a polypeptide mixture categorized as α-, β-, γ-, and δ-zein (MW). In maize, α-zein dominates, accounting for 70–85% of total zein, followed by γ- (10–20%), β- (1–5%), and δ-zein (1–5%). All zein fractions are amphiphilic and include residues of hydrophilic and hydrophobic amino acids soluble in alcohol-water combinations (Esen, 1986; Wlison, 1991). Zein is less significant owing to a shortage of key amino acids like lysine and tryptophan. Zein is well-known for its edible films, coatings, biodegradable packaging, and encapsulation (Shukla & Cheryan, 2001). Anderson and Lamsal (2011b) state that zein may be generated through the dry-grind method and extracted from ground maize by combining it with ethanol, which contains α- and γ-zein. Except for γ-zein, which is rich in cysteine, their amino acid compositions are identical. A high cysteine content may alter zein rheological characteristics, causing gelation of zein solutions (Nonthanum et al., 2012). Due to its great purity, zein is usually sold for USD 10–40/kg. As a result of the low market price of synthetic plastics, zein is uncompetitive until new, cost-effective zein extraction methods are available (Shukla & Cheryan, 2001).

6.4.2 Zein protein characterization

SDS-PAGE is used to separate a mixture of zein proteins and categorize them as α-, β, γ-, and δ-zein (Wilson, 1991; Olsen & Phillips, 2001). Thompson and Larkins (1989) reported α-, β, γ-, and δ-zein with molecular weights of 19 and 22 kDa, 27 and 16 kDa, 14 kDa, and ten kDa. Amphiphilic zein fragments include both hydrophobic and hydrophilic amino acids. This category of hydrophobic proteins is deficient in critical amino acids like lysine and tryptophan but rich in glutamine, leucine, proline, and alanine (Coleman & Larkins, 1999).

α-zein is generally soluble in 60–95% aqueous ethanol, but its solubility in water can be improved through enzymatic modification. Zein has a high amino acid content, accounting for 25% of the amide group and 10% of the proline group. The samples containing α-, β-, γ-, and δ-zein dissolve easily in 60% ethanol (Shukla & Cheryan, 2001). The α-zein dissolves easily in primary solvents such as formic acid and pyridine but not in anhydrous ethanol or pure water (Yao et al., 2009).

γ-zein contains more hydrophilic components than α-zein, and all of these proteins have different solubilities in ethanol. The zein protein may be suspended in 90% isopropanol, 50%–95% ethanol, and water (Wang & Padua, 2010). The high-temperature treatment causes zein to aggregate owing to intermolecular disulphide cross-linking of prolamins (Byaruhanga et al., 2007). Solvents may influence aggregate formation during binary solution evaporation (Govor et al., 2009). This was explained by raising the ethanol level to 90–95% (Anderson et al., 2011; Paraman and Lamsal, 2011).

α-zein is the most significant zein protein discovered in maize. The amino acid sequence of zein includes 20 peptide repeats rich in glutamine, leucine, and alanine. The α-zein peptide comprises ten repetitions of 22 kDa and nine repeats of 19 kDa (Nonthanum, 2013), with various solubility, molecular size, and charge (Lawton, 2002). The lack of lysine and tryptophan amino acids contributes to the poor dietary nitrogen balance of zein protein. Without acidic and necessary amino acids, non-polar amino acid residues dominate, causing zein protein solubility (Deo et al., 2003). There are two significant bands with molecular weights of 24000 and 22000 daltons, respectively (Parris & Dickey, 2001).

6.5 Extraction of zein from maize samples

Corn prolamin proteins provide nitrogen for corn kernel growth during germination (Flint-Garcia et al., 2009). Because of its grease resistance, bright form, and water-insoluble content, zein has many applications, including biodegradable packaging, waterproofing, and fibre. The majority of zein extractions used dried ground corn, which were dissolved and concentrated by adding 80–85% ethanol. After redissolving zein in 90% ethanol, the pigments and lipids were removed by adding absolute ethanol. Due to the yield of 6–7% zein from the total maize, the extraction was not commercially useful (Anderson, 2011). Figure 6.1 depicts a commonly used zein extraction method. To reduce the cost of zein production, zein protein was extracted from whole ground corn rather than corn gluten meal (Hojilla-Evangelista, 2003). Dicky et al. (1998) described the extraction of zein from cornflour using 70% aqueous ethanol. Hojilla-Evangelista et al. (1992) developed a

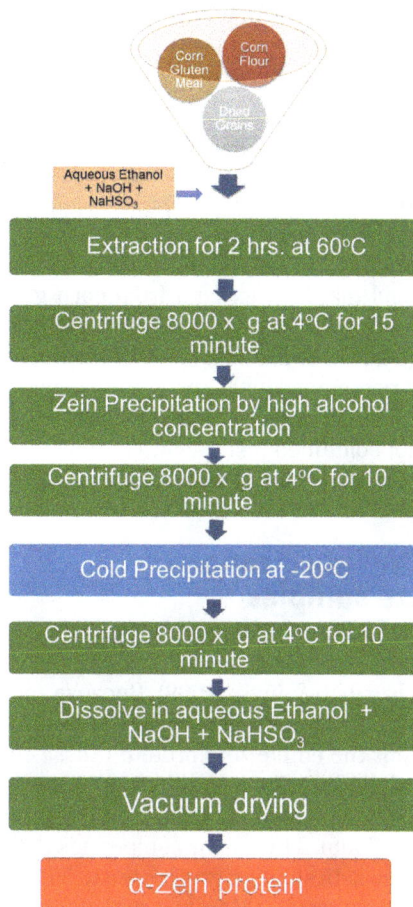

Figure 6.1 Flow chart of zein extraction from corn flour, corn gluten meal and distiller's dried grains

method for dialyzing the extracted protein with water, which improved the extracted protein's purity. The use of NaOH as a solvent also impacts the purity of the final zein product (Coleman & Larkins, 1999). Takahashi and Yanai (1994) demonstrated that in the extraction of zein, aqueous acetone could be replaced with aqueous isopropanol or ethanol and that fats and oils could be extracted from corn gluten meal before zein extraction to improve the extraction method. Cook et al. (1996) demonstrated a method for isolating zein from enzymatically modified gluten meal.

Parris and Dickey (2001) described the zein extraction process using various pre-treatments with a substrate of dry-milled corn, with varying

zein yields, and the recovered zein was also tested for edible film preparations. The nanofiltration and ultrafiltration of zein extracts allow for the
separation of zein and oil and the recovery of solvent throughout the process (Cheryan, 2002). Hojilla-Evangelista and Johnson investigated the zein
extracted from defatted ground corn (2003). Xu et al. reported extracting
zein from ethanol-defatted DDGS under acidic and basic conditions with a
reducing agent (2007).

Size-exclusion chromatography may be used to analyze zein peptides
from oils and pigments (Cheryan, 2009). Selling and Woods (2008) claim
that glacial acetic acid may extract zein from corn gluten meal. The biu-
ret test is the best method for measuring total zein protein in maize flour
alcoholic extracts (Bancila et al., 2016). A 70% isopropanol extract yielded
the most significant percentage yield and protein content. β-zein and α-zein
were discovered in zein extracted with acetic acid, whereas zein extracted
with 70% isopropanol contained just α-zein (Uan-on et al., 2018). Table 6.1
shows the different organic solvent zein extraction methods.

6.6 Isolation and purification of zein from maize samples

Carotenoids like lutein, zeaxanthin, and beta-carotene give zein protein a
yellow colouration (Kurilich & Juvik, 1999). Because zein is used more than
yellow zein, its price is more significant (Sessa et al., 2003). Traditionally,
several ways discolour the edible zein protein (Sessa et al., 2003). Column
chromatography and diafiltration/ultrafiltration successfully removed the
pigments, proving that activated carbon can decolorize zein at different
temperatures (Sessa, 2008). The colour compounds coupled with activated
carbon increased when zein solution was heated to 55 °C, and the lutein-rich
zein was somewhat denatured (Momany et al., 2006). Zein chewing gum
was improved by eliminating smell (Sessa and Palmquist, 2009). Zein fractions were eluted first, followed by non-zein impurities, and xanthophylls
were eluted last with excellent resolution using the LH-20 resin column.
Size-exclusion chromatography may increase zein production and purity
by over 90%, while ultra/nanofiltration reduces zein yield and purity (Kale
& Cheryan, 2007; Kale & Cheryan, 2009).

6.7 Factors affecting maize protein

Several issues threaten farms, reducing yields. These components are
technical, biological, and environmental (Ngoune Tandzi et al., 2018).
Expansion of agricultural land area and intensification of farmland management practices such as irrigation and the use of vast amounts of inputs

Table 6.1 Various processes used for extraction of zein from different types of maize samples

Material	Method of extraction	Process of protein recovery	Final yield	References
Maize	Dry milling, ethanol	ND	Solubilization of zein after 18 h	Russell, 1980
Maize	De-starched, extracted with alcohol in existence of temperature and heat	Filtration and evaporation	First known process for zein recovery	Osborne, 1891
Maize	Wet attrition milling to loosen protein-starch matrix, ethanol	Filtration	82% extraction of both zein and non-zein proteins, purity>90%	Kampen, 1995
Defatted maize	Enzyme hydrolysis of corn starch, extraction with ethanol	ND	95% of zein extracted	Cao et al., 1996
Corn gluten meal	Acetone	Precipitation	30% of zein recovered in white porous granular form	Takahashi & Norimasa, 1994
Corn gluten meal	Enzymatic starch hydrolysis, alkaline treatment, alcohol extraction	Precipitation	High purity (>96%)	Cook et al., 1993, 1996
Corn gluten meal	Ethanol	Precipitation	30% recovery of zein	Takahashi & Yanai, 1996
Corn gluten meal	Isopropyl alcohol	Precipitation	21–32% recovery with 80–87% purity	Wu et al., 1997a, b

(Continued)

Table 6.1 (Continued) Various processes used for extraction of zein from different types of maize samples

Material	Method of extraction	Process of protein recovery	Final yield	References
Distillers dried grains	Ethanol	ND	Low yields 1.5–6.6%. Protein purity was low (37–57%)	Wu et al., 1981; Wu & Stringfellow, 1982; Wolf & Lawton, 1997
Distillers dried grains with soluble	Ethanol	Acidic and basic conditions	α-zein extraction	Xu et al., 2007
Corn gluten meal	Glacial acetic acid	Ethanol	Extract large amount of zein	Selling & Woods, 2008
Corn flours	Ethanol	Ultra-sonication	Total zein proteins	Bancila et al., 2016
Corn gluten meal	Isopropanol	Acetic acid	Extraction of α, β-zein	Uan-on et al., 2018

ND: not determined

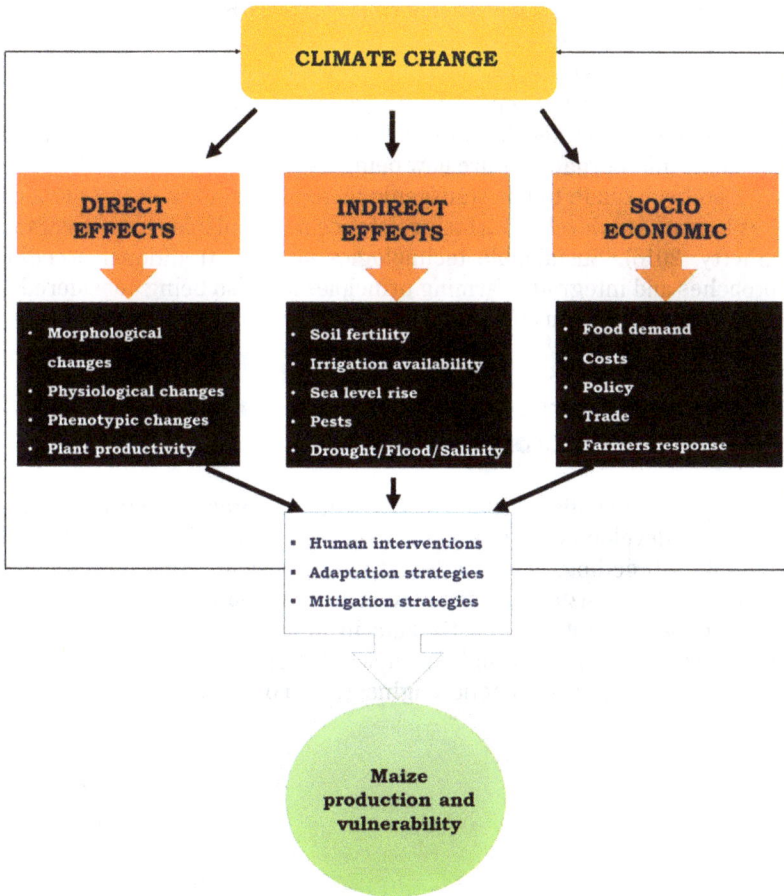

Figure 6.2 General effects of climate change in maize production

such as inorganic fertilizers and synthetic pesticides for pest and weed control have occurred in many nations (Oldfield et al., 2019). Abiotic and biotic limitations affect crop production. Abiotic limitations include soil characteristics (pH, soil components, physicochemical and biological qualities) and climate (drought, cold, flood, heat stress, etc.) (see Figure 6.2). Climate change exacerbates these issues. At the molecular level, abiotic stressors affect plant development and production. Beneficial organisms (pollinators, decomposers, and natural enemies), pests (arthropods, diseases, weeds, vertebrate pests), and anthropogenic evolution are examples of biotic forces. Global warming has a catastrophic impact on plant development and agricultural production, decreasing crop yields by 70% due to climate change (Noya et al., 2018).

Conventional crop improvement methods such as breeding are routinely conducted to increase long-term resistance to specific pests, diseases, and abiotic stresses. However, modern biotechnology techniques such as marker-assisted selection and transgenic approaches that involve genetic modification and high-throughput sequencing of both plant and pathogenic microorganisms are now being used more frequently. Attempts have also been made to use transgenic technologies to modify functional genes in plants to develop intrinsic tolerance mechanisms (Kumaraswamy & Shetty, 2016). Sustainable technologies such as traditional breeding approaches and integrated farming principles are also being considered to develop crop adaptation and improve adaptive mechanisms.

6.8 Quality improvement strategy for multi-nutrient biofortified maize

A variety of conventional and non-conventional breeding strategies are used in the development of multi-nutrient maize. Breeding methods such as mutation breeding, abiotic tolerant varieties, and modern techniques such as marker-assisted and genomic selection could all be used in tandem to boost the rate of genetic gain in multi-nutrient maize breeding (see Figure 6.3). High-throughput molecular breeding techniques, such as genome editing and genetic engineering, could also be useful. QTL

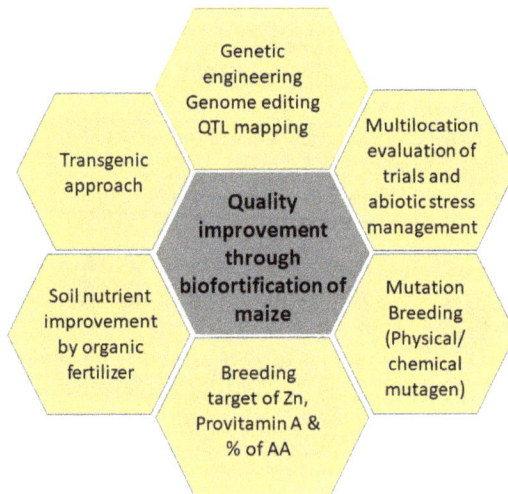

Figure 6.3 Quality improvement strategies for the development of multi-nutrient maize genotypes

mapping and GWAS are two techniques used in molecular breeding to dissect complex traits in maize.

6.8.1 Biofortification through organic fertilizers

The continuous use of chemical fertilizers creates potential polluting effects due to chemicals in the environment, and there is growing interest in the use of organic fertilizer worldwide due to soil fertility depletion (Das et al., 1991). They recycle, and the nutrients in organic manure have been given more thought to ensure long-term land use and agricultural production improvement. The long-term consequences of using organic and inorganic fertilizers together to improve soil fertility and crop yield have been investigated (Liu et al., 1996). Organic and inorganic fertilizers both demonstrated significant benefits in increasing plant N uptake and soil accessible N and improving maize production (Wang et al., 2001).

Organic manure has been widely employed in maize agriculture for its agronomic and environmental benefits (Yang et al., 2014), and it improves maize yields closer to their biological potential while lowering chemical fertilizer input (Dordas et al., 2008). As a result, the additional organic manure input improves soil nutrient storage and crop absorption, resulting in better grain production with lower chemical fertilizer and environmental costs (Masood et al., 2014). However, in semi-arid regions, how soil water and nutrients are transmitted when other organic manures are applied is unknown.

The reuse of organic wastes, which can sustain organic matter, can address the nutritional deficiency. It has long been known to increase soil microbial functioning, aeration, moisture retention, and nutrient availability (Girmay et al., 2008). Many different types of manures (for example, green crop residues, mulch, industrial wastes, animal dung, seaweed sap, and domestic wastes) have been successfully applied to crops, resulting in increased Zn availability to plants through microbial actions, direct donation, or chemical conversion reactions (Maliwal et al., 2007; Quilty & Cattle, 2011).

6.8.2 Mutation breeding technique

Mutation breeding is a popular technique among plant breeders and plant biotechnology researchers. Mutagens are commonly used to induce genetic variation in various plant species, resulting in gene mutation and the expansion of the gene pool for crop improvement. Various mutagen agents and irradiation (both ionizing and non-ionizing) have been successfully used for induced mutation breeding in a variety of crops (Song & Kang, 2003).

Gamma irradiation is one of the most effective methods for removing physical mutagens. Physical mutagens such as X-rays, gamma rays, and thermal and fast neutrons are responsible for approximately 89% of mutant varieties developed worldwide (Kharkwal et al., 2004); gamma irradiation alone is responsible for 70% of mutant varieties (Nagatomi & Degi, 2009). Gamma-ray irradiation of plant material is widely used to induce mutations at the genetic level, altering the number of biochemical processes that lead to the desired changes in the genotype. The primary injury to plant material caused by gamma irradiation is physiological damage, which is primarily limited to growth retardation and higher doses, eventually leading to death in the M1 generation (Yasmin et al., 2020). Gamma irradiation has resulted in many mutant varieties such as plants with resistance to diseases, cold, salt, and plants with desired qualities (Mehlo et al., 2013).

A mutation is a valuable tool for improving one or two easily identifiable traits in a well-adapted variety in plant breeding. While the improved traits are added, the primary genotype of the variety is usually only slightly altered by mutation. Among the physical mutagens, gamma irradiation is the most effective at inducing plant growth and the development of cytological, genetical, biochemical, physiological, and morphological changes. Gamma irradiation increased the number of morphological, viable, and seed mutants in various crops. With the effect of gamma irradiation, the following mutants were observed: high seed protein, small seed, bold seed, tall, bushy, and dwarf. Among mutants, the bold seed mutant has been used in various plant breeding programmes, and it is used as a donor parent for the bold seed trait; this bold seed mutant has also been used successfully in crossbreeding programmes.

Induced mutation via physical and chemical mutagen methods results in genetic variation, which results in new varieties (see Figure 6.4) with improved characteristics (Wongpiyasatid et al., 2000). Mutation breeding can be a valuable supplement to conventional breeding methods and has been used to improve qualitative and quantitative traits in many crops successfully, including sesame (Sharma, 1993), cowpea (Dhanavel et al., 2008; Adekola & Oluleye, 2007; Ronde & Spreeth, 2007; Gnanamurthy et al., 2012), black gram (Thilagavathi & Mullainathan, 2009; Yasmin et al., 2020). Furthermore, mutation breeding takes less time, is easier to manage, and is highly beneficial to developed crop cultivars when compared to other conventional breeding methods (Bolbhat & Dhumal, 2009; Manjaya, 2009). In an updated database of IAEA/MVD (International Atomic Energy Agency)/(Mutant Variety Database), a list of 3,303 varieties released through artificial mutations is maintained; among these, 97 mutant varieties of maize are registered and released by physical mutagens, particularly gamma rays (https://mvd.iaea.org/, accessed 5 June 2021).

The improvement of quality maize protein, both in traditional plant breeding and molecular genetic approaches, has been utilized. Traditional

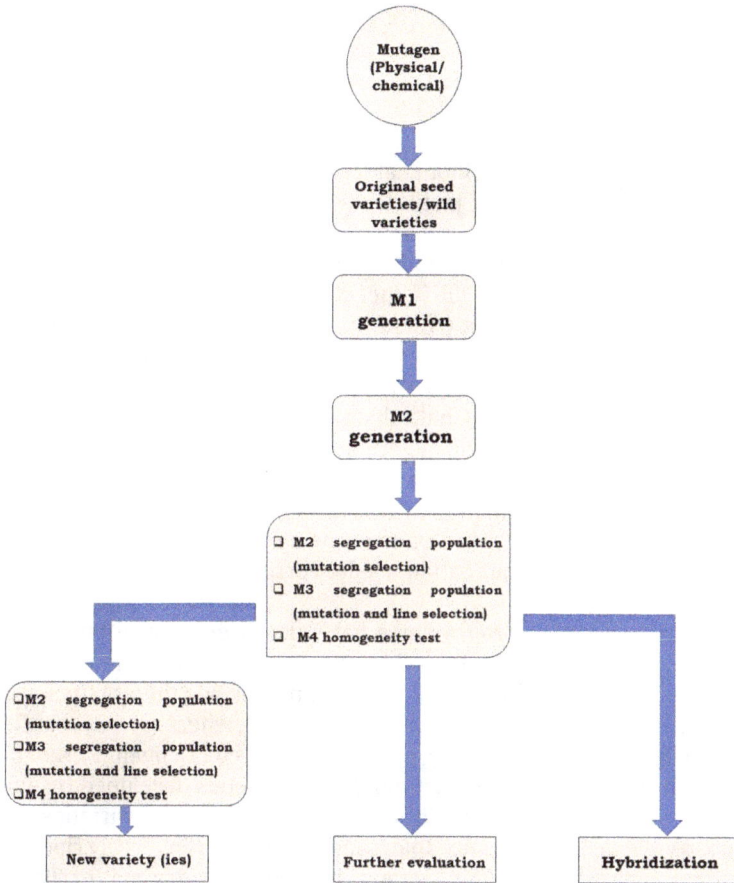

Figure 6.4 Methods of mutation breeding technique

plant breeding will be assumed to include the use of naturally occurring genetic variation via a controlled crossing of maize lines or strains followed by artificial selection and mutagenic agents to increase genetic variation before selection. Molecular genetic approaches are assumed to include methods that require molecular sequence information during development or implementation. Among these methods are the creation of transgenic plants with novel phenotypic characteristics and molecular markers to supplement traditional breeding efforts.

India has created approximately 329 mutant varieties of various crops through direct mutagenesis, including rice, wheat, barley, pearl millet, jute, groundnut, soybean, chickpea, mung bean cowpea, black gram, sugarcane, chrysanthemum, portulaca, tobacco, and dahlia. About 50 of these 329 mutant varieties have been developed using mutant lines in breeding programmes. In the 1950s, the Indian mutation breeding programme achieved success (Bughio et al., 2007). The Indian Agricultural Research Institute (IARI), the Bhabha Atomic Research Centre, Tamil Nadu Agricultural University, and the National Botanical Research Institute were among the primary research centres and institutes involved in developing and releasing various mutants (Wani et al., 2014). These centres' efforts have resulted in a significant breakthrough for India. Since 1957, the IARI has been India's primary research institution for induced mutations, releasing numerous mutant crops and ornamentals (Ahloowalia & Maluszynski, 2001). Several gamma radiation-induced rice mutants were released as high-yielding varieties in India under the "PNR" series. Some of these varieties are short and mature at a young age (Chakrabarti, 1995). Among these varieties, two early ripening and aromatic mutation-derived rice varieties, "PNR381" and "PNR102", are currently popular among farmers in Haryana and Uttar Pradesh. There is no information available on the actual area planted with these varieties. However, based on the rate of new seed replacement by farmers, breeder seeds, foundation seeds, certified seeds, and IARI data are distributed.

Mutation breeding programmes have also been carried out in various European countries. In Bulgaria, for example, induced mutagenesis has resulted in the development of more than 76 new cultivars, including maize (26), durum wheat (9), tomato (6), barley (5), wheat (5), soybean (5), pepper (4), lentil (4), sunflower (3), cotton (2), tobacco (2), bean (2), and pea (2). Maize has the most significant number of varieties developed through mutation breeding, with 26 varieties released thus far. These varieties have high grain yield and productivity, tolerance to dense sowing, early ripening, drought tolerance, high protein content, high biomass dry matter, flowering time shifts, white-coloured grain, strong stems, altered ear length, and increased length number of rows. Kneja 509, a maize hybrid, has become a leading cultivar, accounting for up to 50% of the crop's growing area (Tomlekova, 2010).

Induced mutagenesis and its breeding strategies can improve quantitative and qualitative traits in crops in a fraction of the time that conventional breeding takes. Mutagenic treatment of seeds and other parts remains a valuable tool for isolating desired variants and developing resistance to biotic and abiotic stresses in various crops due to its relative simplicity and low cost. Thus, the released mutant cultivars are already part of the Joint FAO/IAEA Division's overall strategy and commitment to global food security. As a result, the global impact of mutation breeding-derived crop varieties demonstrates mutation breeding's potential as a flexible and practicable approach applicable to any crop, provided appropriate objectives and selection methods are followed.

6.9 Application of zein protein

Zein is used in a variety of industrial products, including food packaging, tablet coatings, cosmetics, adhesives, fibres, food product coatings, plastics, inks, and chewing gum (Shukla & Cheryan, 2001). Zein micro fragments were created and can be used as a replacement for dietary fat residues with a lower calorific protein density than lipids (Stark & Gross, 1991). Corn gluten meal is used to extract zein, which is then used in non-polar coatings for food products and pharmaceutical residues. Zein films have been investigated for use as coatings and as a barrier to prevent contamination in various food applications (Shukla, 1992). Zein films are also used in toilet cleaning blocks (Campbell & Ferrando, 1997) (see Figure 6.5), paper coatings for glossy magazine covers (Trezza & Vergano, 1994), and the production of dry-chemical tests agents.

Zein's most promising and recent application is plastic for packaging and the preparation of biodegradable films. The global demand for such products is estimated to be 15000–250000 tonnes per year. Gennadios et al. (1994) and Baker et al. (1994) published a comprehensive aspect review on coatings and protein-based edible films. Spence et al. (1995) reported using zein in the production of edible films, pharmaceutical tablet coating, bottles, packaging materials, sacks, bags, sheets, films, pipes, rods, laminates, and powders. Unplasticized edible zein protein films, on the other hand, were too brittle for most applications (Parris & Coffin, 1997).

By incorporating food-grade antimicrobial components into the packaging film, the marketability of zein-based plastic films can be increased (Padgett et al., 1998). Wang et al. (1998) demonstrated zein protein

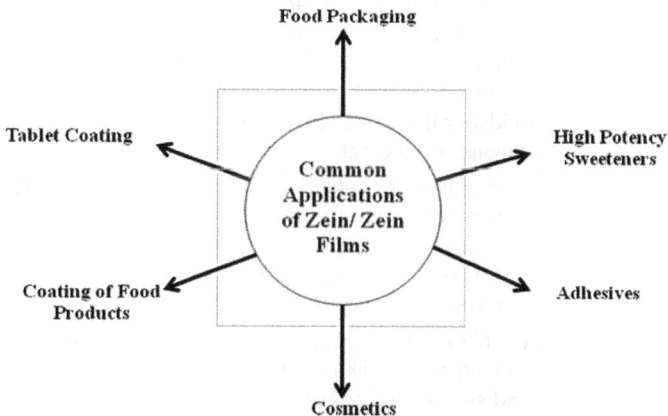

Figure 6.5 Some common applications of zein protein isolated from maize samples

Table 6.2 Various applications of zein protein extracted from different maize samples

Category	Uses	References
Biomedical	Encapsulating lutein into zein nanoparticles improved its aqueous solubility and enabled specific drug delivery in the colon region	Hurtado-López & Murdan, 2006
	The Indole-3-carboxymethyl chitosan (I3C) and 3,3'-diindolylmethane (DIM) degradation rates were improved by zein/carboxymethyl chitosan (zein/CMCS) nanoparticles (both have potential for cancer treatment) resistance to harsh gastric conditions	Luo et al., 2012
	Zein nanoparticles complexed with chitosan containing tocopherol (zein/CS/TOC) improved α-tocopherol stability (TOC)	Luo et al., 2011
	Zein-microparticles increased the stability of rifampicin, isoniazid, and pyrazinamide drugs	Mehta et al., 2011
	Ultrafiltrated α-zein hydrolysates possess activity for reducing blood pressure	Miyoshi et al., 1991
Food packaging	Antioxidant and antibacterial properties on zein-based films were developed after incorporation of different phenolic acids (gallic acid, p-hydroxy benzoic acid or ferulic acids) or flavonoids (catechin, flavone, or quercetin)	Arcan & Yemenicioglu, 2011
	The addition of phenolic and flavonoid compounds also eliminated the classical brittleness and flexibility problems associated with raw zein	Shi & Dumont, 2014
	Zein films can replace commercial coating agents, like carnauba wax and shellac, inside food packets	Lawton, 2002

(Continued)

Table 6.2 (Continued) Various applications of zein protein extracted from different maize samples

Category	Uses	References
	Zein as a protective, impermeable coating for food packaging, wherein different conditions, such as hydrophilic and hydrophobic surfaces, may be required to store edible materials	Subramanian & Sampath, 2007
	Curcumin encapsulated in zein nanofibers possesses improved free radical scavenging activity and sustained release properties	Brahatheeswaran et al., 2012
	To delay ripening/to reduce moisture loss in pear and mango	Luo et al., 2011
	Reduce ripening rate/moisture loss/ microbial contamination in semihard cheese	Pena-Serna & Filho, 2015
Pharmaceutical	Blends of zein with conventional synthetic polymers, such as polyethylene, nylon, and polyvinylpyrrolidone (PVP), and with biodegradable (natural or synthetic) polymers, such as starch, poly(ε-caprolactone) (PCL), etc., retaining the biodegradability character of zein	Shukla & Cheryan, 2011
	Microspheres of poly(D,L-lactide-co-glycolide) (PLA) and zein were prepared by spray drying to release amoxicillin (AMX) for root canal disinfection	Sousa et al., 2010
	Nanofibrous membranes for wound healing were developed by collagen and zein electrospinning in aqueous acetic acid solution. The combination of zein was found to improve the electrospinnability of collagen	Lin et al., 2012

applications as laminated boards, adhesives, and solid colour. Zein peptide has numerous applications, including the formation of rigid, glossy, hydrophobic, greaseproof coatings resistant to microbial attack and exhibits excellent compressibility and flexibility (Shukla & Cheryan, 2001).

Rakotonirainy et al. (2001) described zein resin sheets made from three-ply pressed oleic acid, laminated with tung oil, and used to protect broccoli. Broccoli firmness and colour were preserved using zein films and stored in a refrigerator for six days. The use of zein protein also preserved the integrity of a turkey product. Zein protein's film-forming properties were used to reduce oil consumption during frying (Ilter et al., 2008). Zein coatings have also been used to control seed germination problems (Assis & Leoni, 2009). Zein has a wide range of applications in the food and drug industries, and nanoparticles of zein have been reported as non-polar drug carriers, and zein microspheres produce ciprofloxacin. A purified and decolorized zein has been used in the biomedical and controlled self-assembly fields (Fu et al., 2009).

Dong et al. (2004) used polylactic acid (PLA) and corning microplates as controls to grow mouse fibroblast cells (NIH3T3) and human liver cells (HL 7702) on zein films. Tu et al. (2009) reported using zein for tissue work because it has a high tensile strength that helps hold the cells. Wang et al. (2007) reported three-dimensional porous zein scaffolds with about 80% porosity and 100–380 m diameter pores needed for tissue support. Tu et al. (2009) used bone tissue growth on zein scaffolds to repair critical bone damage to a rabbit's radius bone.

The production of biodegradable materials from natural materials has piqued people's interest all over the world (Tharanathan, 2003). Zein was also investigated for its potential use as a structural material in packaging applications (Lai et al., 1997). Wang and Padua (2010) discovered that zein could produce various microstructures depending on the concentration of zein protein in ethanol solvents. Table 6.2 shows a variety of zein industrial applications.

References

Adekola, O. F., & Oluleye, F. (2007). Influence of mutation induction on the chemical composition of cowpea Vigna unguiculata (L.) Walp. *African Journal of Biotechnology, 6*(18), 2143–2146.

Ahloowalia, B. S., & Maluszynski, M. (2001). Induced mutations a new paradigm in plant breeding. *Euphytica, 118*(2), 167–173.

Anderson, T. J. (2011). Thesis on "extraction of zein from corn co-products". Lowa State University Ames, Iowa.

Anderson, T. J., & Lamsal, B. P. (2011a). Zein extraction from corn, corn products, and coproducts and modifications for various applications: A review. *Cereal Chemistry, 88*(2), 159–173.

Anderson, T. J., & Lamsal, B. P. (2011b). Development of new method for extraction of α-Zein from corn gluten meal using different solvents. *Cereal Chemistry, 88*(4), 356–362.

Arcan, I., & Yemenicioglu, A. (2011). Incorporating phenolic compounds opens a new perspective to use zein as flexible bioactive packaging materials. *Food Research International, 44*(2), 550–556.

Assis, O., & Leoni, A. (2009). Protein hydrophobic dressing on seeds aiming at the delay of undesirable germination. *Scientific Agriculture, 66*(1), 123–126.

Baker, R. A., Baldwin, E. A., & Nisperos-Carriedo, M. O. (1994). Edible coatings and films for processed foods. In J. M. Krochta, E. A. Baldwin, & M. O. Nisperos-Carriedo (Eds.), *Edible coatings and films to improve food quality* (pp. 89–104). Lancaster, PA: Technomic.

Bancila, S., Ciobanu, C. I., Murariu, M., & Drochioiu, G. (2016). Ultrasound-assisted zein extraction and determination in some patented maize flours. *Revue Roumaine de Chimie, 61*(10), 725–731.

Bolbhat, S. N., & Dhumal, K. N. (2009). Induced macromutations in horsegram [Macrotyloma uniflorum (Lam.) Verdc]. *Legume Research, 32*(4), 278–281.

Brahatheeswaran, D., Mathew, A., Aswathy, R. G., Nagaoka, Y., Venugopal, K., Yoshida, Y., ... Sakthikumar, D. (2012). Hybrid fluorescent curcumin loaded zein electrospun nanofibrous scaffold for biomedical applications. *Biomedical Materials*, 7–12.

Bughio, H. R., Asad, M. A., Odhano, I. A., Bughio, M. S., Khan, M A., Mastoi, N. N. (2007). Sustainable rice production through the use of mutation breeding. *Pakistan Journal of Botany, 39*, 2457–2461.

Byaruhanga, Y. B., Erasmus, C., Emmambux, M. N., & Taylor, J. R. N. (2007). Effect of heating cast of kafirin films on the functional properties. *Journal of Science of Food and Agriculture, 87*, 167–175.

Cao, N., Xu, Q., Ni, J., & Chen, L. F. (1996). Enzymatic hydrolysis of corn starch after extraction of corn oil with ethanol. *Applied Biochemistry and Biotechnology, 57*(1), 39–47.

Campbell, S. J., & Ferrando, J. C. (1997). Toilet cleaning compositions. *Patent WO, 97*, 20029.

Chakrabarti, S. N. (1995). Mutation breeding in India with particular reference to PNR rice varieties. *Journal of Nuclear Agriculture and Biology, 24*, 73–82.

Cheryan, M. (2002). Corn oil and protein extraction method. US patent 6,433,146.

Cheryan, M. (2009). *Zein: The industrial biopolymer for the 21st century.*

Coleman, C. E., & Larkins, B. A. (1999). The prolamins of maize. In P. R. Shewry & R. Casey (Eds.), *Seed proteins* (pp. 109–139). Dordrecht: Kluwer Academic Publishers.

Cook, R. B., Mallee, F. M., & Shulman, M. L. (1996). Purification of zein from corn gluten meal. U.S. Patent 5,580-959.

Das, O. P., Ward, K., Ray, S., & Messing, J. (1991). Sequence variation between alleles reveals two types of copy correction at the 27-kDa zein locus of maize. *Genomics, 11*(4), 849–856.

De Ronde, J. A., & Spreeth, M. H. (2007). Development and evaluation of drought resistant mutant germ-plasm of Vigna unguiculata. *Water SA*, *33*(3), 381–386.

Deo, N., Jockusch, S., Turro, N. J., & Somasundaran, P. (2003). Surfactant interactions with zein protein. *Langmuir*, *19*(12), 5083–5088.

Dhanavel, D., Pavadai, P., Mullainathan, L., Mohana, D., Raju, G., Girija, M., & Thilagavathi, C. (2008). Effectiveness and efficiency of chemical mutagens in cowpea (Vigna unguiculata (L.) Walp). *African Journal of Biotechnology*, *7*(22), 4116–4117.

Dickey, L. C., Dallmer, M. F., Radewonuk, E. R., Parris, N., Kurantz, M., & Craig, J. C. (1998). Zein batch extraction from dry-milled corn: Cereal disintegration by dissolving fluid shear. *Cereal Chemistry*, *75*(4), 443–448.

Dong, J., Sun, Q., & Wang, J. Y. (2004). Basic study of corn protein, zein, as a biomaterial in tissue engineering, surface morphology and biocompatibility. *Biomaterials*, *25*(19), 4691–4697.

Dordas, C. A., Lithourgidis, A. S., Matsi, T., & Barbayiannis, N. (2008). Application of liquid cattle manure and inorganic fertilizers affect dry matter, nitrogen accumulation, and partitioning in maize. *Nutrient Cycling in Agroecosystems*, *80*(3), 283–296.

Flint-Garcia, S. A., Bodnar, A. L., & Scott, M. P. (2009). Wide variability in kernel composition, seed characteristics, and zein profiles among diverse maize inbreds, landraces, and teosinte. *Theoretical and Applied Genetics*, *119*(6), 1129–1142.

Fu, J. X., Wang, H. J., Zhou, Y. Q., & Wang, J. Y. (2009). Antibacterial activity of ciprofloxacin-loaded zein microsphere films. *Materials Science and Engineering*, *29*(4), 1161–1166.

Gennadios, A., Weller, C. L., & Gooding, C. H. (1994). Measurement errors in water vapor permeability of highly permeable, hydrophilic edible films. *Journal of Food Engineering*, *21*(4), 395–409.

Girmay, G., Singh, B. R., Mitiku, H., Borresen, T., & Lal, R. (2008). Carbon stocks in Ethiopian soils in relation to land use and soil management. *Land Degradation and Development*, *19*(4), 351–367.

Govor, L. I., Demidov, A. M., Kurkin, V. A., & Mikhailov, I. V. (2009). Multipole mixtures in gamma transitions in the (n, n′ γ) reaction on 160Gd. *Physics of Atomic Nuclei*, *72*(11), 1799–1811.

Hojilla-Evangelista, M. P., & Johnson, L. A. (2003). Sequential extraction processing of high-oil corn. *Cereal Chemistry*, *80*(6), 679–683.

Hojilla-Evangelista, M. P., Johnson, L. A., & Mayers, D. J. (1992). Sequntial extraction processing of flaked whole corn: Alternative corn fractionation technology for ethanol production. *Cereal Chemistry*, 69, 643–647.

Hurtado-López, P., & Murdan, S. (2006). Zein microspheres as drug/antigen carriers: A study of their degradation and erosion, in the presence and absence of enzymes. *Journal of Microencapsulation*, *23*(3), 303–314.

Ilter, S., Dogan, I. S., & Meral, R. (2008). Application of food-grade coatings to turkey buttocks. *Italian Journal of Food Science, 20*, 203–212.

Kale, A., & Cheryan, M. (2007). Rapid analysis of xanthophylls in ethanol extracts of corn by HPLC. *Journal of Liquid Chromatography and Related Technologies, 30*(8), 1093–1104.

Kale, A., Zhu, F., & Cheryan, M. (2007). Separation of high-value products from ethanol extracts of corn by chromatography. *Industrial Crops and Products, 26*(1), 44–53.

Kharkwal, M. C., Pandey, R. N., & Pawar, S. E. (2004). Mutation breeding for crop improvement. In H.K. Jain and M.C. Kharkwal (Eds.), *Plant breeding* (pp. 601–645). Dordrecht: Springer.

Kumaraswamy, S., & Shetty, P. K. (2016). Critical abiotic factors affecting implementation of technological innovations in rice and wheat production: A review. *Agricultural Reviews, 37*(4), 268–278.

Lai, H. M., Padua, G. W., & Wei, L. S. (1997). Properties and microstructure of zein sheets plasticized with palmitic and stearic acids. *Cereal Chemistry, 74*(1), 83–90.

Lawton, J. W. (2002). Zein: A history of processing and use. *Cereal Chemistry, 79*(1), 1–8. https://doi.org/10.1094/cchem.2002.79.1.1

Lin, J., Li, C., Zhao, Y., Hu, J., &Zhang, L. M. (2012). Co-electrospun nanofibrous membranes of collagen and zein for wound healing. *ACS Applied Materials and Interfaces, 4*(2), 1050–1057.

Liu, H., & Setiono, R. (1996). A probabilistic approach to feature selection-a filter solution. *ICML, 96*, 319–327.

Luo, Y., Teng, Z., & Wang, Q. (2012). Development of zein nanoparticles coated with carboxymethyl chitosan for encapsulation and controlled release of vitamin D3. *Journal of Agricultural and Food Chemistry, 60*(3), 836–843.

Luo, Y., Wang, Q., Zhang, B. C., Whent, M., & Yu, L. L. (2011). Preparation and characterization of zein/chitosan complex for encapsulation of α-tocopherol, and its in vitro controlled release study. *Colloid and Surfaces, Part B: Biointerfaces, 85*(2), 145–152.

Maliwal, P. L., Maliwal, P. L., & Mundra, S. L. (2007). *Agronomy at a glance* (Vol. 2). Agrotech Publishing Academy.

Manjaya, J. G. (2009). Genetic improvement of soybean variety VLS-2 through induced mutations. *Small, 38*, 106–109.

Masood, S., Naz, T., Javed, M. T., Ahmed, I., Ullah, H., & Iqbal, M. (2014). Effect of short-term supply of farmyard manure on maize growth and soil parameters in pot culture. *Archives of Agronomy and Soil Science, 60*(3), 337–347.

Mehlo, L., Mbambo, Z., Bado, S., Lin, J., Moagi, S. M., Buthelezi, S., ... Chikwamba, R. (2013). Induced protein polymorphisms and nutritional quality of gamma irradiation mutants of sorghum. *Mutation Research/Fundamental and Molecular Mechanisms of Mutagenesis, 749*(1–2), 66–72.

Mehta, S. K., Kaur, G., & Verma, A. (2011). Fabrication of plant protein micro-spheres for encapsulation, stabilization and in vitro release of multiple anti-tuberculosis drugs. *Colloids and Surfaces A: Physicochemical and Engineering Aspects, 375*(1–3), 219–230.

Miyoshi, S., Kaneko, T., Yoshizawa, Y., Fukui, F., Tanaka, H., & Maruyama, S. (1991). Hypotensive activity of enzymatic alpha-zein hydrolysate. *Agricultural and Biological Chemistry, 55*, 1407–1408.

Momany, F. A., Sessa, D. J., Lawton, J. W., Selling, G. W., Hamaker, S. A., & Willett, J. L. (2006). Structural characterization of α-zein. *Journal of Agricultural and Food Chemistry, 54*(2), 543–547.

Nagatomi, S., & Degi, K. (2009). *Mutation breeding of chrysanthemum by gamma field irradiation and in vitro culture* (pp. 258–261). Rome: Food and Agriculture Organization of United Nations.

Ngoune Tandzi, L., Mutengwa, C. S., Ngonkeu, E. L. M., & Gracen, V. (2018). Breeding maize for tolerance to acidic soils: A review. *Agronomy, 8*(6), 84.

Nonthanum, P., Lee, Y., & Padua, G. W. (2012). Effect of γ-zein on the rheo-logical behavior of concentrated zein solutions. *Journal of Agricultural and Food Chemistry, 60*(7), 1742–1747.

Noya, I., González-García, S., Bacenetti, J., Fiala, M., & Moreira, M. T. (2018). Environmental impacts of the cultivation-phase associ-ated with agricultural crops for feed production. *Journal of Cleaner Production, 172*, 3721–3733.

Oldfield, E. E., Bradford, M. A., & Wood, S. A. (2019). Global meta-analysis of the relationship between soil organic matter and crop yields. *Soil, 5*(1), 15–32.

Olsen, M. S., & Phillips, R. L. (2001). Molecular genetics improvement of protein quality in maize. *Impact of Agriculture on Human and Nutrition, II*, 1–9.

Padgett, T., Han, I. Y., & Dawson, P. L. (1998). Incorporation of food-grade anti-microbial compounds into biodegradable packing films. *Journal of Food Protection, 61*(10), 1330–1335.

Parris, N., & Coffin, D. R. (1997). Composition factors affecting the water vapor permeability and tensile properties of hydrophilic zein films. *Journal of Agricultural and Food Chemistry, 45*(5), 1596–1599.

Parris, N., & Dickey, L. C. (2001). Extraction and solubility characteristics of zein proteins from dry-milled corn. *Journal of Agriculture and Food Chemistry, 49*(8), 3757–3760.

Quilty, J. R., & Cattle, S. R. (2011). Use and understanding of organic amend-ments in Australian agriculture: A review. *Soil Research, 49*(1), 1–26.

Rakotonirainy, A. M., Wang, Q., & Padua, G. W. (2001). Evaluation of zein films as modified atmosphere packaging for fresh broccoli. *Journal of Food Science, 66*(8), 1108–1111.

Selling, G. W., & Woods, K. K. (2008). Improved isolation of zein from corn gluten meal using acetic acid and isolate characterization as solvent. *Cereal Chemistry, 85*(2), 202–206.

Sentayehu, A. (2008). Protein, tryptophan and lysine contents in quality protein maize. *North Indian Ethiopian Journal of Health Sciences*, *18*(2), 9–15.

Sessa, D. J. (2011). U.S. Patent No. 7,939,633. Washington, DC: U.S. Patent and Trademark Office.

Sessa, D. J., Eller, F. J., Palmquist, D. E., & Lawton, J. W. (2003). Improved methods for decolorizing corn zein. *Industrial Crops and Products*, *18*(1), 55–65.

Sessa, D. J., & Palmquist, D. E. (2009). Decolorization/deodorization of zein via activated carbons and molecular sieves. *Industrial Crops and Products*, *30*(1), 162–164.

Sharma, S. M. (1993). Utilization of national collection of sesame in India. ICAR-IBPGR Reg. In Workshop on sesame evaluation and improve, 28–30th, September, Nagpur, India.

Shi, W., & Dumont, M. J. (2014). Review: Bio-based films from zein, keratin, pea, and rapeseed protein feedstocks. *Journal of Materials Science*, *49*(5), 1915–1930.

Shukla, R., & Cheryan, M. (2001). Zein: The industrial protein from corn. *Industrial Crops and Products*, *13*(3), 171–192.

Shukla, T. P. (1992). Trends in zein research and utilization. *Cereal Foods World (USA)*, *37*, 225.

Sousa, F. F. O., Luzardo-Alvarez, A., Perez-Estevez, A., Seoane-Prado, R., & Blanco-Mendez, J. (2010). Development of a novel AMX-loaded PLGA/zein microsphere for root canal disinfection. *Biomedical Materials*, 5–9.

Spence, K. E., Jane, J. L., & Pometto, A. L. (1995). Dialdehyde starch and zein plastic: Mechanical properties and biodegradability. *Journal of Environmental Polymer Degradation*, *3*(2), 69–74.

Stark, L. E., & Gross, A. T. (1990). Hydrophobic protein microparticles and preparation thereof. *Patent WO, 90*, 03123.

Subramanian, S., & Sampath, S. (2007). Adsorption of zein on surfaces with controlled wettability and thermal stability of adsorbed zein films. *Biomacromolecules*, *8*(7), 2120–2128.

Tanakahashi, H., & Yanai, N. (1994). Process for refining zein. U.S. Patent 5342923.

Teklewold, A., Gissa, D. W., Tadesse, A., Tadesse, B., Bantte, K., Friesen, D., & Prasanna, B. M. (2015). *Quality Protein Maize (QPM): A guide to the technology and its promotion in Ethiopia*. Ethiopia: Addis Ababa.

Tharanathan, R. N. (2003). Biodegradable films and composite coatings: Past, present and future. *Trends in Food Science and Technology*, *14*(3), 71–78.

Thilagavathi, C., & Mullainathan, L. (2009). Isolation of macro mutants and mutagenic effectiveness, efficiency in black gram [Vigna mungo (L) Hepper]. *Global Journal of Molecular Sciences*, *4*(2), 76–79.

Tomlekova, N. B. (2010). Induced mutagenesis for crop improvement in Bulgaria. *Plant Mutation Reports*, *2*(2), 4–27.

Trezza, T. A., & Vergano, P. J. (1994). Grease resistance of corn zein coated paper. *Journal of Food Science, 59*(4), 912–915.

Tu, J., Wang, H., Li, H., Kerong, D., Zhang, X., & Wang, J. (2009). The in vivo bone formation by mesenchymal stem cells in zein scaffolds. *Biomaterials, 26*, 4369–4376.

Uan-On, T., Baibang, C., & Shuwisitkul, D. (2018). Extraction and characterization of zein protein from corn for controlled drug release. *Current Applied Science and Technology, 18*(3), 167–178.

Villegas, E., Vasal, S. K., & Bjarnason, M. (1992). Quality protein maize-What it is and how was it developed. In E. T. Mertz (Ed.), *Quality protein maize* (pp. 27–48). St. Paul, MN: American Association Cereal Chemists.

Vivek, B. S., Krivanek, A. F., Palacios-Rojas, N., Twumasi-Afriyie, S., & Diallo, A. O. (2008). *Breeding quality protein maize (QPM): Protocols for developing QPM cultivars*. Mexico: CIMMYT.

Wang, X., Woo, Y. M., Kim, C. S., & Larkins, B. A. (2001). Quantitative trait locus mapping of loci influencing elongation factor 1α content in maize endosperm. *Plant Physiology, 125*(3), 1271–1282.

Wang, X., Yan, J., Zhang, X., Zhang, S., & Chen, Y. (2020). Organic manure input improves soil water and nutrients use for sustainable maize (Zea mays. L) productivity on the Loess Plateau. *PLoS ONE, 15*(8), e0238042.

Wang, Y., & Padua, G. W. (2010). Formation of zein microphases in ethanol-water. *Langmuir, 26*(15), 12897–12901.

Wang, Y., Smith, D. E., Fant, A. B., & Muehlbauer, J. L. (1999). U.S. Patent No. 5,858,634. Washington, DC: U.S. Patent and Trademark Office.

Wang, Z., Chen, X., Wang, J., Liu, T., Liu, Y., Zhao, L., & Wang, G. (2007). Increasing maize seed weight by enhancing the cytoplasmic ADP-glucose pyrophosphorylase activity in transgenic maize plants. *Plant Cell, Tissue and Organ Culture, 88*(1), 83–92.

Wani, M. R., Kozgar, M. I., Tomlekova, N., Khan, S., Kazi, A. G., Sheikh, S. A., Ahmad, P. (2014). Mutation breeding: A novel technique for genetic improvement of pulse crops particularly Chickpea (*Cicer arietinum* L.). In A. Parvaiz, M. R. Wani, M. M. Azooz, & P.T. Lam-Son (Eds.), *Improvement of crops in the era of climatic changes* (pp. 217–248). New York: Springer.

Wilson, C. M. (1991). Multiple zeins from maize endosperms characterized by reversed-phase high performance liquid chromatography. *Plant Physiology, 95*(3), 777–786.

Wu, S., Myers, D. J., & Johnson, L. A. (1997a). Effect of maize hybrid and meal drying conditions on yield and quality of extracted zein. *Cereal Chemistry, 74*, 268–273.

Wu, S., Myers, D. J., & Johnson, L. A. (1997b). Factors affects yield and composition of zein extracted from commercial corn gluten meal. *Cereal Chemistry, 74*(3), 258–263.

Xu, W., Reddy, N., & Yang, Y. (2007). An acidic method of zein extraction from DDGS. *Journal of Agriculture and Food Chemistry, 55*(15), 6279–6284.

Yang, L., Wang, W., Yang, W., & Wang, M. (2013). Marker-assisted selection for pyramiding the waxy and opaque-16 genes in maize using cross and backcross schemes. *Molecular Breeding, 31*(4), 767–775.

Yao, C., Li, X., Song, T., Li, Y., & Pu, Y. (2009). Biodegradable nanofibrous membrane of zein/silk fibroin by electrospinning. *Polymer International, 58*(4), 396–402.

Yasmin, K., Arulbalachandran, D., Dilipan, E., & Vanmathi, S. (2020). Characterization of ^{60}CO γ-ray induced pod trait of blackgram-A promising yield mutants. *International Journal of Radiation Biology, 96*(7), 929–936.

Chapter 7

Bioactive compounds, antioxidant properties, and health benefits of whole maize and its components

Shweta Suri and Anuradha Dutta

DOI: 10.1201/9781003245230-7

7.1 Introduction

Maize, botanically known as *Zea mays* L., is an Indo-American term used for corn, and means that it sustains life. Maize has its origins in Mexico and is considered one of the most widely cultivated cereal grains due to its diverse adaptability to the soil and climatic conditions. Several types of maize, for example normal yellow/white grain, popcorn, baby corn, sweet corn, waxy corn, high amylase corn, high oil variety of corn, quality protein maize, etc., are cultivated in various regions around the world. As per the Food and Agriculture Organization, the annual production of maize has amply grown worldwide in past years, increasing from around 820.80 million tonnes in 2009 to 1148.4 million tonnes in 2019. The United States is the leading producer of maize, with 347.04 million tonnes of production per year, contributing to 35% of total world production. The other major countries producing maize are China (mainland), Brazil, Argentina, Ukraine, Indonesia, India, Mexico, and Romania (FAO STAT, 2019). Because of the high yield among other cereal crops, maize is referred to globally as the queen of cereal grains. India stands in seventh position with an annual production of 27.71 million tonnes. In the context of India, the main maize-producing states are Uttar Pradesh, Punjab, Bihar, Himachal Pradesh, Haryana, Madhya Pradesh, Maharashtra, Rajasthan, West Bengal, Karnataka, Andhra Pradesh, and Jammu and Kashmir, together representing more than 95% of the overall maize production in India (Milind & Isha, 2013).

Maize is regarded as an important cereal crop, having a bounty of nutrients required to meet the needs of humans. Maize is utilized as feed, food, and natural resource for a number of purposes. Maize is a staple food of over 200 million people in Asia, Africa, and Latin America. It is also utilized in the making of traditional cuisines, for example, tortillas, porridge, and couscous (Rooney & Serna-Saldivar, 2003). Maize kernels are used in the form of cob, roasted, dried, fried, and fermented forms. Maize is also used in the preparation of cakes, porridge, bread, gruel, and alcoholic drinks. Moreover, the utilization of maize for food applications has risen in industrialized nations in the form of breakfast cereals, snack items, dietetic items, and specifically, for gluten-free food sources (Blandino et al., 2017). In addition, maize grains are utilized as a source of starch, protein, oil, sweetener, and sometimes for fuel (Kaur et al., 2021). Industries utilize 12–15% of India's maize production. Corn oil, cornmeal, popcorn, corn syrup, rice corn, corn flakes, and corn soap are a few common corn-based items. Corn oil is a light yellow oil extracted from corn kernels. Corn seed oil is known for its high essential oil content, required for brain development. Unsaturated fatty acids, viz. PUFA and MUFA, are present in high concentrations in corn oil (Ambika Rajendran et al., 2012).

Different maize varieties (flour corn, flint corn, waxy corn, sweet corn, dent corn, popcorn, pod corn, striped corn, and amylo corn) as well as colours (white, yellow, red, and purple) of maize exhibit diverse antioxidant properties. Different bioactive constituents, like carotenoids, anthocyanins, phytosterol, and polyphenolic compounds are abundantly found in maize (Luo & Wang, 2012). These secondary metabolites are usually present in the whole maize kernel either in the form of soluble, free, or conjugated/insoluble bound forms (Prior et al., 2006).

Maize also contains high resistant starch, primarily amylose, that helps in the management of chronic illnesses, for instance atherosclerosis, cancer, as well as obesity-related complications. Several varieties of maize contain varying amounts of bioactive components like phenolics, flavonoids, and starch content. Hence, the combined utilization of different varieties of maize provides a blend of all phytochemicals with optimum nutritional properties (Siyuan et al., 2018). Maize is well known for its medicinal and therapeutic potential due to its bioactive components. The major bioactive complexes usually observed in maize are polyphenols, carotenoids, xanthophylls, vitamins, as well as dietary fibre. These bioactive compounds, especially polyphenols, contribute towards maize's high total antioxidant activity. Evidence suggests that consuming foods that have high antioxidant potential leads to the counteraction of different oxidative stress-related illnesses, like cancer and cardiovascular disease (Adom & Liu, 2002). In contrast to other staple cereal crops like wheat, rice, and oats, maize holds high total antioxidant activity. Also, maize-based foods comprise of a high vitamin concentration and dietary fibre. Dietary fibre, including cellulose, hemicellulose, inulin, resistant starch, etc., is present in abundance in maize grain and helps in maintaining gut health, controlling body weight, and cholesterol levels (Blandino et al., 2017).

This chapter covers recent work done on various aspects related to the bioactive nature of maize grain along with its health benefits. In addition, the effect of different processing techniques on the bioactive composition of maize is also discussed.

7.2 Structure of maize grain

The main parts of the maize grain are the endosperm (82%), embryo/germ (12%), pericarp or seed coat (5%), and tip cap (1%). Bioactive compounds found in maize are usually present in the bran, outer layer, or germ part of the grain (Singh et al., 2019). The longitudinal section of maize grain comprises the aleurone layer, pericarp, endosperm, scutellum, epithelium, coleoptile, and radicle as shown in Figure 7.1. Starch is the major macronutrient found in the maize kernel. The maize kernel is comprised primarily of starch, viz. amylopectin and amylose (72–73% of the total weight), followed by proteins (approx. 8–12% of the whole maize kernel) (Diaz-Gomez et al., 2017). In context to the amino acids present in maize, glutamic acid is one found in greater quantities (Tang et al., 2013). The four storage proteins (prolamins, albumins, globulins, and glutelin) are abundantly observed in maize kernels. Maize protein, particularly prolamin, usually recognized as zein, is located in the protein bodies of the rough endoplasmic reticulum. Zein constitutes about 44–79% of the protein found in maize endosperm (Giuberti et al., 2012). Maize protein lacks lysine and tryptophan. Basically, four different fractions of zein protein (α, γ, β, and δ) are present in maize, with α-zein being the most valued, as it stores a lot of nitrogen (Momany et al., 2006). Micronutrients in maize grain are traced predominantly to the germ portion, which comprises 78% of the whole kernel (Suri & Tanumihardjo, 2016).

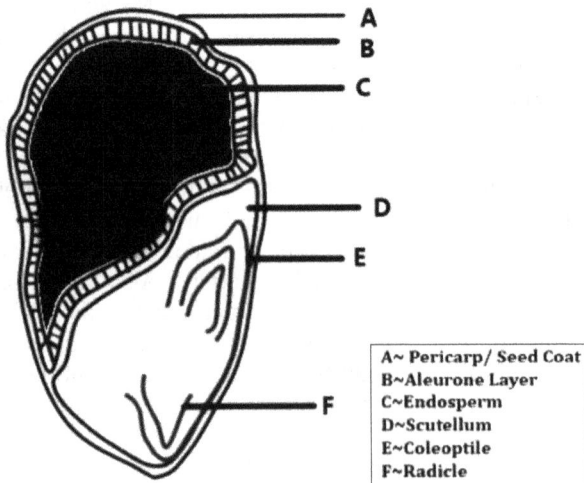

A~ Pericarp/ Seed Coat
B~Aleurone Layer
C~Endosperm
D~Scutellum
E~Coleoptile
F~Radicle

Figure 7.1 Longitudinal section of maize grain

7.3 Nutritional characterization of maize

The nutritional composition of maize varies based on the variety, colour, and type. Geographical area and climatic conditions also affect the morphology of maize. Maize is known as a good source of protein, dietary fibre, vitamins, as well as minerals. Whole maize and sweet variety i.e. sweet corn, contain a varied range of vitamins (vitamin B-complex, carotenoids, and ascorbic acid), minerals (calcium, potassium, magnesium, sodium, zinc, and phosphorus), along with resistant starch. Also, the fatty acid profile of maize suggests the presence of monounsaturated fats, polyunsaturated fats, and smaller quantities of saturated fats (Siyuan et al., 2018). Besides this, maize contains vitamins K and E, folic acid, selenium, *N-ferrulyltryptamine* and *N-p-coumaryltryptamine* (Rouf Shah et al., 2016). Minerals, such as potassium, are present in good amounts in maize to meet human dietary requirements (Kumar & Jhariya, 2013). Among the staple cereals like rice and wheat, maize is considered to contain great amounts of vitamins, minerals, and essential fats. The nutritional composition of some of the staple cereal grains is presented in Table 7.1 (Longvah et al., 2017).

7.4 Bioactive components present in maize

Several bioactive compounds like carotenoids, anthocyanins, and polyphenolic compounds have been found in maize. Bioactive compounds present in maize also differ depending on the type and colour of the maize grain. The anthocyanin content of blue maize grains are derived from cyanidin and malvidin, while red maize contains the phytochemicals, viz. pelargonidin, cyanidin, and malvidin, that contribute to its anthocyanin content. In addition, maize exhibits high antioxidant activity (181.4 μmol vitamin C equivalent/g) in contrast to staple cereals such as wheat, rice, and oat. Bioactive components contribute to the high antioxidant content of maize. In the context of the polyphenols found in maize, the ferulic acid and flavonoids add to the total phenol content (Adom & Liu, 2002). Major bioactive components existing in the maize grain are illustrated in Figure 7.2.

7.4.1 Polyphenolic compounds

Polyphenols are found in abundance in maize, particularly in the bran or outer covering of the grain (Zhao et al., 2005). The key phenolic components present in maize are ferulic acid (4-hydroxy-3-methoxycinnamic acid) plus anthocyanin. The refined bran of maize comprises higher amounts of ferulic acid (Zhao & Moghadasian, 2008). The total phenols, total ferulic

Table 7.1 Nutritional profile of maize (*Zea mays* L.), wheat (*Triticumaestivum*), and rice (*Oryza sativa*) per 100 g

Nutrients	Maize	Wheat	Rice
Macronutrients			
Protein (g)	8.80	10.59	7.94
Ash (g)	1.17	1.42	0.56
Total fat (g)	3.77	1.47	0.52
Dietary fibre (g)	12.24	11.23	2.81
Carbohydrate (g)	64.77	64.72	78.24
Energy (K Joule)	1398	1347	1491
Vitamins			
Vitamin B1 (mg)	0.33	0.46	0.05
Vitamin B2 (mg)	0.09	0.15	0.05
Vitamin B3 (mg)	2.69	2.68	1.69
Vitamin B6 (mg)	0.34	0.26	0.12
Folate (μg)	25.81	30.09	9.32
Carotenoids			
Lutein (μg)	186	52.56	1.49
Zeaxanthin (μg)	42.4	1.47	-
β-carotene (μg)	186	3.03	-
Total carotenoids (μg)	893	287	16.87
Minerals			
Magnesium (mg)	145	125	19.30
Phosphorus (mg)	279	315	96.00
Potassium (mg)	291	366	108
Sodium (mg)	4.44	2.50	2.34
Iron (mg)	2.49	3.97	0.65
Zinc (mg)	2.27	2.85	1.21
Fatty acids			
Palmitic acid (mg)	363	176	143
Stearic acid (mg)	42.45	14.83	14.50
Arachidic acid (mg)	7.14	-	1.46
Oleic acid (mg)	700	141	109
Linoleic acid (mg)	1565	616	234
α-linolenic acid (mg)	40.76	38.51	9.51

(Continued)

Table 7.1 (Continued)   Nutritional profile of maize (*Zea mays* L.), wheat (*Triticumaestivum*), and rice (*Oryza sativa*) per 100 g

Nutrients	Maize	Wheat	Rice
Amino acids			
Histidine (g)	2.70	2.65	2.45
Methionine (g)	2.10	1.75	2.60
Isoleucine (g)	3.67	3.83	4.29
Leucine (g)	12.24	6.81	8.09
Phenylalanine (g)	5.14	4.75	5.36
Tryptophan (g)	0.57	1.40	1.27
Arginine (g)	4.20	5.13	7.72
Proline (g)	7.88	10.25	4.31
Tyrosine (g)	3.71	3.12	4.36
Alanine (g)	7.73	3.64	5.51

Source: Indian food composition tables (Longvah et al., 2017).

acid, and total anthocyanin content found in different varieties of maize are shown in Figure 7.3.

Maize polyphenols display antioxidant and other beneficial properties like cell differentiation, deoxyribonucleic acid (DNA) repair, procarcinogen deactivation, prevention of formation of N-nitrosamine, and modification of oestrogen metabolism (Shahidi, 2004). Several polyphenolic components, viz. vanillic, protocatechuic, ferulic, isoferulic acid, syringic acid, sinapic acid, *p-hydroxybenzoic*, cyanidin-3-O-glucoside, caffeic, *p-coumaric*, quercetin, and kaempferol, are present in different varieties of maize (whole corn, baby corn, waxy corn, popcorn, as well as sweet corn) (Kandil et al., 2012; Das & Singh, 2016). These polyphenolic compounds are predominantly observed in the pericarp and embryo/germ of maize grains, where the pericarp region of maize grain comprises the highest amount of polyphenolics. The phenolic compounds also contribute to the grain colour. Quercetin and catechin are the major phenols that add colour to the grain. Corn leaf and silk are enriched with phenolic compounds, for example *p* -hydroxybezoic acid, *p*-coumaric, protocatechuic, vanillic, syringic, sinapic, and chlorogenic acids (And & Shahidi, 2005).

7.4.1.1 Ferulic acid

Among phenolic components, ferulic acid is the main bioactive phenolic component found in whole maize. Ferulic acid/4-hydroxyl-3-methoxycinnamic

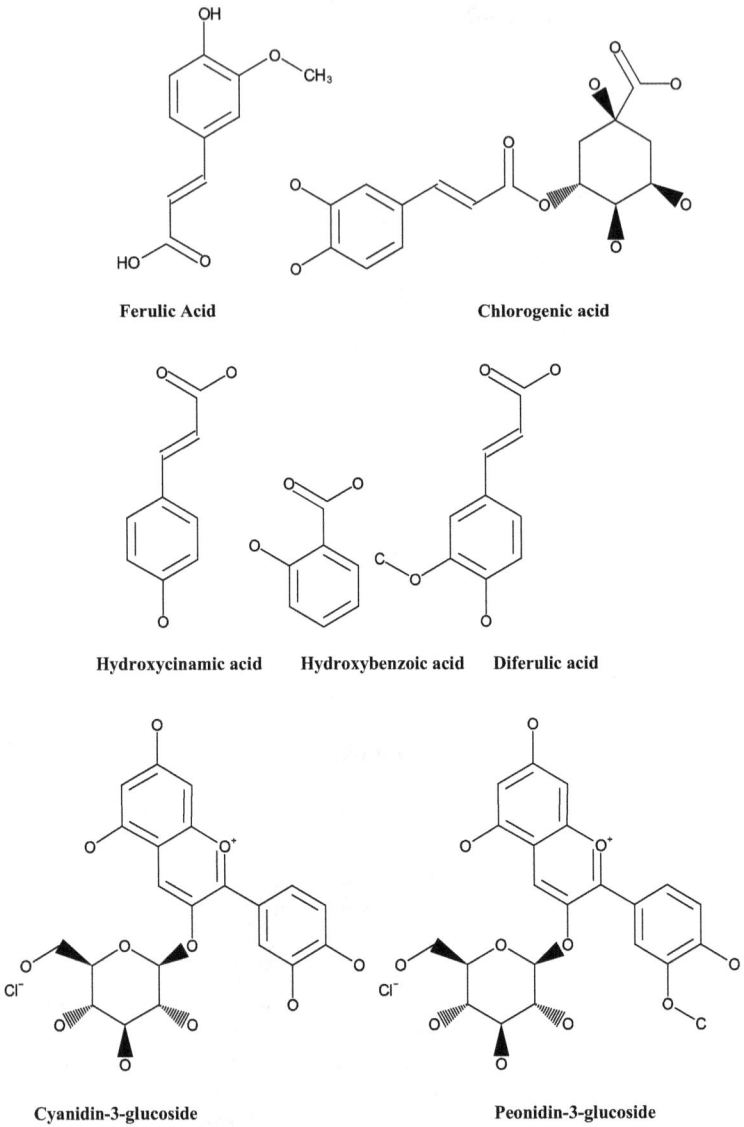

Figure 7.2 Illustration showing major polyphenolic compounds found in the maize grain. Source: Pubchem https://pubchem.ncbi.nlm.nih.gov/

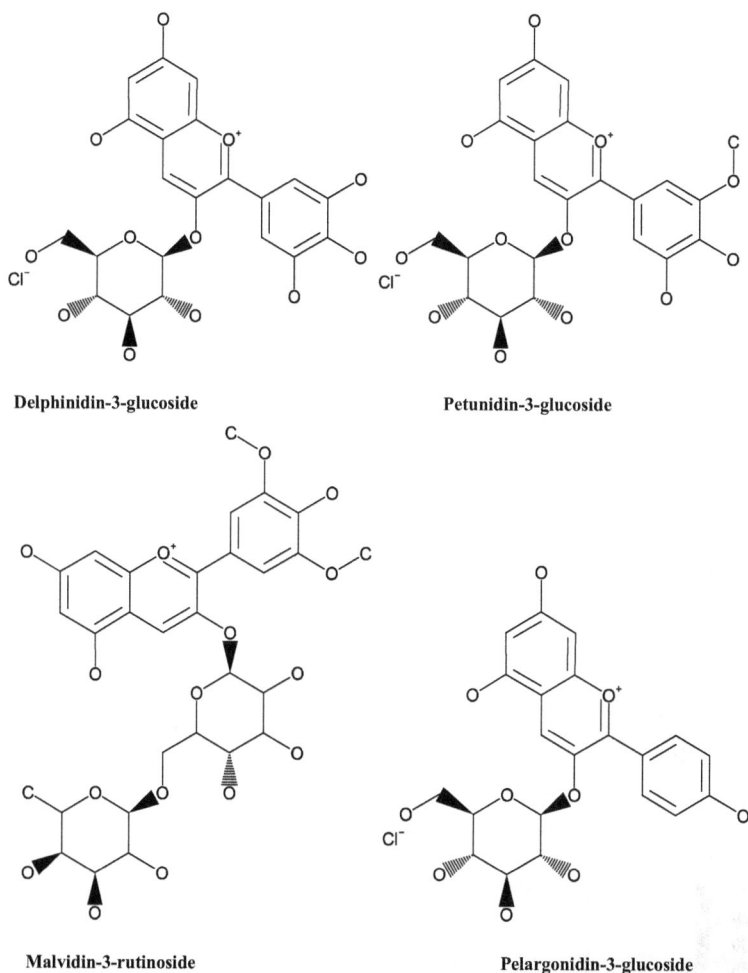

Delphinidin-3-glucoside

Petunidin-3-glucoside

Malvidin-3-rutinoside

Pelargonidin-3-glucoside

Figure 7.2 (Continued)

acid is an organic compound that represents approximately 3.1% to 4.0% of maize grain on a dry weight basis (Saulnier et al., 1995; Santiago & Malvar, 2010). Ferulic acid content ranging from 216 to 3400 mg/100 g is found in maize (Zhao et al., 2005; Vitaglione et al., 2008). Ferulic acid present in maize grains is no less than ten times greater than the other staple cereal grains (Vitaglione et al., 2008). Ferulic acid is mainly observed in free as well as conjugated forms that are covalently allied with lignin or other biopolymers. The summation of free and conjugated forms relates to the total ferulic acid

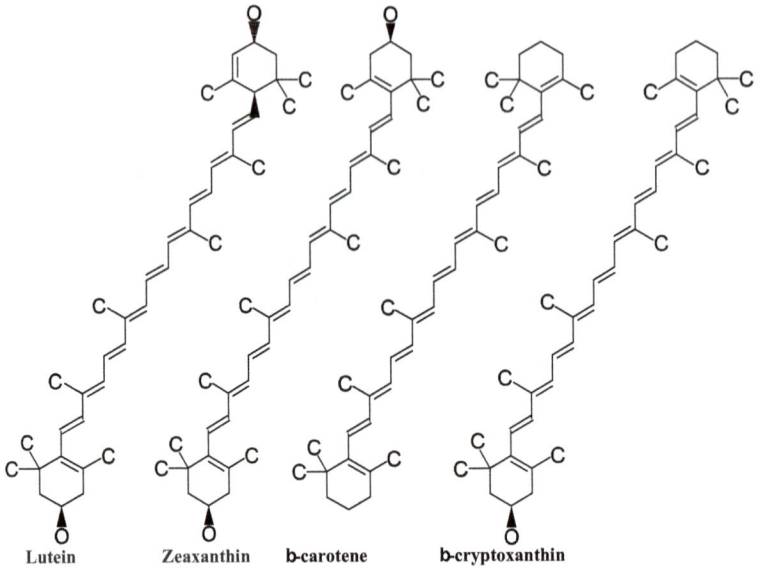

Lutein   Zeaxanthin   b-carotene   b-cryptoxanthin

Figure 7.2 (Continued)

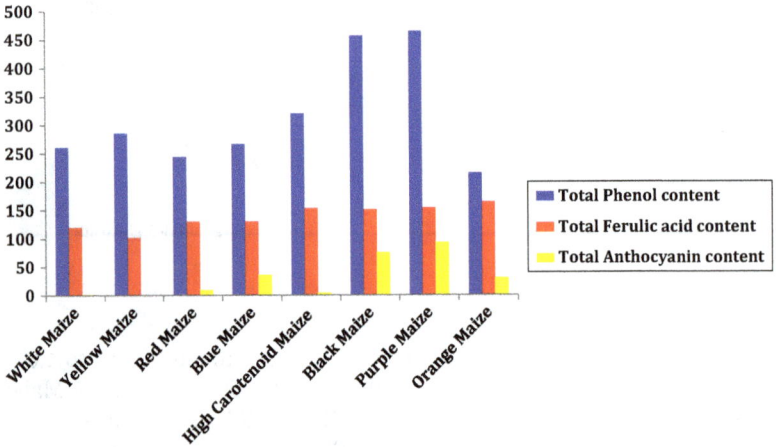

Figure 7.3 Polyphenol content (mg/100 g of dry weight) of different varieties of maize #Total phenols measured as mg Gallic Acid Equivalent/100 g and total anthocyanin content as mg cyanidin-3-glucoside/100 g. Source: Lopez-Martinez et al. (2009); de la Parra et al. (2007)

composition in maize (Zhao et al., 2005). However, the content of free ferulic acid in maize grain is relatively low (1–6% of the total ferulic acid) (Lopez-Martinez et al., 2009). The bioavailability of ferulic acid recovered from maize is reliant upon several factors, for example, the form of ferulic acid plus the daily dietary intake; these two factors would additionally influence the profile and concentration of ferulic acid in the human gastrointestinal tract (Van der Logt et al., 2003). In addition, the bioavailability of ferulic acid obtained from the maize is lower than the free ferulic acid. This is as a consequence of the conjugation with arabinose or arabinoxylan from maize bran (Rondini et al., 2004). In cereals, ferulic acid is mainly seen in the cell wall region (Zhao et al., 2005). Recently, the bound form of ferulic acid has been utilized in the enzymatic synthesis of vanillin. Also, vanillin yield obtained from the maize bran's bound ferulic acid is perceived to be higher in contrast to that obtained from wheat bran and flax shrives (Buranov & Mazza, 2009). Ferulic acid shows various beneficial properties, viz. anti-inflammatory, anti-carcinogenic, anti-diabetic, neuro-protective, and cardiovascular properties (Bento-Silva et al., 2018), though the therapeutic effects of ferulic acid are influenced by the concentration of ferulic acid and its metabolite in the body. Hence it is very important to conduct a bioavailability study of the phenolic components both before and after processing.

7.4.1.2 Anthocyanins

Anthocyanins – which come from the Greek word *anthos* meaning flower and *kyaneos* meaning blue – are polyphenolic phytochemicals that belong to the flavonoid group. These anthocyanins are hydrophilic glycosides of polyhydroxy/polymethoxy derivatives of 2-phenylbenzopyrylium or flavylium salts (Luo & Wang, 2012). Anthocyanins impart red-orange to blue-violet colours to fruits, vegetables, and grains. Anthocyanins are well known for their therapeutic effects leading to a healthy body (Wallace & Giusti, 2015).

Anthocyanidins are the principal pigment compound present in coloured grains, amongst which maize has been demonstrated to contain the second-highest amount, as well as varying forms of anthocyanins (Abdel-Aal et al., 2006). Commonly occurring anthocyanins in the maize are peonidin-3-glucoside, pelargonidin-3-glucoside, pelargonidin-3-(6″-malonylglucoside), cyanidin-3-(6″-malonylglucoside), cyanidin-3-glucoside, and cyanidin-3-(3″, 6″-dimalonylglucoside). Among all the types of anthocyanin found in maize, cyanidin derivatives are the main anthocyanin, contributing to above 70% in the kernel (both pericarp as well as aleurone layer), seed, as well as cob of purple maize (Aoki et al., 2002). The amount and type of anthocyanin present in maize depend on the variety of maize. The deeper the colour of maize, the more is its anthocyanin content. Particularly purple maize comprises the highest anthocyanin content, being many times more prominent than the red variety of maize, which contains the second high anthocyanin amount. Because of the high anthocyanins present in purple and red maize, these two varieties are studied most (Luo & Wang, 2012). The purple, black, and red varieties of maize contain greater quantities of anthocyanins in contrast to the yellow, orange, and white phenotypes. Blue, red, and purple varieties of maize contain good amounts of anthocyanidin (approximately 325 mg/100 g DW) constituting 75–90% cyanidin, 15–20% peonidin, and 5–10% pelargonidin (Siyuan et al., 2018).

In addition to imparting colour, anthocyanin has additionally been notable for its disease-preventing/health-promoting and nutraceutical properties, for example it is anti-diabetic, anti-carcinogenic, anti-inflammatory, anti-microbial, anti-neoplastic, anti-atherogenic, lipid-lowering, and has the ability to prevent platelet aggregation and to stimulate the immune system (Ghosh & Konishi, 2007). Anthocyanin present in colourful varieties of maize prevents colorectal carcinogenesis (Hagiwara et al., 2001) and chances of obesity, ameliorates hyperglycaemia (Tsuda et al., 2003), and assists in gastro-protection (Matsumoto et al., 2006).

7.4.2 Carotenoids

Carotenoids are phytochemicals belonging to the family of yellow, orange, and red pigments. The carotenoid content of maize varies based on the genotype or the type of maize. Carotenoids found in different varieties of maize are shown in Figure 7.4. An enormous amount of carotenoids are found in

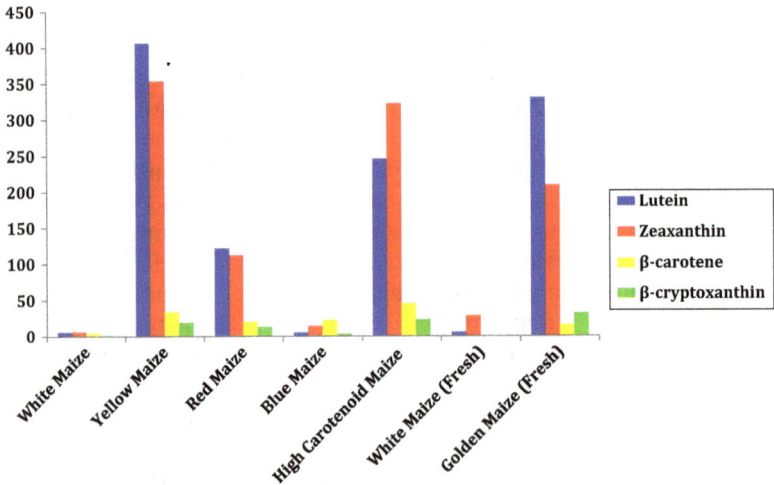

Figure 7.4 Carotenoids content (μg/100 g DW) of different varieties of maize. Source: Lopez-Martinez et al. (2009); de la Parra et al. (2007)

yellow varieties of maize, particularly in the horny as well as floury endo-sperm (Liu, 2007). Carotenoids, viz. lutein and zeaxanthin, are present in major quantities; however β-carotene, α-cryptoxanthin, β-cryptoxanthin, and α-carotene are observed in lesser amounts in maize. The yellow and high carotenoid varieties of maize contain greater quantities of lutein and zeaxanthin, whereas the blue and white varieties of maize contain lesser amounts of these pigments (de la Parra et al., 2007). The oxygenated carot-enoid derivatives are known as xanthophylls and contribute to the yellow colour of maize. The yellow maize contains high amounts of carotenoids (approximately 823 µg/100 g DW) together with lutein (50%), zeaxanthin (40%), β -carotene (4%), β -cryptoxanthin (3%), and α-carotene (2%) (Siyuan et al., 2018). Carotenoid pigments are unstable with heat, light, pH, and aerobic conditions. The lutein pigment exhibits a higher stability to heat as compared to other carotenoid pigments. The carotenoids present in maize exhibit health-promoting effects; for instance lutein showed antitumor-pro-moting effects and tumour suppression in mice. It has been observed that lutein and zeaxanthin assist in preventing age-related ocular degeneration which is a human disorder similar to cataracts and leads to blindness (Park et al., 1998).

7.4.2.1 Carotene (α-carotene and β-carotene)

The yellow variety of maize, silage, and stalklage contains a high provita-min A content. Carotene, like α-carotene and β-carotene, exhibits provita-min activity, which indicates that it can be effectively churned in the gut and tissues to the active form of vitamin A (Luo & Wang, 2012). β-carotene exhibits disease prevention activities such as anti-cancer properties. It helps in inducing apoptosis of cancer cells, leukaemia cells, melanoma cancer cells, and gastric cancer cells; hence the carotene shows effective chemo-preventive effect (Palozza et al., 2001).

7.4.2.2 Xanthophylls

Two classes of xanthophylls, viz. lutein and zeaxanthin, are found in maize grain. Mostly yellow varieties of maize comprise high levels of xanthophyll. Maize grains contain xanthophylls (2.07 mg/100 g), lutein (1.50 mg/100 g), and zeaxanthin (0.57 mg/100 g) (Moros et al., 2002). Recent research has emphasized the impact of lutein and zeaxanthin on human health and wellbeing. Anti-tumour properties of hydroxyl groups of lutein and zea-xanthin extracted from maize were examined. The variation in the anti-proliferative properties on the human mouth epithelial cancer line KB cell amongst the altered di-acetylation lutein/zeaxanthin and unaltered one was evaluated. The study reported the role of hydroxyl groups present in lutein and zeaxanthin in their effective anti-tumour activities (Sun & Yao, 2007). Furthermore, lutein also acts as a chemo-preventive agent, leading to its inhibitory activities during the advancement of hepato-carcinogenesis (Moreno et al., 2007).

7.4.3 Phytosterols

Phytosterols are plant-based sterols that are crucial parts of plant cell walls and membranes. Maize oil contains a promising amount of phytosterols. The key phytosterols found in maize grains are stigmasterol, sitosterol, and campesterol (Verleyen et al., 2002). Phytosterol exhibits hypocholesterolemic action (Piironen et al., 2000). It helps in lowering serum cholesterol levels especially total LDL-cholesterols (Jiang & Wang, 2005). Also, it inhibits dietary cholesterol absorption and biosynthesis (Ostlund et al., 2002). Phytosterols are distributed in varying sections of the maize kernel, for instance in endosperm, pericarp, as well as germ (Harrabi et al., 2008).

7.5 Extraction of bioactive components from maize

Different researchers analyzed the polyphenol content of the maize and its parts by using diverse extraction methods, solvents, and extraction parameters as presented in Table 7.2. Studies utilized both conventional and green extraction methods such as ultrasound extraction techniques, for obtaining bioactive compounds from maize. Fuentealba et al. (2017) utilized solvent extraction involving ethanol, aqueous acetone, and methanol for extraction as well as quantification of polyphenols from Chilean Cristalino corn (*Zea mays* L.), where high quantities of phenolic acid (ranging from 27.5 to 63.7 mg GAE/100 g) was observed in utilizing aqueous acetone as solvent. Additionally, ultrasound-assisted extraction and conventional extraction technology were employed for obtaining polyphenols from corn (*Zea mays* subsp. Mays L.) by-products. The study reported the presence of greater concentrations of chorogenic acid (42.4 µg/g DW), caffeic acid (13.7 µg/g DW), ferulic acid (48.1 µg/g DW), apigenin (7.9 µg/g DW), and pelargonidin (2.6 µg/g DW) in extracts obtained by ultrasound extraction at a frequency of 40 kHz, methanol solvent (70%), and time (15 minutes) (Aires & Carvalho, 2016). This shows that ultrasound technology is an efficient and sustainable extraction technique. Solvent extraction techniques for extraction of both soluble and bound phenols from Bolivian purple corn (*Zea mays* L.) variety resulted in recovery of soluble phenols (68.1 to 309.7 mg GAE/100 g DW) and bound phenol (242.9 to 529.6 mg GAE/100 g DW) (Cuevas Montilla et al., 2011). The above-mentioned studies clearly explain the potential recovery of bioactive compounds especially polyphenols through utilizing conventional as well as energy-saving methods.

Table 7.2 Extraction of polyphenols from maize grain

Maize cultivars	Extraction method	Solvent type	Extraction conditions				Total phenol content	References
			Solvent conc.	Temp.	Pressure	Time	Extraction yield		

Maize cultivars	Extraction method	Solvent type	Solvent conc.	Temp.	Pressure	Time	Extraction yield	Total phenol content	References
Purple corn cob (Zea mays L.)	Sequential extraction	Supercritical carbon dioxide	–	–	–	–	–	–	Monroy et al. (2016)
		Ethanol	–	68 °C	480bar	–	7.9±2.0%	282.61 mg GAE/g extract	
		Water	–	65 °C	450bar	–	13.1±2.2%	389.49 mg GAE/g extract	
Chilean Cristalino corn (Zea mays L.)	Solvent extraction	Aqueous acetone	67.9–92.1%	–	–	15.8–64.2 min	–	Free phenolic acid 27.5 to 63.7 mg GAE/100 g	Fuentealba et al. (2017)
		Ethanol						Free phenolic acid 42.8 to 53.3 mg GAE/100 g	
		Methanol						Free phenolic acid 39.3 to 53.2 mg GAE/100 g	
Corn (Zea mays subsp. Mays L.) by-product	Conventional extraction	Methanol	70%	700 °C	–	15 min	–	Chorogenic acid: 37.3±0.6 µg/g DW Caffeic acid: 12.0±0.5 µg/g DW Ferulic acid: 42.3±0.6 µg/g DW Apigenin: 6.9±0.6 µg/g DW Pelargonidin: 2.1±0.1 µg/g DW	Aires & Carvalho (2016)

(Continued)

225

Table 7.2 (Continued) Extraction of polyphenols from maize grain

Maize cultivars	Extraction method	Extraction conditions					Total phenol content	References
		Solvent type	Solvent conc.	Temp.	Pressure	Time	Extraction yield		
	Ultrasound extraction	Methanol	70%	–	–	15 min	–	Chorogenic acid: 42.4±0.7 µg/g DW Caffeic acid: 13.7±0.5 µg/g DW Ferulic acid: 48.1±0.6 µg/g DW Apigenin: 7.9±0.6 µg/g DW Pelargonidin: 2.6±0.1 µg/g DW	
Zea mays L filaments	Ultrasound-assisted extraction	Ethanol	61.08%	–	–	–	–	Polyphenol content: 7.1±0.015 mg/g	Lingzhu et al. (2015)
Bolivian purple corn (Zea mays L.)	Solvent extraction	Ethanol	80%	–	–	–	–	Soluble phenols 68.1 to 309.7 mg GAE/100 g DW Bound phenol 242.9 to 529.6 mg GAE/100 g DW.	Cuevas Montillaet al. (2011)

(Continued)

Table 7.2 (Continued) Extraction of polyphenols from maize grain

Maize cultivars	Extraction method	Extraction conditions					Extraction yield	Total phenol content	References
		Solvent type	Solvent conc.	Temp.	Pressure	Time			
Corn (Zea mays L.) grain	Solvent extraction	Methanol	80%	–	–	–	–	Soluble Phenol • Pericarp (Free phenol: 107.3 to 181.7 µg/g GAE/100 g Glycosylated phenol: 14.9 to 28.7 µg/g MS Esterified phenol: 86.6 to 137.1 µg/g MS) • Endosperm (Free phenol: 45.3 to 94.2 µg/g MS Glycosylated phenol: 12.0 to 20.7 µg/g MS Esterified phenol: 49.0 to 80.6 µg/g MS) • Germ (Free phenol: 375.0 to 480.7 µg/g MS Glycosylated phenol: 17.9 to 45.0 µg/g MS Esterified phenol: 105.1 to 202.2 µg/g MS) Insoluble Phenol • Pericarp: 872.2 to 1037.8 µg/g MS • Endosperm: 445.0 to 662.2 µg/g MS • Germ: 766.4 to 1092.7 µg/g MS	Cabrera-Soto et al. (2009)

7.6 Health benefits of whole maize grain

Maize contains plentiful amounts of nutrients and bioactive components comprising fibre, micronutrients, vitamins and minerals, and phytochemicals that aid in the management of diseases. Water-soluble vitamins, for example vitamin B-complex commonly found in maize, are advantageous for the heart, hair, skin, and brain. Also, these water-soluble vitamins in maize help in strengthening joint mobility. The occurrence of lipid-soluble vitamins such as A, K, and C along with β-carotene and selenium assists with the functioning of thyroid organ and immune functioning. Among minerals, the potassium found in maize exhibits diuretic effects. Moreover, maize grain contains essential fatty acid such as linoleic acid which helps in the maintenance of blood pressure and blood cholesterol (Rouf Shah et al., 2016). The more recent research proposes that the consistent ingestion of whole maize grain decreases the risk of chronic illnesses, viz. cardiovascular, non-insulin-dependent diabetes mellitus, obesity, and digestion-related issues (Siyuan et al., 2018).

The nutritional components of maize grain endow it with numerous health benefits. The high fibre content of maize grain prevents digestion-related problems such as constipation and colorectal cancer (Adom & Liu, 2002). Antioxidants found in whole maize averts lung cancer, and lutein pigment forestalls age-related vision loss. Besides this, investigations have revealed that the intake of whole grains is significantly linked with low BMI and decreased risk of obesity. Whole grains appear to have a high impact on the visceral fat tissues. At the point when visceral fat tissue mass is reduced, there is substantial improvement in cardiovascular risk factors, even with just weight reduction (Harris & Kris-Etherton, 2010).

Maize bran assists in reducing the gain in weight and obesity by imparting the feeling of satiety. In a study, the effect of incorporation of maize bran on satiety was checked, where high fibre muffins were fed to 20 healthy men and women. The study revealed that the muffins comprising of resistant starch and maize bran had a high influence on satiety as compared to the control muffins (Willis et al., 2009).

Furthermore, the consumption of phytosterol found in maize helps in lowering cholesterol absorption, both serum total as well as low-density lipo-protein cholesterol. The dietary phytosterol works through the mechanism of hindering the cholesterol absorption from the intestine and stimulating the synthesis of cholesterol leading to greater cholesterol elimination in the stool. Maize oil consumption for an extended duration of time leads to a reduction of cholesterol concentration and prevents atherosclerosis. Also, the cholesterol absorption was found to be 38% higher among subjects having phytosterol-removed commercial maize oil as compared to the subjects having the original maize oil for two weeks of administration (Ostlund Jr et al., 2002).

Resistant starch present in the maize grain is extremely nutritious, and its intake helps in the prevention of obesity and lowering cholesterol levels. It is also helpful in skin problems and cancers. Resistant starch is a starch type that can't be processed in the small intestine, and it directly passes to the colon where fermentation takes place by the microflora (Perera et al., 2010). The physiological impacts of resistant starch include an enhanced glycaemic response and improved colon health, lower intake of calories, modulated fat metabolism, and cardiovascular disease prevention (Liu, 2007).

7.7 Therapeutic properties of maize

Maize grain induces several bioactive effects such as antioxidant, anti-cancer, anti-tumour, anti-microbial, and hepato-protective effects. Maize polyphenols are helpful in preventing diseases and chronic illness. This is mostly because of their anti-oxidative properties, although the inversion of epigenetic changes can have great impacts too (Wu et al., 2016). In addition, extensive research on the therapeutic nature of polyphenols affirmed that they not only forestall different disease conditions but also effect the propagation of disease, its suppression, and also add to the healing process (Ramos, 2008). Nowadays, polyphenols are widely researched for forestalling and treating different malignancies like cancer.

Moreover, the anthocyanin present in maize also exhibits therapeutic properties. The cancer prevention action of anthocyanins relies basically upon the aglycone molecule, however, might be influenced by covalently bound sugars' varying solubility and permeability of the membrane. The chemical structure of anthocyanin allows it to donate hydrogen atoms/electrons to free radicals, snare them in their aromatic structure, and thereby exhibit potent antioxidant effects (Magaia-Cerino et al., 2020). Intervention studies state that *in-vivo* intake of anthocyanins increases the plasma and serum antioxidant activity in golden Syrian hamsters as well as in human participants, respectively (Mazza et al., 2002; Auger et al., 2004). A long-term intervention trial on rats revealed that the consumption of anthocyanin obtained from the pigmented variety of maize led to protection of rat hearts from ischemia-reperfusion injury (Toufektsian et al., 2008). Some of the common therapeutic properties of maize are discussed below.

7.7.1 Antioxidant activity

Antioxidants are natural or synthetic substances that avert or delay cellular damage. Maize contains phytochemicals such as phenolic acids and flavonoids that contribute to its antioxidant activity. Ferulic acid, diferulic acid, and chlorogenic acid are some of the potent phenols that exhibit antioxidant properties. Flavonoids present in maize, viz. naringenin glucoside, maysin, kaempferol, morin, hyperoside, and rutin, also add to its antioxidant potential. These antioxidants lessen the formation of free radicals and eliminate reactive oxygen species (ROS) (Navarro et al., 2018). Besides this, flavonoid pigments found in maize ought to be explored primarily for their antioxidant and neuro-protective activities. These also can be viewed as inductors of apoptosis and lipolysis of adipocytes.

Furthermore, the antioxidant activity of maize is due to maize peptides. Antioxidant peptides have acquired attention lately due to their significance to the food sector as food colour preservatives and as a defence against deterioration from the creation of ROS and RNS (Rizzello et al., 2016). Other than their use in the food industry, maize peptide might be helpful in preventing oxidative stress which is linked with degenerative disorders like cancer as well as atherosclerosis (Zhu et al., 2019). Maize peptides are made up of explicit amino acids like phenylalanine, lysine, proline, leucine, tyrosine, alanine, cysteine, tryptophan, and histidine, which can scavenge free radicals (Diaz-Gomez et al., 2017; Zhu et al., 2019). Peptides having a low molecular weight (<5 kDa) exhibit high antioxidant properties. Maize peptides display potential hydroxyl radical and free radical-scavenging activity, lipid peroxidation inhibition, plus ion chelating capacity (Tang et al., 2013). In addition, antioxidant activity does not change in peptides recovered from total maize protein or the zein portion of maize. An additional factor influencing antioxidant action is the pH, for instance; the native α-zein exhibits greater antioxidant potential as compared to the α-zein obtained in the mild acidic/basic conditions (Zheng et al., 2012).

The total antioxidant content of maize has been analyzed by using different methods, viz. DPPH, ABTS, hydroxyl and superoxide anion with radical-scavenging properties, Fe(II) and Cu(II)-chelating capacities as well as ORAC assays (Zhu et al., 2019). A study showed that maize peptides having a molecular weight ranging from 0.5 to 1.5 kDa display higher antioxidant effects in contrast to the peptides under 500 Da, which usually form a defensive layer around oxidative inhibitors (Hu et al., 2020).

Additionally, investigations have shown the positive influence of thermal processing on the anti-oxidative nature of maize. This is due to the release of bound forms of phytochemicals into the matrix. Higher antioxidant activity of the processed sweet variety of maize was reported in contrast to the unprocessed form (Dewanto et al., 2002).

7.7.2 Anti-inflammatory activity

Inflammation is a biological response mediated through inflammatory cells because of cellular infections, injuries, and toxins. It is the body's process of healing. Maize peptides exhibit strong anti-inflammatory properties by regulating different pro-inflammatory proteins, for example TNF-α-induced pathways and nuclear factor kappa B (NF-κB) (Liang et al., 2020). Three maize peptides having sequences PPYLSP, FLPPVTSMQ, and IIGGAL were recently recognized. These peptides displayed *in-vitro* inflammatory activity (Liang et al., 2018). Consequently, *in-vitro* research on the maize peptides to check their potential as a nutraceutical in averting cardiovascular disorders was proposed. The study showed the ability of peptides to forestall both TNF-α-induced overexpression of two adhesion particles (TNFR1 and TNFR2) and oxidative stress in EA.hy926 cells (Liang et al., 2020), thereby proving their anti-inflammatory nature.

Recently, the immune-modulatory function of maize protein zein hydrolysates, as well as maize germ digested using thermolysin, was studied in human macrophage-like U937 cells (Liu et al., 2020). Five peptides obtained from the α-zein fraction of maize were recognized. Specifically, the peptides FLPFNQL and PFNQL displayed prominent *in-vitro* inhibitory action on IL-6 production. The FLPFNQL peptide was observed to have the highest inhibitory action by decreasing IL-6 levels by 58%. Looking at the findings of these studies, it could be said that these maize peptides exhibit potent anti-inflammatory effects. However, most of these studies are conducted in *in-vitro* models; therefore some *in-vivo* studies in animals are required to prove the anti-inflammatory properties of these maize peptides (Trinidad-Calderon et al., 2021).

Moreover, the anthocyanin present in the pigmented maize also presents anti-inflammatory activity. For instance, in an *in-vitro* examination of mono- or co-culture of macrophage or/and adipocytes with anthocyanin from extracts obtained from purple and red maize varieties, an inhibition of NF-κB and JNK pathways through regulating IκBα and JNK phosphorylation was exhibited. Further, the administration of maize extracts led to a lessening of pro-inflammatory cytokine production as well as lipolysis, in addition to enhancement of the glucose transporter-4 membrane (GLUT4) translocation (Zhang et al., 2019). Moreover, purple varieties of maize having high anthocyanin content increase the free fatty acid receptor-1 (FFAR1) and glucokinase (GK) action in Caco-2, INS-1E, and HepG2 cells' culture (Luna-Vital & Gonzalez de Mejia, 2018).

7.7.3 Anti-cancer activity

Maize exhibits potential anti-cancer activity because of the peptides present in it. These peptides exert chemo-preventive effects via various mechanisms such as regulation of apoptosis, angiogenesis, immune system stimulation, and decreases in hepatic tumour growth. Many investigations have revealed the anti-cancer properties of amino acids/peptides/dietary proteins in regulating apoptosis and angiogenesis, which is significant in monitoring tumour metastasis (De Mejia & Dia, 2010).

In the context of the structure of peptides, both glycine and arginine amino acids have been viewed as necessary against malignant growth action. In particular, glycine residues fundamentally play a structural role due to their cyclization activity and their impact on the development of β-turns. The arginine residues are accepted to play a part in cancer prevention. Other amino acids, for instance cysteine, isoleucine, lysine, and tryptophan, are observed at different locations in anti-cancer peptides; notwithstanding, their effects stay obscure (Li & Yu, 2015).

In addition, anthocyanin present in coloured varieties of maize also exhibits anti-cancer activity (Stintzing & Carle, 2004). The administration of anthocyanins obtained from purple maize at 5% dietary level for a period of 36 weeks showed a strong inhibitory effect on colorectal carcinogenesis in male rodents. Also, it showed that lesion formation in the colon was essentially suppressed (Hagiwara et al., 2001). Besides this, the studies demonstrate that the anthocyanin obtained from the acid-treated extracts of the hybrid blue variety of maize displays effective anti-proliferative action against breast carcinoma, hepatic carcinoma, colon carcinoma, as well as prostate carcinoma cell lines (Urias-Lugo et al., 2015).

7.7.4 Hepato-protective activity

The liver is the main organ that has the ability to detoxify harmful substances. Liver damage caused due to hepato-toxic agents leads to serious consequences. Therefore, there is an urgent need for specialists that could shield it from such harm. Maize-based peptides exhibit protective effects on the liver (Yu et al., 2012). *In-vivo* studies utilizing rodents revealed that the maize bioactive peptides ensure the protection of liver tissue and also help in metabolism of alcohol (Lv et al., 2013).

Some reports on the hepato-protective activity of maize suggested that the maize gluten meal-derived peptides decrease the degree of cellular damage indicators in animal models. Subsequently, it was reported that pentapeptide QLLPF decreases the alcohol levels in the blood of mice (Ma et al., 2012). Inhibitory action on the hepatocyte apoptosis was seen in another study done on mice consuming the same pentapeptide and a mix of maize peptides (Ma et al., 2015). As per the previous *in-vivo* studies done on maize peptide, it could be said that the maize peptides are helpful in exhibiting a hepato-protective effect. However, the specific mechanism of maize peptide on hepato-protection needs to be explored.

7.8 Effects of processing techniques on the bioactive components of maize grain

Maize grain is either consumed in whole form or processed into various products. In the whole form, maize is either eaten as boiled, roasted, or frozen sweetcorn and also as popcorn. Several processing techniques have various impacts on the biochemistry and nutritional activity of maize. The impact of the processing technique varies based on the type and variety of maize. For instance, a decline in the resistant starch content of diverse varieties of maize was noticed after cooking. A steep decrease from 14% to 3.2% resistant starch was observed in blue maize flour while a decline from 11.5% to 1.5% was seen in white variety of maize (Camelo-Mendez et al., 2017). Some of the common processing techniques utilized for maize processing are milling, germination, fermentation, baking, and extrusion (Suri & Tanumihardjo, 2016). The effect of processing on the bioactive content of maize is displayed in Table 7.3.

By definition, germination or sprouting includes those processes which start with the uptake of water by the seed crop and complete with the growth of embryonic axis (Rifna et al., 2019). It is one of the techniques proposed for improving the nutritional composition and functional attributes of maize grain. Germination of maize is advantageous in speeding the biosynthesis of tocopherols, niacin, and phenols which thereby lead to enhancement of the *in-vitro* antioxidant activity of maize flour (Zilic et al., 2015). In addition, germination of maize improves the digestibility of maize-based proteins (Nkhata et al., 2018). In a study done by Gong et al. (2018), it was detected that germination of maize leads to a 169% and 230% increase in free and bound phenolic acid content, respectively. Germination enhances the polyphenol content from 12 to 15 types. Also, a 311% increment in the flavonoid content of maize was observed upon germination. Besides this, germination also leads to an increase in the phytase activity. A study showed a six-fold upsurge in phytase activity and a decline in the phytic acid level to 65% of the original phytic acid value with 72 hours of germination (Egli et al., 2002). However, there is limited data in regards to the profile and bioactivity of germinated maize-based peptides. *In-vivo* trials on the bioactive properties of constituents of maize need to be explored.

Milling is another common method for the processing of grains. It helps in the grinding of whole maize grain into smaller pieces and fine flour, thereby eliminating bran and germ (Suri & Tanumihardjo, 2016). However, milling could bring about the loss of the vast majority of the nutrients. The manner by which maize is milled and taken differs incredibly from one nation to another (Blandino et al., 2017). Milling can be done in two forms – dry and wet. The end product of dry milling of maize consists of maize flour, grits, and meal, while the wet milling yields starch, syrup, and dextrose

Table 7.3 Effect of processing on bioactive content (µg/g DW) of maize grain

Maize cultivar	Type of bioactive compound	Bioactive compound (in raw maize grain)	Bioactive content after germination	Bioactive content after extrusion	Bioactive content after sequential extrusion and germination
Yellow corn (Zea mays L.)	Vanillic acid	–	28.95±1.26	–	23.71±0.67
	Syringic acid	–	10.52±0.73	–	12.39±0.57
	Protocatechuic acid	12.85±0.49	12.12±0.77	11.82±0.61	42.79±0.82
	Benzoic acid (bound form)	55.46±1.17	45.21±2.66	50.92±1.37	9.17±0.37
	Salicylic acid (bound form)	13.48±0.67	9.38±0.44	12.24±1.03	–
	Chlorogenic acid	15.77±0.18	–	14.76±0.62	7.96±0.34
	Caffeic acid	–	8.98±0.22	–	43.13±2.77
	Sinapic acid	–	–	40.23±1.54	27.56±0.6
	p-coumaric acid	18.69±0.72	28.84±1.02	18.10±0.55	11.06±0.39
	Ferulic acid	–	11.98±0.37	–	43.13±2.77
	Sinapic acid	42.27±1.12	42.26±2.07	–	38.87±2.21
	Quercetin (bound form)	10.73±0.71b	33.72±1.86	9.92±1.04	79.99±2.64
	Quercetin-3-O-galactoside (bound form)	104.52±3.32	89.73±4.81	95.14±2.86	
	Kaempferol (bound form)	14.58±0.73	–	10.96±1.17	9.96±0.69
	Catechin	12.85±0.59	–	12.33±0.71	26.11±2.23
	Lutin	10.92±0.79	–	10.20±0.35	11.12±0.17
	Naringenin	11.23±1.17	–	8.82±0.26	–
	Epicatechin	–	9.22±0.73	–	8.46±0.49

Source: Adapted with permission from Gong et al. (2018). Copyright 2018, Elsevier.

(Slavin et al., 2000). Dry milling of maize leads to a reduction in particle size, mechanical separation, as well as the removal of bran and germ from the endosperm portion of maize, leading to a shelf-stable product. Dry milling causes an increase in the thiamin and riboflavin content and bioavailability of maize. In addition, wet milling is commonly used in industries for the separation of maize into starch, oil, protein, and fibre. Maize can also be milled in a traditional stone mill leading to retention of maize germ (Suri & Tanumihardjo, 2016). Milled maize can be utilized for the formulation of different food items like porridge, polenta, corn meal, grits, and baked foods.

A recent study analyzed the bioactive components present in maize fractions obtained from dual dry-milling operations, as well as the effects of extraction on the yield, as affected by maize kernel hardness and the distribution of the bioactive components in products and by-products of the milling processes. The study reported that the milling fraction of maize grain has an uneven distribution of bioactive compounds. Also, the bioactive components present in different maize fractions relate to the milling process utilized, as indicated by its effectiveness in removing the germ and bran residuals from degerminated endosperm segments. In contrast, the yield of extraction which is interlinked with the kernel hardness, doesn't appear to influence the bioactive nature/composition of maize to a great extent (Blandino et al., 2017).

Other maize-based processing techniques involve soaking, fermentation, and nixtamalization utilized during the preparation of food products. Fermentation of maize is done in preparation of porridge such as *ogi* (a fermented porridge usually consumed in Nigeria). The effect of fermentation on the retention of carotenoids of high β-carotene in maize porridge was examined in a study, where they found that the fermentation of maize before cooking diminishes the carotenoid loss that occurs during cooking (Li et al., 2007). In addition, fermentation, as well as germination of maize, can expand its nutritional value. One investigation indicated that fermentation of maize done after germination leads to retention of the riboflavin and niacin levels (Lay & Fields, 1981). Further, fermentation is useful for reduction of phytic acid levels of maize compared to cooking. Around 66% of reduction in phytic acid was reported in maize by fermentation, while it was 16–17% by cooking. Besides this, soaking of maize flour for a duration of 24 to 48 hours before the fermentation process resulted in around 96% reduction in phytic acid levels. An increase in the soluble iron content (4–6%) was reported in maize following the fermentation and soaking process. In another experiment, fermentation of maize flour along with germination or other fermentation along with 10 mg phytase showed 9% soluble iron. However, fermentation plus 50 mg phytase caused a 99% decline in phytic acid content in addition to 43% increment in soluble iron (Sandberg & Svanberg, 1991). In another study, lactic acid fermentation of maize flour brought about a decrease of phytic acid to about 88%, and much low phytic acid levels are reported when a starter culture (61%) or germinated maize flour (71%) is used (Hotz & Gibson, 2001), hence such approaches could be collectively utilized for better reduction of phytic acid and nutrient enrichment in maize flour. In addition, a number of studies reported the use of microbial fermentation to obtain maize peptides. Microbes such as *Aspergillus niger* exhibited a role in hydrolyzing the yellow maize filtrate (Sheriff et al., 2012); *Bacillus natto* helped in fermenting the corn gluten meal and additionally enabling the hydrolysates to have antioxidant activity (Zheng et al., 2012); and *Enterococcus faecium* and *Bacillus subtilis* improved the digestibility of corn and soybean meal following the two-step fermentation process (Shi et al., 2017). These are some favourable examples showing the effects of microbial fermentation of maize; however, some *in-vivo* studies are still needed.

Nixtamalization is one of the processing techniques utilized for the formulation of maize-based tortillas in which maize is boiled with alkaline lime solution (usually calcium hydroxide at 1% to 5%), heated followed by soaking overnight, subsequently leading to chemically modified dough (Suri & Tanumihardjo, 2016; Kamau et al., 2020). The nixtamalization process can cause numerous physical as well as biochemical changes, for example, improved bioavailability of vitamins and minerals as well as reduction of mycotoxins (Schaarschmidt & Fauhl-Hassek, 2019). This process decreases phytic acid content in maize by about 20%, sufficient to possibly enhance iron absorption. Also, removal of phytic acid through nixtamalization improves the zinc bioavailability (Bressani et al., 2004). Correspondingly, improvement in the nutrient content was reported after the nixtamalization process. An increase in magnesium content of maize from 165 mg/100 g to 180 mg/100 g was reported after nixtamalization and development into dough and tortillas (Mendoza et al., 1998). It also leads to retention of thiamine (35% to 40%), riboflavin (48% to 68%), and niacin (70%) (Bressani et al., 2002). Nixtamalization leads to about an 18-fold increase in calcium, while for iron 70 to 100% retention and zinc 90 to 100% retention were reported. The nixtamalization process marginally increases the protein content from 8% to 9% as a consequence of the concentration effect caused by the loss of water. Also, the treatment of maize with lime can expand the balance of essential amino acids (Bressani et al., 2004).

Extrusion is another processing method used for maize. It is a high-temperature short-time procedure that leads to a change in the chemical and nutritional values of maize and maize-based items (Athar et al., 2006). Extrusion can lower the anthocyanin content of maize. A study showed a 57% decrease in the anthocyanin content of blue maize (Mora-Rochin et al., 2010). When heated at 98 °C, the purple maize extract exhibited high degradation of anthocyanins (Cevallos-Casals & Cisneros-Zevallos, 2004). Also, a wide reduction in the nutritional content of maize occurs at high temperatures. Extrusion of degermed maize grits leads to a reduction in

B-vitamin content, viz. thiamine (50%), riboflavin (16%), and niacin (25%) (Bressani et al., 2002). Meanwhile, the high temperature of extrusion processing helps to lower the phytic acid content thereby improving the nutrient bioavailability (Singh et al., 2007). Also in terms of retention of nutrients present in maize grain, extrusion processing is helpful causing the retention of iron (67%) (Hazell & Johnson, 1989), anthocyanin and polyphenols (43%) (Mora-Rochin et al., 2010), thiamine (50%), riboflavin (84%), pyridoxine (100%), and niacin (75%) (Bressani et al., 2002). In addition, an upsurge in the total phenols and flavonoid content of the yellow variety of maize flour was reported after extrusion processing (Gong et al., 2018). However, additional research is needed to decide the impact of extrusion on the content and bioavailability of the maize. Therefore, different processing techniques have a substantial influence on the nutritional quality of the maize grain. Most of the water-soluble vitamins especially vitamin B-complex is lost during storage and long-term milling processes (degermination), as well as during the soaking and cooking process. However, other processing techniques, viz. fermentation, germination, and nixtamalization, can lead to an upsurge in the bioavailability of vitamin B_2 (riboflavin) and vitamin B_3 (niacin). The content of carotenoids can be increased by eliminating germs because they are mainly found in the endosperm. Also, the bioavailability of the minerals may be boosted through processing techniques that help in reducing the phytic acid content.

7.9 Conclusion

The current chapter highlights the bioactive compounds present in different parts of maize and also summarizes their potential anti-oxidative properties and health benefits to human. The chapter sums up several recent studies on the bioactive nature of maize and the impact of processing conditions on bioactive components present in maize. It could be said that maize is a vital source of both macro- and micronutrients in the human diet, predominantly for people who use it as a common food in Latin America, Asia, and Africa. Maize is eaten in cob, roasted, dried, fried, and fermented forms and used in traditional foods. Maize contains various biologically active phytochemicals, including polyphenolic compounds (ferulic acid and anthocyanin), carotenoids (carotene and xanthophyll), and plant sterols. These phytochemicals in maize play a noteworthy part in the management of chronic illness. Maize protein – zein – is usually found in the endosperm portion and can be used in the pharmaceutical and nutraceutical industries. Maize grain exhibits several bioactive actions including antioxidant, anti-cancer, anti-tumour, anti-microbial, and hepato-protective effects. All this indicates maize's great usefulness in the defence and prevention of various diseases and its health-promoting properties.

References

Abdel-Aal, E. S. M., Young, J. C., & Rabalski, I. (2006). Anthocyanin composition in black, blue, pink, purple, and red cereal grains. *Journal of Agricultural and Food Chemistry, 54*(13), 4696–4704.

Adom, K. K., & Liu, R. H. (2002). Antioxidant activity of grains. *Journal of Agricultural and Food Chemistry, 50*(21), 6182–6187.

Aires, A., & Carvalho, R. (2016). Compositional study and antioxidant potential of polyphenol extracted from corn by-product, using ultrasound extraction method. *Austin Chromatography, 3*(1), 1–5.

AmbikaRajendran, R., Nirupma, S., VinayMahajan, D. P., Chaudhary, S., & Sai, R. K. (2012). Corn oil: An emerging industrial product. Directorate of maize research, New Delhi, Technical Bulletin, No. 8: 36 p. Retrieved from https://iimr.icar.gov.in/wp-content/uploads/2020/03/CORN-OIL-An-emerging-industrial-product.pdf

And, T. M., & Shahidi, F. (2005). Antioxidant potential of pea beans (*Phaseolus vulgaris* L.). *Journal of Food Science, 70*(1), S85–S90.

Aoki, H., Kuze, N., Kato, Y., & Gen, S. E. (2002). Anthocyanins isolated from purple corn (*Zea mays* L.). *Foods and Food Ingredients Journal of Japan, 199*, 41–45.

Athar, N., Hardacre, A., Taylor, G., Clark, S., Harding, R., & McLaughlin, J. (2006). Vitamin retention in extruded food products. *Journal of Food Composition and Analysis, 19*(4), 379–383.

Auger, C., Laurent, N., Laurent, C., Besançon, P., Caporiccio, B., Teissedre, P. L., & Rouanet, J. M. (2004). Hydroxycinnamic acids do not prevent aortic atherosclerosis in hypercholesterolemic golden Syrian hamsters. *Life Sciences, 74*(19), 2365–2377.

Bento-Silva, A., Patto, M. C. V., & do Rosário Bronze, M. (2018). Relevance, structure and analysis of ferulic acid in maize cell walls. *Food Chemistry, 246*, 360–378.

Blandino, M., Alfieri, M., Giordano, D., Vanara, F., & Redaelli, R. (2017). Distribution of bioactive compounds in maize fractions obtained in two different types of large scale milling processes. *Journal of Cereal Science, 77*, 251–258.

Bressani, R., Turcios, J. C., Colmenares de Ruiz, A. S., & de Palomo, P. P. (2004). Effect of processing conditions on phytic acid, calcium, iron, and zinc contents of lime-cooked maize. *Journal of Agricultural and Food Chemistry, 52*(5), 1157–1162.

Bressani, R., Turcios, J. C., & De Ruiz, A. S. C. (2002). Nixtamalization effects on the contents of phytic acid, calcium, iron and zinc in the whole grain, endosperm and germ of maize. *Food Science and Technology International, 8*(2), 81–86.

Buranov, A. U., & Mazza, G. (2009). Extraction and purification of ferulic acid from flax shives, wheat and corn bran by alkaline hydrolysis and pressurised solvents. *Food Chemistry, 115*(4), 1542–1548.

Cabrera-Soto, M. L., Salinas-Moreno, Y., Velazquez-Cardelas, G. A., & Espinosa Trujillo, E. (2009). Content of soluble and insoluble phenols in the structures of corn grain and their relationship with physical properties. *Agrociencia, 43*(8), 827–839.

Camelo-Mendez, G. A., Agama-Acevedo, E., Tovar, J., & Bello-Pérez, L. A. (2017). Functional study of raw and cooked blue maize flour: Starch digestibility, total phenolic content and antioxidant activity. *Journal of Cereal Science, 76*, 179–185.

Cevallos-Casals, B. A., & Cisneros-Zevallos, L. (2004). Stability of anthocyanin-based aqueous extracts of Andean purple corn and red-fleshed sweet potato compared to synthetic and natural colorants. *Food Chemistry, 86*(1), 69–77.

Cuevas Montilla, E., Hillebrand, S., Antezana, A., & Winterhalter, P. (2011). Soluble and bound phenolic compounds in different Bolivian purple corn (*Zea mays* L.) cultivars. *Journal of Agricultural and Food Chemistry, 59*(13), 7068–7074.

Das, A. K., & Singh, V. (2016). Antioxidative free and bound phenolic constituents in botanical fractions of Indian specialty maize (*Zea mays* L.) genotypes. *Food Chemistry, 201*, 298–306.

de la Parra, C., Serna Saldivar, S. O., & Liu, R. H. (2007). Effect of processing on the phytochemical profiles and antioxidant activity of corn for production of masa, tortillas, and tortilla chips. *Journal of Agricultural and Food Chemistry, 55*(10), 4177–4183.

De Mejia, E. G., & Dia, V. P. (2010). The role of nutraceutical proteins and peptides in apoptosis, angiogenesis, and metastasis of cancer cells. *Cancer and Metastasis Reviews, 29*(3), 511–528.

Dewanto, V., Wu, X., & Liu, R. H. (2002). Processed sweet corn has higher antioxidant activity. *Journal of Agricultural and Food Chemistry, 50*(17), 4959–4964.

Diaz-Gomez, J. L., Castorena-Torres, F., Preciado-Ortiz, R. E., & Garcia-Lara, S. (2017). Anti-cancer activity of maize bioactive peptides. *Frontiers in Chemistry, 5*, 44. https://doi.org/10.3389/fchem.2017.00044

Egli, I., Davidsson, L., Juillerat, M. A., Barclay, D., & Hurrell, R. F. (2002). The influence of soaking and germination on the phytase activity and phytic acid content of grains and seeds potentially useful for complementary feedin. *Journal of Food Science, 67*(9), 3484–3488.

FAO STAT. (2019). *Food and agriculture data. Food and agriculture organization*. Rome: FAO. Retrieved from http://www.fao.org/faostat/en/?#data/QC

Fuentealba, C., Quesille-Villalobos, A. M., González-Muñoz, A., SaavedraTorrico, J., Shetty, K., & GálvezRanilla, L. (2017). Optimized methodology for the extraction of free and bound phenolic acids from Chilean Cristalino corn (*Zea mays* L.) accession. *CyTA: Journal of Food, 15*(1), 91–98.

Ghosh, D., & Konishi, T. (2007). Anthocyanins and anthocyanin-rich extracts: Role in diabetes and eye function. *Asia Pacific Journal of Clinical Nutrition, 16*(2), 200–208.

Giuberti, G., Gallo, A., & Masoero, F. (2012). Quantification of zeins from corn, high-moisture corn, and corn silage using a turbidimetric method: Comparative efficiencies of isopropyl and tert-butyl alcohols. *Journal of Dairy Science, 95*(6), 3384–3389.

Gong, K., Chen, L., Li, X., Sun, L., & Liu, K. (2018). Effects of germination combined with extrusion on the nutritional composition, functional properties and polyphenol profile and related in vitro hypoglycemic effect of whole grain corn. *Journal of Cereal Science, 83*, 1–8.

Hagiwara, A., Miyashita, K., Nakanishi, T., Sano, M., Tamano, S., Kadota, T., … Shirai, T. (2001). Pronounced inhibition by a natural anthocyanin, purple corn color, of 2-amino-1-methyl-6-phenylimidazo [4, 5-b] pyridine (PhIP)-associated colorectal carcinogenesis in male F344 rats pretreated with 1, 2-dimethylhydrazine. *Cancer Letters, 171*(1), 17–25.

Harrabi, S., St-Amand, A., Sakouhi, F., Sebei, K., Kallel, H., Mayer, P. M., & Boukhchina, S. (2008). Phytostanols and phytosterols distributions in corn kernel. *Food Chemistry, 111*(1), 115–120.

Harris, K. A., & Kris-Etherton, P. M. (2010). Effects of whole grains on coronary heart disease risk. *Current Atherosclerosis Reports, 12*(6), 368–376.

Hazell, T., & Johnson, I. T. (1989). Influence of food processing on iron availability in vitro from extruded maize-based snack foods. *Journal of the Science of Food and Agriculture, 46*(3), 365–374.

Hotz, C., & Gibson, R. S. (2001). Assessment of home-based processing methods to reduce the phytate content and phytate/zinc molar ratio of white maize (*Zea mays*). *Journal of Agricultural and Food Chemistry, 49*(2), 692–698.

Hu, R., Chen, G., & Li, Y. (2020). Production and characterization of antioxidativehydrolysates and peptides from corn gluten meal using papain, ficin, and bromelain. *Molecules, 25*(18), 4091.

Jiang, Y., & Wang, T. (2005). Phytosterols in cereal by-products. *Journal of the American Oil Chemists' Society, 82*(6), 439–444.

Kamau, E. H., Nkhata, S. G., & Ayua, E. O. (2020). Extrusion and nixtamalization conditions influence the magnitude of change in the nutrients and bioactive components of cereals and legumes. *Food Science and Nutrition, 8*(4), 1753–1765.

Kandil, A., Li, J., Vasanthan, T., & Bressler, D. C. (2012). Phenolic acids in some cereal grains and their inhibitory effect on starch liquefaction and saccharification. *Journal of Agricultural and Food Chemistry, 60*(34), 8444–8449.

Kaur, R., Pahwa, A., & Bhise, S. (2021). Maize: A potential grain for functional and nutritional properties. *Cereals and Cereal-Based Foods: Functional Benefits and Technological Advances for Nutrition and Healthcare, 65*, 65–80.

Kumar, D., & Jhariya, A. N. (2013). Nutritional, medicinal and economical importance of corn: A mini review. *Research Journal of Pharmaceutical Sciences, 2319*, 555X.

Lay, M. M. G., & Fields, M. L. (1981). Nutritive value of germinated corn and corn fermented after germination. *Journal of Food Science, 46*(4), 1069–1073.

Li, S., Tayie, F. A., Young, M. F., Rocheford, T., & White, W. S. (2007). Retention of provitaminA carotenoids in high β-carotene maize (*Zea mays*) during traditional African household processing. *Journal of Agricultural and Food Chemistry, 55*(26), 10744–10750.

Li, Y., & Yu, J. (2015). Research progress in structure-activity relationship of bioactive peptides. *Journal of Medicinal Food, 18*(2), 147–156.

Liang, Q., Chalamaiah, M., Liao, W., Ren, X., Ma, H., & Wu, J. (2020). Zeinhydrolysate and its peptides exert anti-inflammatory activity on endothelial cells by preventing TNF-α-induced NF-κB activation. *Journal of Functional Foods, 64*, 103598.

Liang, Q., Chalamaiah, M., Ren, X., Ma, H., & Wu, J. (2018). Identification of new anti-inflammatory peptides from zeinhydrolysate after simulated gastrointestinal digestion and transport in Caco-2 cells. *Journal of Agricultural and Food Chemistry, 66*(5), 1114–1120.

Lingzhu, L. V., Lu, W., Dongyan, C., Jingbo, L., Songyi, L., Haiqing, Y., & Yuan, Y. (2015). Optimization of ultrasound-assisted extraction of polyphenols from maize filaments by response surface methodology and its identification. *Journal of Applied Botany and Food Quality, 88*, 152–163.

Liu, P., Liao, W., Xingpu, Q., Wenlin, Y., & Wu, J. (2020). Identification of immunomodulatory peptides from zein hydrolysates. *European Food Research and Technology, 246*(5), 931–937.

Liu, R. H. (2007). Whole grain phytochemicals and health. *Journal of Cereal Science, 46*(3), 207–219.

Longvah, T., Ananthan, R., Bhaskarachary, K., Venkaiah, K., & Longvah, T. (2017). *Indian food composition tables*. Hyderabad: National Institute of Nutrition (pp. xii–ixx).

Lopez-Martinez, L. X., Oliart-Ros, R. M., Valerio-Alfaro, G., Lee, C. H., Parkin, K. L., & Garcia, H. S. (2009). Antioxidant activity, phenolic compounds and anthocyanins content of eighteen strains of Mexican maize. *LWT - Food Science and Technology, 42*(6), 1187–1192.

Luna-Vital, D. A., & Gonzalez de Mejia, E. (2018). Anthocyanins from purple corn activate free fatty acid-receptor 1 and glucokinase enhancing in vitro insulin secretion and hepatic glucose uptake. *PLOS ONE, 13*(7), e0200449.

Luo, Y., & Wang, Q. (2012). Bioactive compounds in corn. In Yu Liangli (Lucy), Rong Tsao, Fereidoon Shahidi (Eds.), *Cereals and Pulses: Nutraceutical Properties and Health Benefits*, 85–103.

Lv, J., Nie, Z. K., Zhang, J. L., Liu, F. Y., Wang, Z. Z., Ma, Z. L., & He, H. (2013). Corn peptides protect against thioacetamide-induced hepatic fibrosis in rats. *Journal of Medicinal Food, 16*(10), 912–919.

Ma, Z. L., Zhang, W. J., Yu, G. C., He, H., & Zhang, Y. (2012). The primary structure identification of a corn peptide facilitating alcohol metabolism by HPLC–MS/MS. *Peptides*, *37*(1), 138–143.

Ma, Z., Hou, T., Shi, W., Liu, W., & He, H. (2015). Inhibition of hepatocyte apoptosis: An important mechanism of corn peptides attenuating liver injury induced by ethanol. *International Journal of Molecular Sciences*, *16*(9), 22062–22080.

Magaia-Cerino, J. M., Peniche-Pavía, H. A., Tiessen, A., & Gurrola-Díaz, C. M. (2020). Pigmented maize (*Zea mays* L.) contains anthocyanins with potential therapeutic action against oxidative stress-a review. *Polish Journal of Food and Nutrition Sciences*, *70*(2), 85–99.

Matsumoto, H., Nakamura, Y., Iida, H., Ito, K., & Ohguro, H. (2006). Comparative assessment of distribution of blackcurrant anthocyanins in rabbit and rat ocular tissues. *Experimental Eye Research*, *83*(2), 348–356.

Mazza, G., Kay, C. D., Cottrell, T., & Holub, B. J. (2002). Absorption of anthocyanins from blueberries and serum antioxidant status in human subjects. *Journal of Agricultural and Food Chemistry*, *50*(26), 7731–7737.

Mendoza, C., Viteri, F. E., Lönnerdal, B., Young, K. A., Raboy, V., & Brown, K. H. (1998). Effect of genetically modified, low-phytic acid maize on absorption of iron from tortillas. *American Journal of Clinical Nutrition*, *68*(5), 1123–1127.

Milind, P., & Isha, D. (2013). Zea maize: A modern craze. *International Research Journal of Pharmacology*, *4*(6), 39–43.

Momany, F. A., Sessa, D. J., Lawton, J. W., Selling, G. W., Hamaker, S. A., & Willett, J. L. (2006). Structural characterization of α-zein. *Journal of Agricultural and Food Chemistry*, *54*(2), 543–547.

Monroy, Y. M., Rodrigues, R. A. F., Sartoratto, A., & Cabral, F. A. (2016). Optimization of the extraction of phenolic compounds from purple corn cob (*Zea mays* L.) by sequential extraction using supercritical carbon dioxide, ethanol and water as solvents. *Journal of Supercritical Fluids*, *116*, 10–19.

Mora-Rochin, S., Gutiérrez-Uribe, J. A., Serna-Saldivar, S. O., Sánchez-Peña, P., Reyes-Moreno, C., & Milán-Carrillo, J. (2010). Phenolic content and antioxidant activity of tortillas produced from pigmented maize processed by conventional nixtamalization or extrusion cooking. *Journal of Cereal Science*, *52*(3), 502–508.

Moreno, F. S., Toledo, L. P., de Conti, A., Heidor, R., Jordao Jr, A., Vannucchi, H., … Ong, T. P. (2007). Lutein presents suppressing but not blocking chemopreventive activity during diethylnitrosamine-induced hepatocarcinogenesis and this involves inhibition of DNA damage. *Chemico-Biological Interactions*, *168*(3), 221–228.

Moros, E. E., Darnoko, D., Cheryan, M., Perkins, E. G., & Jerrell, J. (2002). Analysis of xanthophylls in corn by HPLC. *Journal of Agricultural and Food Chemistry*, *50*(21), 5787–5790.

Navarro, A., Torres, A., Fernandez-Aulis, F., & Pena, C. (2018). Bioactive compounds in pigmented maize. In S. Fahad (Ed.), *Corn- production and human health in changing climate* (pp. 69–91). London: InTech.

Nkhata, S. G., Ayua, E., Kamau, E. H., & Shingiro, J. B. (2018). Fermentation and germination improve nutritional value of cereals and legumes through activation of endogenous enzymes. *Food Science and Nutrition, 6*(8), 2446–2458.

OstlundJr, R. E., Racette, S. B., Okeke, A., & Stenson, W. F. (2002). Phytosterols that are naturally present in commercial corn oil signifi- cantly reduce cholesterol absorption in humans. *American Journal of Clinical Nutrition, 75*(6), 1000–1004.

Palozza, P., Calviello, G., Serini, S., Maggiano, N., Lanza, P., Ranelletti, F. O., & Bartoli, G. M. (2001). β-Carotene at high concentrations induces apoptosis by enhancing oxy-radical production in human adenocarcinoma cells. *Free Radical Biology and Medicine, 30*(9), 1000–1007.

Park, J. S., Chew, B. P., & Wong, T. S. (1998). Dietary lutein from mari- gold extract inhibits mammary tumor development in BALB/c mice. *Journal of Nutrition, 128*(10), 1650–1656.

Perera, A., Meda, V., & Tyler, R. T. (2010). Resistant starch: A review of ana- lytical protocols for determining resistant starch and of factors affect- ing the resistant starch content of foods. *Food Research International, 43*(8), 1959–1974.

Piironen, V., Lindsay, D. G., Miettinen, T. A., Toivo, J., & Lampi, A. M. (2000). Plant sterols: Biosynthesis, biological function and their importance to human nutrition. *Journal of the Science of Food and Agriculture, 80*(7), 939–966.

Prior, R. L., Wu, X., & Gu, L. (2006). Flavonoid metabolism and challenges to understanding mechanisms of health effects. *Journal of the science of food and agriculture, 86*(15), 2487–2491.

Ramos, S. (2008). Cancer chemoprevention and chemotherapy: Dietary polyphenols and signalling pathways. *Molecular Nutrition and Food Research, 52*(5), 507–526.

Rifna, E. J., Ramanan, K. R., & Mahendran, R. (2019). Emerging technology applications for improving seed germination. *Trends in Food Science and Technology, 86*, 95–108.

Rizzello, C. G., Tagliazucchi, D., Babini, E., Rutella, G. S., Saa, D. L. T., & Gianotti, A. (2016). Bioactive peptides from vegetable food matrices: Research trends and novel biotechnologies for synthesis and recovery. *Journal of Functional Foods, 27*, 549–569.

Rondini, L., Peyrat-Maillard, M. N., Marsset-Baglieri, A., Fromentin, G., Durand, P., Tomé, D., ... Berset, C. (2004). Bound ferulic acid from bran is more bioavailable than the free compound in rat. *Journal of Agricultural and Food Chemistry, 52*(13), 4338–4343.

Rooney, L. W., & Serna-Saldivar, S. O. (2003). Food use of whole corn and dry-milled fractions. In P. J. White and L. A. Johnson (Eds.), *Corn: Chemistry and technology* (2nd ed., pp. 495–535). St Paul: American Association of Cereal Chemists.

Rouf Shah, T., Prasad, K., & Kumar, P. (2016). Maize—A potential source of human nutrition and health: A review. *Cogent Food and Agriculture, 2*(1), 1166995.

Sandberg, A. S., & Svanberg, U. (1991). Phytate hydrolysis by phytase in cereals; effects on in vitro estimation of iron availability. *Journal of Food Science, 56*(5), 1330–1333.

Santiago, R., & Malvar, R. A. (2010). Role of dehydrodiferulates in maize resistance to pests and diseases. *International Journal of Molecular Sciences, 11*(2), 691–703.

Saulnier, L., Vigouroux, J., & Thibault, J. F. (1995). Isolation and partial characterization of feruloylated oligosaccharides from maize bran. *Carbohydrate Research, 272*(2), 241–253.

Schaarschmidt, S., & Fauhl-Hassek, C. (2019). Mycotoxins during the processes of nixtamalization and tortilla production. *Toxins, 11*(4), 227.

Shahidi, F. (2004). Functional foods: Their role in health promotion and disease prevention. *Journal of Food Science, 69*(5), R146–R149.

Sheriff, M., Zainab, K. A., Laminu, H. G., & Maisaratu, A. (2012). Hydrolysis of gelatinized maize, millet and sorghum starch by amylases of Aspergillusniger. *Bio-Science Research, 9*(2), 92–93.

Shi, C., Zhang, Y., Lu, Z., & Wang, Y. (2017). Solid-state fermentation of corn-soybean meal mixed feed with Bacillus subtilis and Enterococcus faecium for degrading antinutritional factors and enhancing nutritional value. *Journal of Animal Science and Biotechnology, 8*(1), 1–9.

Singh, N., Singh, S., & Shevkani, K. (2019). Maize: Composition, bioactive constituents, and unleavened bread. In *Flour and breads and their fortification in health and disease prevention* (pp. 111–121). Cambridge: Academic Press.

Singh, S., Gamlath, S., & Wakeling, L. (2007). Nutritional aspects of food extrusion: A review. *International Journal of Food Science and Technology, 42*(8), 916–929.

Siyuan, S., Tong, L., & Liu, R. (2018). Corn phytochemicals and their health benefits. *Food Science and Human Wellness, 7*(3), 185–195.

Slavin, J. L., Jacobs, D., & Marquart, L. (2000). Grain processing and nutrition. *Critical Reviews in Food Science and Nutrition, 40*(4), 309–326.

Stintzing, F. C., & Carle, R. (2004). Functional properties of anthocyanins and betalains in plants, food, and in human nutrition. *Trends in Food Science and Technology, 15*(1), 19–38.

Sun, Z., & Yao, H. (2007). The influence of di-acetylation of the hydroxyl groups on the anti-tumor-proliferation activity of lutein and zeaxanthin. *Asia Pacific Journal of Clinical Nutrition, 16*(1), 447–452.

Suri, D. J., & Tanumihardjo, S. A. (2016). Effects of different processing methods on the micronutrient and phytochemical contents of maize: From A to Z. *Comprehensive Reviews in Food Science and Food Safety, 15*(5), 912–926.

Tang, M., He, X., Luo, Y., Ma, L., Tang, X., & Huang, K. (2013). Nutritional assessment of transgenic lysine-rich maize compared with conventional quality protein maize. *Journal of the Science of Food and Agriculture, 93*(5), 1049–1054.

Toufektsian, M. C., de Longeril, M., Nagy, N., Salen, P., Donati, M. B., Giordano, L., ... Martins, C. (2008). Chronic dietary intake of plant-derived anthocyanins protects the rat heart against ischemia-reperfusion injury. *Journal of Nutrition, 138*(4), 747–752.

Trinidad-Calderon, P. A., Acosta-Cruz, E., Rivero-Masante, M. N., Diaz-Gomez, J. L., Garcia-Lara, S., & Lopez-Castillo, L. M. (2021). Maize bioactive peptides: From structure to human health. *Journal of Cereal Science, 100*, 103232.

Tsuda, T., Horio, F., Uchida, K., Aoki, H., & Osawa, T. (2003). Dietary cyanidin 3-O-β-D-glucoside-rich purple corn color prevents obesity and ameliorates hyperglycemia in mice. *Journal of Nutrition, 133*(7), 2125–2130.

Urias-Lugo, D. A., Heredia, J. B., Muy-Rangel, M. D., Valdez-Torres, J. B., Serna-Saldivar, S. O., & Gutiérrez-Uribe, J. A. (2015). Anthocyanins and phenolic acids of hybrid and native blue maize (*Zea mays* L.) extracts and their antiproliferative activity in mammary (MCF7), liver (HepG2), colon (Caco2 and HT29) and prostate (PC3) cancer cells. *Plant Foods for Human Nutrition, 70*(2), 193–199.

Van der Logt, E. M. J., Roelofs, H. M. J., Nagengast, F. M., & Peters, W. H. M. (2003). Induction of rat hepatic and intestinal UDP-glucuronosyltransferases by naturally occurring dietary anticarcinogens. *Carcinogenesis, 24*(10), 1651–1656.

Verleyen, T., Forcades, M., Verhé, R., Dewettinck, K., Huyghebaert, A., & De Greyt, W. (2002). Analysis of free and esterified sterols in vegetable oils. *Journal of the American Oil Chemists' Society, 79*(2), 117–122.

Vitaglione, P., Napolitano, A., & Fogliano, V. (2008). Cereal dietary fibre: A natural functional ingredient to deliver phenolic compounds into the gut. *Trends in Food Science and Technology, 19*(9), 451–463.

Wallace, T. C., & Giusti, M. M. (2015). Anthocyanins. *Advances in Nutrition, 6*(5), 620–622.

Willis, H. J., Eldridge, A. L., Beiseigel, J., Thomas, W., & Slavin, J. L. (2009). Greater satiety response with resistant starch and corn bran in human subjects. *Nutrition Research, 29*(2), 100–105.

Wu, J. C., Lai, C. S., Lee, P. S., Ho, C. T., Liou, W. S., Wang, Y. J., & Pan, M. H. (2016). Anti-cancer efficacy of dietary polyphenols is mediated through epigenetic modifications. *Current Opinion in Food Science, 8*, 1–7.

Yu, G. C., Lv, J., He, H., Huang, W., & Han, Y. (2012). Hepatoprotective effects of corn peptides against carbon tetrachloride-induced liver injury in mice. *Journal of Food Biochemistry, 36*(4), 458–464.

Zhang, Q., Luna-Vital, D., & de Mejia, E. G. (2019). Anthocyanins from colored maize ameliorated the inflammatory paracrine interplay between macrophages and adipocytes through regulation of NF-κB and JNK-dependent MAPK pathways. *Journal of Functional Foods, 54*, 175–186.

Zhao, Z., Egashira, Y., & Sanada, H. (2005). Phenolic antioxidants richly contained in corn bran are slightly bioavailable in rats. *Journal of Agricultural and Food Chemistry, 53*(12), 5030–5035.

Zhao, Z., & Moghadasian, M. H. (2008). Chemistry, natural sources, dietary intake and pharmacokinetic properties of ferulic acid: A review. *Food Chemistry, 109*(4), 691–702.

Zheng, X. Q., Liu, X. L., & Liu, Z. S. (2012). Production of fermentative hydrolysate with antioxidative activity of extruded corn gluten meal by Bacillus natto. In *Applied mechanics and materials* (Vol. 138, pp. 1142–1148). Trans Tech Publications Ltd.

Zhu, B., He, H., & Hou, T. (2019). A comprehensive review of corn protein-derived bioactive peptides: Production, characterization, bioactivities, and transport pathways. *Comprehensive Reviews in Food Science and Food Safety, 18*(1), 329–345.

Zilic, S., Delic, N., Basic, Z., Ignjatovic-Micic, D. R. A. G. A. N. A., Jankovic, M., & Vancetovic, J. (2015). Effects of alkaline cooking and sprouting on bioactive compounds, their bioavailability and relation to antioxidant capacity of maize flour. *Journal of Food and Nutrition Research, 54*(2), 155–164.

Chapter 8

Coloured maize and its unique features

Ramandeep Kaur, Ramandeep Kaur, Urvashi, and Arashdeep Singh

DOI: 10.1201/9781003245230-8

8.1 Introduction

Starch, proteins, lipids, carbohydrates, anthocyanins, salicylic acid, resins, saponins, potassium, sodium salts, sulphur, phosphorus, and other phenolic chemicals are all abundant in pigmented maize. The aleurone and pericarp monolayers of the grain contain high amounts of anthocyanins and phenolic chemicals, which give maize cultivars their distinctive blue, red, purple, and black colour (Espinosa Trujillo et al., 2009). The most prevalent beads are red and blue/purple and at least 59 races have been identified, many of which correlate to coloured grain variants (Magaa-Cerino et al., 2020). The amount of anthocyanins in different coloured maize cultivars ranged from 2.50 to 696.07 mg CGE/kg d.m (Žilić et al., 2012). Anthocyanins in coloured grains have the potential to be beneficial dietary elements due to their antioxidative properties, anti-inflammatory potential, as phase II enzyme catalysts and anti-cell proliferating, and having hypoglycaemic effects (Tsuda et al., 2003; Urias-Peraldi et al., 2013). In general, the aleurone layer and, to a lesser extent, the starchy endosperm of the seeds contain a significant amount of colours (Betrán et al., 2000).

Physical qualities such as grain size, density, hardness, and chemical composition, as well as environmental factors such as climate, soil type, production practices, and lastly, individual genetic elements of each variety, all contribute to the variety of coloured maize types. Each variety of pigmented maize has a distinct observable property, such as colouring, as well as a distinct biological activity based on the quantity and profile of secondary metabolites, which increases its functional food potential. Pelargonidin, for example, is the most prevalent anthocyanidin in light red grains (Abdel-Aal et al., 2006), whereas cyanidin derivatives predominate in blue grains which also includes grains with blue, purple, and black colour (Pedreschi & Cisneros-Zeballos, 2007; Zhao et al., 2009). Cyanidin derivatives predominate in magenta red kernels, but they also contain derivatives of pelargonidin and peonidin (Salinas-Moreno et al., 2005). The primary anthocyanins identified in maize are cyanidin 3-O-glucoside and pelargonidin 3-O-glucoside, which have been described as malonylated in a reasonably high proportion (approximately 40%) as cyanidin 3-O-(6''-malonyl-glucoside) and cyanidin 3-O-(3'', 6''-dimalonyl-glucoside). Some landraces' cobs might have up to three different coloured kernels (Barrientos-Ramrez et al., 2018). The pigment is found in the aleurone layer in bluegrain types, in the pericarp in light red grain varieties, and in both the pericarp and aleurone layer in magenta/red grain kinds (Salinas-Moreno et al., 2012). Pigmented maize grains include both acylated and non-acylated anthocyanins. In comparison to other cereals, pigmented maize has a high quantity of acylated anthocyanins, which boosts its chemical stability under extreme pH and temperature circumstances (Salinas-Moreno et al., 2005). Cyanidin-3-O-glucoside, pelargonidin-3-O-glucoside, and peonidin 3-O-glucoside were the acylated anthocyanins found in maize grain. One or more acyls may bond to the sugar moiety (aliphatic acids such as malic, malonic, or succinic acid). Cinnamic acids (p-coumaric, caffeic, ferulic, or synaptic) acting on glucose and rhamnose can also produce acyl radicals (Wang et al., 1997).

8.2 Pigment accumulation and biosynthesis in pigmented maize

Colour development is a crucial stage in maize kernel growth. Maize kernels lose their colours with their adaptation to new growth settings, primarily due to artificial selection by humans, and as a result, most farmed maize is yellow (Hu et al., 2016). Because of a growing knowledge of the nutritional relevance of pigments in the prevention of chronic diseases, there has been an increase in demand for dark-coloured maize kernels recently. As a result, efforts to understand the colour creation in a variety of crops, including maize kernels, have been focused on historic flavonoid-rich landraces (Petroni et al., 2014; Casas et al., 2014). Flavonoids, primarily anthocyanins and specifically phlobaphenes, are responsible for the colourful pigments found in maize kernels. Blue, purple, red, and black colours are caused by anthocyanin compounds, whereas red and purple hues are caused by phlobaphenes (Zilić et al., 2012; Morohashi et al., 2012).

Colours in coloured maize, as well as white and yellow grains, have been related to genes involved in the manufacturing of anthocyanins. The r1(red1)/b1(booster1) family, which belongs to the bHLH (basic helix-loop-helix) transcription factor class, the c1(colourless1)/pl1(purple plant1)/p1(pericarp colour1) family, which belongs to the MYB-like transcription factors, and the WD40 factor PAC1 all play a role in this pathway (pale aleurone colour 1). Each member of this family manifests itself in a tissue-specific or plant-wide fashion (Sharma et al., 2011; Hu et al., 2016). The allelic combination of the MYB and bHLH loci influences maize pigmentation patterns to a great extent (Shen & Petonilo, 2006). The allelic combination of the regulatory pl1/ B1 genes, for example, produces the dark blue hue seen in some maize grains grown in tropical climes (Lago et al., 2014). This allelic combination can trigger cyanidin-3-O-glucoside production primarily in the maize pericarp and certain tissues. The buildup of pelargonidin-3-O-glucoside in the aleurone layer requires the r1/c1 combination (Sharma et al., 2011; Li et al., 2019).

A ternary complex of transcription factors, comprising the MYB factor C1 (colourless 1) or PL1 (purple leaf 1), the bHLH factor R1 (red colour 1) or B1 (booster 1), and the WD40 factor PAC1, regulates the anthocyanin pathway in maize (pale aleurone colour 1). Although each transcription factor appears to be controlled individually, anthocyanin pigmentation requires the expression of all three. In the absence of R1 or B1, however, the MYB member P1 (pericarp colour1) controls the production of phlobaphene. These transcriptional factors create precursors of different anthocyanins and phlobaphenes by regulating three common flavonoid biosynthetic structural genes, chalcone synthase (CHS), chalcone isomerase (CHI), and dihydroflavonol 4-reductase (DFR). Furthermore, epigenetics play an essential role in the regulation of the complex spatial and temporal pigmentation in maize, due to the notably uncompact structure of the promoters of R1, B1, and P1 genes (Hu et al., 2016; Petroni et al., 2014). Maize gets its blue hue colour expression from anthocyanins in the aleurone layer which is controlled by anthocyaninless-1 (a1), anthocyaninless-2 (a2), bronze-1 (bz1), bronze-2 (bz2), colourless-1 (c1), colourless-2 (c2), defective kernel-1 (dek1), red aleurone (pr), colourless (r), and viviparous-1 (v1) (vp1). Every factor requires the presence of a dominant allele for deep purple colour. The presence of dilution elements such as C2-Idf can lessen the blue colour. The R locus alleles (R and r) have dosage-dependent expression. R/R/r genotypes are purple, but R/r/r genotypes are mottled purple (Betrán et al., 2000).

8.3 Bioactive compounds in coloured maize

Pigmented maize is a major source of carbohydrates, minerals, and phenolic compounds (Pedreschi & Cisneros-Zevallos, 2007). The colour parameters in 15 Mexican varieties of coloured maize varied from 25.13 to 63.64. The percentage of carbohydrates, protein, and moisture ranged from 71.30–74.88, 9.72–12.57, and 7.98–9.67 respectively. The presence of phosphorous, potassium, magnesium, zinc, and iron were also reported (Rodriguez-Salinas et al., 2020). The presence of a maximum amount of anthocyanins and phenolic compounds in aleurone and pericarp layer of grains is mainly responsible for providing specific colour (black, blue, red, and purple) to corn varieties (Trujillo et al., 2009). Among differently coloured grains, maximum numbers of anthocyanins were reported in coloured maize as compared to coloured rice, barley, and wheat grains. The predominant anthocyanins were cyanidin 3-glucoside in black and red rice and in bluish, purplish, and reddish corns, pelargonidin 3-glucoside in pinkish corn, and delphinidin 3-glucoside in blue-coloured wheat (Abdel-Aal et al., 2006). The amount of anthocyanin in several varieties of coloured maize ranged from 2.50 to 696.07 mg CGE/kg (Žilić et al., 2012) while in a few varieties, it can reach up to 1052 mg Cy3glu/kg cyanidin-3-O-glucoside for each kg of sample (Urias-Lugo et al., 2015).

Variation in geographical regions and environmental factors have led to diversification in maize. The colour variation might be due to the presence of different anthocyanin derivatives in different concentrations. The cropping location, accessions, and location-accession also significantly affect the total flavonoid and phenolic content of corn in 57 native maize accessions and three commercial varieties grown in different cities of Mexico, and Central and South America. The anthocyanin content in blue maize was higher than that in red maize. Significantly, decreasing differences were observed among the different grain colours (i.e., blue red>yellow in total phenolic, total flavonoids, and antioxidant activity), and significant interactions were observed between the localities and grain colours (Martínez-Martínez et al., 2019). Cyanidin and pelargonidin has been reported as major anthocyanin in coloured maize. These are mainly present in their glycosylated form as cyanidin 3-O-(6"-malonyl-glucoside) and cyanidin 3-O-(3",6"-dimalonyl-glucoside). The cobs of some landraces might contain kernels having three colours (Barrientos-Ramírez et al., 2018). The coloured pigment could be present in the aleurone layer, pericarp, or both (Salinas-Moreno et al., 2012). Pelargonidin and cyanidin derivatives were reported as predominant anthocyaninin light-coloured and blue-coloured corn, respectively (Zhao et al., 2009; Abdel-Aal et al., 2006). Magenta red-coloured kernels mostly contained cyanidin analogues, however in few cases, pelargonidin and peonidin analogues were also identified (Salinas-Moreno et al., 2005).

Acylated and non-acylated forms of anthocyanins have been reported in coloured corn. The high concentration of acylated anthocyanins has been found in Mexican pigmented maize. The acylated anthocyanins detected in the maize grain were cyanidin-3-O-glucoside, pelargonidin-3-O-glucoside, and peonidin 3-O- -glucoside (de Pascual-Teresa et al., 2002; Salinas-Moreno et al., 2005).

Soluble and insoluble-bound fraction of phenolic compounds were analyzed in nine Bolivian purple corn (*Z. mays* L.) varieties using HPLC-DAD and HPLC-ESI-MSn in Bolivian purple corn varieties. All the varieties revealed the presence of anthocyanin in varying amounts. The concentration of ferulic acid and p-coumaric acid (non-anthocyanin phenolics) varied from 132.9 to 298.4 mg/100 g and 607.5 mg/100 g dry weight (DW), respectively. Cyanidin-3-glucoside and its malonated derivative were detected as major anthocyanins. The total amount of phenolic compounds varied from 311.0 to 817.6 mg gallic acid equivalents (GAE)/ 100 g DW while the total monomeric anthocyanin content varied from 1.9 to 71.7 mg cyanidin-3-glucoside equivalents/100 g DW. Several dimalonylated monoglucosides of cyanidin, peonidin, and pelargonidin were present as minor constituents (Montilla et al., 2011). The phytochemical analysis of whole kernels of yellow, red, blue, and purple maize varieties showed a clear discrimination in their chemical profile. The concentration of acylated anthocyanins was higher as compared to glycosilated anthocyanins. The purple kernel maize variety was enriched in total phenolics, flavonoids and pro-anthocyanidin, whereas yellow and red maize has more of carotenoids and total tocols content (Suriano et al., 2021). Phytochemical analysis of blue maize (BM) (*Z. mays* L.) flour revealed that total phenolic content of blue maize flour (164 ± 14 mg gallic acid/g of dry matter) was higher than white maize (127 ± 7 mg gallic acid/g of dry matter). Amount of anthocyanin determined was 2.0 ± 0.5 mg cyanidin 3-glucoside/100 g (Camelo-Méndez et al., 2017).

Flours prepared from four local Italian coloured (orange to red and dark red) maize revealed that insoluble-bound phenols and flavonoids contribute about 70–80%. The amount of anthocyanins was lower whereas the red-brick phlobaphenes were reported in larger quantities. Zeaxanthin and β-carotene were present in higher quantities whereas lutein was reported in a lower quantity in local varieties as compared to commercial varieties. Out of all local varieties, maximum amount of zeaxanthin, β-carotene, and β-cryptoxanthin were detected in the nano variety (Capocchi et al., 2017).

HPLC-DAD-ESI/ MS analysis of kernels from 12 varieties of waxy corn revealed that the acylated derivatives contributed about 67.1–88.2% and 46.2–83.6% of the total anthocyanin contents at the milk and mature stages, respectively. Cyanidin-3-glucoside and its derivatives were identified as being most prominent. The amount of monomeric anthocyanin increased during maturation of all varieties of corn. Maximum amount of anthocyanins were reported in the kernels of a purplish black waxy corn (KKU-WX111031) variety (Harakotr et al., 2014b). The determination of anthocyanins composition during grain filling in purple corn kernels revealed that amount of anthocyanins increased with darkening of grain colour. Quantitatively, cyanidin-3-β-O glucoside was the major compound with concentration ranging from 57.0–409.1 mg kg^{-1} fresh weight in raw kernels and 1027.6 mg kg^{-1} in dry seeds followed by pelargonidin-3-β-O-glucoside and malvidin-3-β-O-glucoside. The physicochemical properties and amino acid content determined the building up of anthocyanins in corn. The amount of anthocyanin increased in a stepwise rather than linear fashion (Kim et al., 2020).

Differential spectrophotometry revealed that the total anthocyanin content in the white grain of the inbred line of sweetcorn CE401 and purple grain of the population F4(CE401×Chornosteblova) was 1174.5 and 2951.4 mg/kg, respectively. The concentration of anthocyanins in grains increased linearly with increase in colour intensity. In both varieties, the amount of glucosidic anthocyanins was double in quantity as compared to non-glycosidic in white and purple grains. Peanidin-3-glucoside and pelargonidin were dominant in the glucosidic and non-glucosidic forms, respectively (Psolova et al., 2019).

The acylated anthocyanins cyanidin-3-(6″-succinylglucoside) (Cy-Suc-Glu) and cyanidin-3-(6″-disuccinylglucoside) (Cy-diSuc-Glu) were found to be the most abundant in 15 Mexican blue-coloured maizes, accounting for 52.1% and 15.6% of total anthocyanin content, respectively.

Additional significant anthocyanins are cyanidin-3-glucoside, pelargonidin-3-glucoside, pelargonidin-3-(6"-malonylglucoside), and cyanidin-3-(6"-malonyglucoside). The raw blue maize also showed a similar anthocyanins profile (Mora-Rochín et al., 2016). The concentration of anthocyanins, a group of flavonoids, varied from 1.5% and 6.0% in the purple variety of maize. In the purple maize, the percentage of carbohydrate, proteins, and minerals and vitamins (Band C) was 80, 11, and 2, respectively (Guillén-Sánchez et al., 2014). Eight anthocyanin compounds were isolated from blue corn extract (BCE) and its yellow variant. The principal anthocyanins are pelargonidin, cyanidin-3-glucoside, and various glycoside derivatives. In comparison to saturated fatty acids, the proportion of unsaturated fatty acids (linoleic acid and oleic acid) (stearic and palmitic acid) was lower. Among sterols, β-sitosterol and campesterol were present in larger amounts than stigmastanol and stigmasterol (Smorowska et al., 2021).

A considerable variation in the genetic variation in 192 Turkish maize landraces for phytic acid and total phenolic compounds was observed. The phytic acid content was found between 0.82–4.87 mg g^{-1} and the total phenolic content was between 0.03% and 1.99% (Kahriman et al., 2021). Phytochemical evaluation of pericarp, aleurone layer, kernel, and corncob of 52 corn populations with different grades of pigmentation of Mexican purple corn revealed that total anthocyanin (TA) content of purple corn was higher than blue and red corn. TA ranged from 0.0044 to 0.0523 g, 0.2529 to 2.6452 g, and 0.2529 to 2.6452 g of TA per 100 g aleurone layer, pericarp, and kernel biomass, respectively (Mendoza-Mendoza1 et al., 2020). A study revealed that anthocyanin content remained unchanged with mild cooking of standard sugary sweetcorn (*Z. mays saccharate* Sturt.) whereas stronger treatment resulted in change in amount of anthocyanins without structural changes in molecules (Lago et al., 2020).

In the chromatographic studies, seven anthocyanins were reported in blue maize extracts. Cyanidin-3-(600-malonylglucoside) was the major one (Camelo-Méndez et al., 2016). In another similar study, cyanidin, peonidin, and pelargonidin anthocyanin glycosides were isolated from the rose corn. HPLC profile revealed that the most abundant anthocyanin was Cyanidin-3-(6-malonyl-glucoside), a malonic derivative of cyanidin (Ramírez et al., 2018).

Four native maize varieties (Arrocillo, Cónico, Peruano, and Purepecha) analyzed using HPLC revealed that the anthocyanin profile was similar in all the samples and total anthocyanin content varied from 54 to 115 mg/100 g of sample. Cyanidin-3-glucoside, pelargonidin- 3-glucoside, peonidin-3-glucoside, cyanidin-3-(6" malonylglucoside), and cyanidin-3-(3",6"-dimalonylglucoside) were the main anthocyanins identified (Moreno et al., 2005). High-performance liquid chromatography (HPLC) analysis revealed the existence of ferulic acid, quercetin, sinapic acid, gallic acid, and protocatechuic acid in white, yellow, and purple-coloured maize.

Total phenolic content (TPC) varied from 903 to 1843 lg GAE/g. Maximum amount of phenolics were reported in purple accessions. Ferulic acid was the major phenolic acid present in all accessions (Trehan et al., 2018). In 15 coloured corn varieties of maize, total phenolic, total flavonoid, total anthocyanin, and condensed tannin content varied from 349.39 to 485.71 mgGAE/100 g, 24.00 to 105.75 mgCE/100 g, 1.38 to 74.52 mgC3GE/100 g, and 33.70 to 158.55 mgCE/100 g, respectively. The percentage composition of bound fractions was higher in all cases. Ferulic acid and coumaric acid were the main compounds identified (Rodriguez-Salinas et al., 2020).

HPLC profiling of the hydrolyzed and non-hydrolyzed extracts of pigmented Mexican corn showed differences in minor composition only. Major anthocyanins detected by MALDI-ToF MS were cyanidin-3-glucoside ($1 \times 10{-}6$) and pelargonidin-3-glucoside ($1 \times 10{-}7$ M) in the hydrolyzed and non-hydrolyzed extracts, respectively (Castañeda-Ovando et al., 2010). Nixtamalization process led to reduction in the anthocyanin content in creole corn races before and after tortilla formation. Before processing, the amount of anthocyanins was higher in blue and red varieties than white and yellow varieties. The percentage loss of 64 and 83 in anthocyanin content was observed in blue corn when processed to tortillas and masa (Mendoza-Díaz et al., 2012), and another study reported the detection of more anthocyanins in the blue corn (cyanidin 3-O- -glucoside, equivalent to 620.9 mg/kg) than in other colours via the extrusion process. It also resulted in formation of anthocyanin derivatives and thus changed the anthocyanin content (Escalante-Aburto et al., 2013a, 2013b). Another study found that uncooked maize had a greater proportion of glycosylated anthocyanins and a smaller proportion of acylated anthocyanins (Mora-Rochín et al., 2016). The variation in the anthocyanins and total phenolic content was observed during high-energy milling (HEM) as compared to traditional nixtamalization (TN) in four creole corn flours obtained from maize varieties grown under rainy temporal conditions in the semidesert of Mexico. Total anthocyanins and phenolics were more in the HEM as compared to traditional nixtamalization (Amador-Rodríguez et al., 2019).

8.4 Pigment extraction from coloured maize

The coloured pigments present in plants are generally carotenoids, anthocyanins, flavonoids, and cholorophylls etc. The major pigments present in coloured maize are anthocyanins along with small quantities of flavonoids and carotenoids. Several extraction methods like solvent extraction (stirred vessel, soxhlet, and ultrasonication), pressurized liquid extraction (PLE), supercritical fluid extraction, and subcritical water extraction etc. have been developed over the past few years to extract these pigments from coloured maize varieties. The total content of each pigment (anthocyanin, phenols, flavonoids etc.) is usually carried out with the help of spectroscopic method, and total content is measured by taking some standard compound and reporting as an equivalent of that standard compound. However the identification and confirmation of these individual pigments can be done by high-performance liquid chromatography (HPLC) coupled with mass spectrometry (MS). In brief, reverse phase chromatography is performed using non-polar C-18 column as stationary phase, and for mobile phase polar solvents most commonly acetonitrile and dilute trifluroacetic acid (0.1%) are used. The detection is generally carried out with the help of UV/DAD detectors. The identification of individual components can also be carried out with the help of FT-ICR (Fourier transform ion cyclotron resonance) and LC/MS (liquid chromatography and mass spectrometry). Nankar et al. (2016) reported the presence of previously unidentified anthocyanin (cyanidin 3-disuccinylglucoside) in blue corn using these methods.

Most of these coloured pigments, especially anthocyanins, are water soluble, but the mixture of water and organic solvent is recommended to increase the extraction efficiency, as most of these molecules are bound to carbohydrates due to ionic interactions. The conventional extraction procedure is solvent extraction which includes extraction of coloured pigments with mixture of water and polar organic solvents, such as methanol, acetone, and ethanol, in certain proportions (40–100% of solvent) as extracting solvents. The solution is acidified with acids such as HCl (0.01 to 1 %) or citric acid (1%) to increase the ionization of extracting solution thereby breaking ionic interactions between the anthocyanins and carbohydrates (Zhao et al., 2009). As these pigments are very sensitive to high temperature, pH, and light exposure (Haggard et al., 2018), so care should be taken of these parameters during the extraction procedure. The extraction of these pigments can be increased by increasing the extraction time and temperature, but being unstable the anthocyanins are more prone to degradation under these conditions. In general the temperature of 50–60 °C is maintained during the solvent extraction with varying time intervals for different matrices. After extraction the extracts are concentrated under vacuum at low temperature (45 °C or below) and stored in the dark in a freezer (-4 to -20 °C) until further analysis. In one previous study the extraction of anthocyanins was carried out both from seed and cob (Yang & Zhai, 2010) by conventional extraction methods, and anthocyanin content was found to be higher for cob than seeds thereby depicting their utilization as antioxidants and natural food colourants in industry.

The extraction of carotenoids and anthocyanins along with their monomeric content was reported by Harakotr et al. (2015) in 49 different genotypes of waxy corn kernels. For extraction, 2.0 g of corn samples were added to 25 ml of acetone: water (70:30) mixture which was acidified with HCl (0.01%). The flasks were shaken for about 2 hours at room temperature. After shaking, the samples were filtered, evaporated, diluted again with

70% acetone, and further extracted with chloroform (15 ml). The acetone layer was recovered and stored for further analysis of monomeric anthocyanin content by pH differential method (Jing et al., 2007).

Yang et al. (2009) evaluated the anthocyanin content in purple corn with the help of tristimulus colourimetry. The highest yield of anthocyanins was obtained at 70 °C with 1:25 solid:liquid ratio and extraction time of 73 min. The presence and confirmation of three anthocyanins (pelargonidin-3-glucoside, cyanidin-3-glucoside, and peonidin-3-glucoside) was carried out with the help of HPLC-MS. One of the studies (de la Para et al., 2007) also reported the effect of nixtamalization on anthocyanin, total phenolic, and carotenoid content in five different varieties of corn (white, yellow, red, blue, and high carotenoid). The highest content of anthocyanins was reported in blue corn as compared to red corn. Overall it was observed that lime cooking significantly reduced the pigment concentrations (phenols, anthocyanin, and carotenoids) (Hu & Xu, 2011) extracted the carotenoids, anthocyanins, and phenolics from three types of waxy corn (black, white, and yellow). The extraction was carried out by solvent extraction method using 80% acetone. The extraction procedure was repeated three times to get maximum yield. For the extraction of bound pigments, soluble free residues were flushed with nitrogen and treated with 4 N NaOH for one hour with continuous shaking. After shaking the mixture was acidified with hydrochloric acid up to pH 2 and extracted with ethyl acetate. The supernatant was used for analysis of carotenoid and phenolic content. The extract was further acidified and heated to break anthocyanin-glycoside bond for anthocyanin determination. The highest carotenoid content was found in yellow corn (22.5–39.61 µg/g of dry weight) whereas anthocyanins were major constituents (0.09–276.11 mg of cyanidin 3-glucoside equivalents/ 100 of dry weight) in black corn while both the pigments were present in the lowest quantity in white corn. Also the carotenoid and anthocyanin content decreases with maturity in yellow and white corn while it increases in case of black corn. In the recent reports the identification of individual anthocyanins in the mixture have become possible with the help of HPLC coupled with MS. Harakotr et al. (2014) reported the anthocyanin content in coloured varieties of waxy corn at matured stage. The identification and quantification of individual anthocyanins was carried out with the help of HPLC-DAD-ESI/ MS analysis. The highest quantity of anthocyanins detected were cyanidin-3-glucoside and its derivatives. Furthermore, acylated anthocyanins present in milk and mature stage were 67.1–88.2% and 46.2–83.6% of the total contents, respectively. The positive correlation was also observed in anthocyanin content and antioxidant activity.

Sprouts of corn contain health benefits, and their consumption is promoted these days due to their increased nutritional value, which is basically due to the increased phytochemical content of the seeds on sprouting or in their seedling stage. A study was conducted to compare the phytochemical content of yellow, white, and violet waxy and field varieties of corn. The extracts of seeds, sprouts, and seedlings of corn were prepared via the solvent extraction method using methanol as solvent. The extracts were further acidified with HCl (1%) and shaken for about two hours. The filtered extracts were concentrated and used for determination of pigments (phenols, anthocyanins, carotenoids etc.). It was found that highest content of these pigments was observed in seedlings as compared to seeds and sprouts (Chalorcharoenying et al., 2017).

The utilization of waste is always encouraged, and in maize crop, cob, silk, and husk are waste materials and can be utilized to increase the production efficiency of the crop. In one such study (Simla et al., 2016), anthocyanin content was determined in cob, kernel, silk, and husk of three varieties of purple waxy corn. For extraction, the sample (0.5 g) was treated with 80% methanol containing 1% citric acid. All the contents were mixed well and stored for 24 h at 4 °C in the dark. After extraction, the mixture was filtered and centrifuged. The supernatant obtained was stored at -20 °C for further analysis. Total anthocyanin content was determined using pH differential method. It was found that seed stage possessed the higher anthocyanin content than edible stage. Although seeds possessed the highest anthocyanin content, still significant amount of anthocyanins are present in corn silk and corn cob which can be utilized as an anthocyanin source in the food industry. Lao and Giusti (2016) compared the already reported methods of identification (HPLC-PDA-MS) and quantification (pH differential, total anthocyanin contents, and HPLC methods) of anthocyanins. The experiment was carried out on 14 varieties of purple corn cob taking intact and hydrolyzed samples. The quantification of anthocyanins was found to be strongly dependent on the method employed during extraction as well as quantification of pigments (anthocyanins and other pigments). Further, Li et al. (2017) compared the anthocyanin distribution in blue and purple maize co-products via three fractionation methods. Total monomeric anthocyanin contents were found to be ten times more in purple maize as compared to blue maize with maximum concentration in pericarp of purple variety and aleurone layer of blue maize.

In conventional extraction techniques of plant pigments, organic solvents were used. With the increasing concern to be environment friendly and more economic, organic solvents are getting replaced with environmentally friendly solvents. In one such previous report (Monroy et al., 2016) new solvents such as supercritical carbon dioxide, ethanol, and water were used for sequential extraction of plant pigments from purple corn (cob and pericarp). Higher yields of anthocyanins were observed in ethanol whereas a higher phenolic and flavonoid content was obtained via water extraction. Adding to that, sequential extraction techniques proved to be better at increasing the extraction yield, leading to extracts with different chemical compositions.

With the advancement of extraction techniques, many studies have been conducted to compare the yield and extract composition of coloured cob by using conventional and the latest extraction procedures. In one such study, the extracts of Peruvian purple corn pericarp (*Zea mays* L.) were prepared via four methods with two conventional low pressure methods (stirred vessel and soxhlet) and two new high pressure methods (supercritical fluid extraction using CO_2 as solvent, along with EtOH-H_2O (70:30, v/v) as a co-solvent) and pressurized liquid extraction (PLE) with EtOH-H_2O (70:30, v/v) as a solvent. It was found that extraction yields of phenolic compounds and anthocyanins were better with high pressure methods at a higher temperature (60 °C) as compared to low pressure methods (Monroy et al., 2020).

Coloured corn cobs are a rich source of anthocyanin pigments, adding to their health benefits. Anthocyanins are highly unstable molecules and are prone to degradation post-extraction, and most of them are difficult to extract and remain bound to the corn matrix. Several methods have been developed over the years to extract anthocyanins from different corn varieties which have been discussed above. The identification and quantification of anthocyanins have also become possible with the help of HPLC/MS but still some bound and unstable anthocyanins were not identified with this technique. Hong et al. (2020b) recently developed and optimized ultra-high-performance liquid chromatography–diode array detector–mass spectrometry method for efficient extraction of anthocyanins, which had previously remained unexplored due to degradation at high temperature and interconversions under acidic conditions. In total, 18 anthocyanins were quantified from five different varieties of coloured corn. Purple sweetcorn pericarp and blue aleurone maize varieties mainly constitute cyanidin-based glucosides as the main anthocyanin with 75.5% and 91.6% of total anthocyanin content, respectively, whereas pelargonidin-based glucosides composed the main anthocyanins of reddish-purple-pericarp sweetcorn (61.1%) and cherry-aleurone maize (74.6%). Importantly, in the previous reports, the anthocyanins were acetylated and succinylated during the extraction processes and the actual quantification of anthocyanin was not achieved. So using this technique, an actual quantification of anthocyanins was reported.

8.5 Coloured maize-based specialty products

Anthocyanins are natural pigments safe for human consumption that can be used as food additives. Interest in anthocyanins as a promising alternative food colouring has increased because of their colouring properties and potential health benefits (Giusti & Wrolstad, 2003). Colour is the most essential sensory quality of foods from the standpoint of the market. In recent years, the food industry has used a growing number of natural plant-based colourants with both colouring and antioxidant characteristics. Natural colourants, in contrast to chemical additions, have sparked substantial interest due to their perceived safety and potential health benefits. However, they are difficult to incorporate into food systems because of their limited stability when exposed to elements such as light, oxygen, temperature, and pH (Ba.kowska-Barczak, 2005). Flavonoids/anthocyanins, particularly tandem to carotenoids and betalains, are the principal family of photosynthetic pigments which have found use in nutritional and functional systems due to their antioxidant qualities and possible health benefits (Tanaka et al., 2008). Different coloured maize variants are very well tolerated in the manufacturing of popcorn, tortilla chips, snacks, extruded snacks, expanded products, semolina, pasta, bread, and confections such as cakes and muffins, not only for their appealing colour but rather because of their significant nutritional and medicinal characteristics due to their high anthocyanin and fibre content.

8.5.1 Tortillas and tortilla chips

Tortillas are the most important source of nutrients and calories for many people. Tortillas are traditionally prepared from nixtamalized maize grains, which involves heating whole kernels in a hot, alkaline solution (usually 1.5–2% lime, calcium oxide) and subsequently grinding the drained and rinsed corn (in this form named nixtamal) into a fresh dough called masa. This is then sheeted and rolled, followed by grilling on hot pan on both sides to produce tortillas, which can be further fried to prepare tortilla chips. Traditionally, tortillas and tortilla chips were prepared from normal maize. Nowadays, maize tortillas are a very common food in many countries where they provide about 50% of human energy intake (Campas-Baypoli et al., 1999). Nowadays, anthocyanins are of great interest from the nutraceutical or functional nutrition viewpoint owing to their perceived potential health benefits; they are considered natural antioxidants in view of their ability to trap free radicals, which produce molecular and cell damage.

The nixtamalization procedure is the initial step to making maize-based foodstuffs like tortillas; nevertheless, white grains are favoured in both indigenous and industrial methods. Caribbean maize breeds, primarily coloured variations, are gaining popularity because of their high levels of anthocyanins and carotenoids (Mendoza-Díaz et al., 2012). The term nixtamalization refers to the alkaline cooking process of converting maize into foodstuffs such as tortillas and snacks (maize chips, tortillas chips, and tacos). In addition, lime-cooking extrusion is indeed an alternate method for producing dough suited to enchiladas with comparable properties to those created using standard methods (Milan-Carrillo et al., 2004).

Masa is used in many Central American recipes and is generated through the nixtamalization procedure that involves an alkaline thermal decomposition of maize to produce dough or flour (nixtamalized maize flour) after a drying procedure that decreases moisture content to 8–10%. The thermal, physicochemical, sensorial, and nutritive value of nixtamalized products are all significantly influenced by this procedure (Mariscal Moreno et al., 2015). Tortillas, tortilla chips, and other Hispanic foods made with masa and nixtamalized flour have high protein, calcium, and dietary fibre. In Mexican red-coloured maize, the conventional nixtamalization and lime-cooking extrusion technique reduced carotenoid content and radical scavenging capacity more than in yellow ones. Only the proportion of lutein in conventional and extruded tortillas was less influenced (Corrales-Bañuelos et al., 2016). For artisanal handcrafted and commercial/industrial tortilla made with blue and white maize, unique nutrient values and bioactive chemical concentrations were recorded. Industrially manufactured white tortillas exhibited lower antioxidant potential, fibre, calcium, and total polyphenol concentration than blue handcrafted tortillas; in addition, blue homemade tortillas had 4.5 times the ferulic acid concentration of commercially produced white tortillas. As a result, when compared to standard manufactured tortillas, artisanal fresh tortillas exhibited higher nutritional-nutraceutical qualities (Colín-Chávez et al., 2020).

Customarily, pigmented maize has been used for tlacoyos, quesadillas, and fajitas, however as the demand for pigmented maize products grows, production companies can make nixtamalized maize flours from coloured maize, resulting in a coloured nixtamalized masa with functional ingredients and nutraceutical properties, from which new products with improved visual qualities can be made (Cortes-Gomez et al., 2005). Blue maize has been widely studied for the preparation of tortillas and tortilla chips (Cortes-Gomez et al., 2005; Hernandez-Uribe et al., 2007; Del Pozo-Insfran et al., 2006; De La Parra et al., 2007; Aguayo-Rojas et al., 2012; Mendoza-Díaz et al., 2012; López-Martínez et al., 2012; Bello-Perez et al., 2015; Hernández-Martínez et al., 2016; Mendez-Lagunas et al., 2020). In an alternative to the traditional nixtamalization process, Cortes-Gomez et al. (2005) prepared the tortillas from fractionated blue colour maize in which pericarp and tip separated from the endosperm were nixtamalized first separately and then with added endosperm. It was reported that tortillas made from fractioned nixtamalized flour prepared with calcium hydroxide and a nixtamalization time of 45 min showed functional characteristics similar to the traditional blue tortilla with several advantages over traditional nixtamalization process such as lower processing time and water consumption, not generating polluting effluents, and the use of the whole grain. Bello-Perez et al. (2015) in their study reported that tortillas crafted from blue-coloured maize by nixtamalization with calcium salts (NCS) rather than conventional nixtamalization generated a fresh blue-coloured tortilla with relatively high indigestible starch and less RS resistant starch than one prepared with the conventional nixtamalization method. NCS tortillas also had a pretty low expected glycemic load and relatively high antioxidant properties indicating a potential health benefit. The ingestion of a blue-coloured tortilla made with the NCS technique can have some health beneficial properties. Hernandez-Uribe et al. (2007) in their study reported that tortillas prepared from blue colour maize showed elevated fat and protein contents than white tortilla, and they also exhibit softer texture due to blue tortillas' lower total starch content which decreased interactions among starch chains. The tortillas prepared from coloured maize also have moderate resistant starch content and also lower glycaemic index than in white tortillas, showing a reduced starch retrogradation tendency.

Red colour maize is also used for the preparation of tortillas owing to their higher content of anthocyanins and carotenoids in comparison to blue maize (Corrales-Banuelos et al., 2016; De La Parra et al., 2007; Aguayo-Rojas et al., 2012; Mendoza-Díaz et al., 2012; López-Martínez et al., 2012). The tortillas can be made successfully from red colour pigmented maize using conventional nixtamalization or extrusion lime-cooking process. Nowadays, consumers are interested in these products owing to the higher concentration of anthocyanins, carotenoids, and other compounds that could have an added or synergistic effect on biological activity in the human body, and tortillas prepared from red maize can serve this purpose well (Mendoza-Díaz et al., 2012).

Cereal products made from pigmented grains are attractive to consumers. Black and purple pigmented maize can also be used successfully for the preparation of pigmented maize tortillas, and compared to white or yellow tortillas, they have a smooth texture and a sweet taste (López-Martínez et al., 2012; Hernández-Martínez et al., 2016). Tortillas made from black maize retained 34.1% of anthocyanin content while those made from purple retained 24.9, also the tortilla chips made from black maize retain higher anthocyanins (28.6%) in comparison to those made from purple maize (López-Martínez et al., 2012).

8.5.2 Bread, muffins, and cake

Bread is a popular wheat-based food product around the globe, and because bread is generally high in carbohydrates, it is an ideal food for fortifying with bioactive components, necessary micro- and macronutrients, and also a vehicle for their delivery (El-Megeid et al., 2009). Anthocyanins are indeed the principal antioxidants found in coloured maize grains, and substituting wholegrain anthocyanin-rich maize flour for white wheat flour can considerably increase bread's functional characteristics and also nutritional worth (Žilić et al., 2012). The functional characteristics of the finished product were enhanced by utilizing 30% of wheat flour with wholegrain blue and dark-red colour popped maize flour, which contains polyphenolic compounds with bioactive compounds, enhancing the content of total phenolics, anthocyanins, and phenolic acids in maize-mix bread (Simic et al., 2018). Anthocyanins from maize flour may make a significant contribution to the beneficial properties of bread, including its appealing colour, overcoming thermal deterioration during baking. Anthocyanin-rich maize flour had no influence on the enhancement in bread antioxidant properties attributed to the synergistic and antagonistic interactions that emerge from the coexistence of multiple antioxidant components in food. The Maillard reaction was found to have a significant impact on the increase in antioxidative properties in wheat and maize-mix breads.

Muffins are popularly consumed at breakfast as well as being an evening snack food, and coloured maize can also be utilized to develop functional gluten-free muffins, as zein protein, which is not able to form viscoelastic networks upon hydration as wheat gluten does, can be modified functionally at higher temperatures; zein-based doughs had properties similar to wheat dough (Bugusu et al., 2001; Erickson et al., 2012). Several research findings on the rheological characteristics of doughs and the variables that influence bread dough, which are crucial for the shortlisting of technological conditions, dramatically upped the use of maize flour in the prototype of gluten-free product lines. Purple corn flours, as well as white and yellow germplasm, were shown to have the possibility to be used in the creation of gluten-free eggless muffins even without use of hydrocolloids. Flour from different accessions of purple maize can be used to produce gluten-free eggless muffins without any hydrocolloids exhibiting good height and low specific volume and higher levels of phenolic compounds and antioxidant activities (Trehan et al., 2018). In contrast to muffins made from yellow maize accessions, sensory examination of purple maize muffins demonstrated that they were not as appealing. These formulations look very promising, particularly because they are eggless, and their use might help to mitigate the rise in egg-food allergies.

A blend of purple maize and rice flour was also tested as a wheat flour alternative for bakery products. The purple maize cakes with maize silk had the maximum anthocyanin content and high antioxidant activity, as well as the best textural score, with a ratio of 71.1:28.9. Furthermore, its acidification with 3% fumaric acid retained the colour and prevented phenolic degradation, making this a high-nutritional food (Kokkaew & Pitirit, 2016). A further fascinating application of coloured maize has been in the making of polvorones, a traditional Mexican bread treat that is both energetic and nutritious. This dish was more appealing in terms of colour, flavour, and acceptance than meals made with wheat flour. Furthermore, the availability of anthocyanins emphasized healthful positive characteristics, since the only deterioration observed during uncooked flour processing is indeed the de-acylation of certain acylated anthocyanins (Vázquez-Carrillo et al., 2018).

8.5.3 Cookies and crackers

Among bakery products, cookies could easily be fortified with polyphenols to promote health and reduce disease risk, and refined wheat flour can be partially replaced anthocyanin-rich ingredients for such products (Mildner-Szkudlarz et al., 2009; Pasqualone et al., 2014). Considering that the food industry's primary goal is indeed the replacement of synthetic colourants, this can be accomplished with acylated anthocyanins. Zilic et al. (2016) used red and blue maize flour to make functional cookies, demonstrating that adding citric acid to coloured maize dough boosted the phenolic compounds in coloured maize cookies by stabilizing anthocyanins. Citric acid lowered the pH of total anthocyanins in maize cookie dough and final cookies, causing acylation of sugar residues or flavylium cation. Citric acid is a useful organic acid ingredient of bakery food as it stabilizes the anthocyanins in maize cookies. Cookies were also prepared from blue, dark blue, and dark red popcorn maize varieties (Kocadağı et al., 2016) and the availability of multiple polyphenolic compounds in these coloured maize cookies had a very significant effect on dicarbonyl creation and eradication, as organic polyphenols compounds could reduce glyoxal, methylglyoxal, and diacetyl content all through thermal treatment of baking, particularly those items containing ammonium bicarbonate, which creates an alkaline environment. Due to the ability of their phenolic components to trap C2, C3, or C4-dicarbonyl molecules, coloured maize flour can be a source of natural dietary antiglycation agents. Sales et al. (2018) showed that in developing nations like the Philippines, there is increased interest in adopting indigenous crops as alternative food sources to combat food and nutrition shortages, and in this manner indigenous pigmented maize (*Zea mays* L.) called *camotes* can be successfully used for the development of crackers by replacing all-purpose flour. They discovered that crackers made with an 80:20 camotes:all-purpose flour mix look good compared to those made with only all-purpose flour and that incorporating camotes flour enhances the nutritional content to all-purpose flour crackers in terms of protein, dietary fibre, lysine, tryptophan, zinc, antioxidant properties, phenolic compounds, and flavones. Such crackers can be used as a nutritionally dense alternative source of food in impoverished countries to combat food and nutrition scarcity.

8.5.4 Pasta

Pasta, a wheat-based foodstuff, is a widely consumed meal that can be purchased ready to eat or prepared quickly at home. As there is a growing demand for gluten-free products, it is in both celiac patients' and the general public's best interests to buy gluten-free items, especially those that have organoleptic and nutritional features that are comparable to traditional gluten-containing foods (Camelo-Méndez et al., 2018b). The creation of gluten-free pasta with bioactive components is an important strategic area for the food manufacturing industry, as there has been a considerable growth in the production of food aimed at celiac disease (CD) sufferers in recent years (Matthias et al., 2011). Clients with CD appreciate gluten-free pasta, although there are only a few gluten-free items that include bioactive constituents like phytonutrients and soluble dietary fibre (especially resistant starch), which significantly contribute to the possible health beneficial properties of gluten-free eating (Camelo-Méndez et al., 2016).

The antioxidant capacity of blue maize and its putative anti-cancer characteristics (related to its anthocyanin concentration) can be employed to boost the nutraceutical value of gluten-free pasta (Aguayo-Rojas et al., 2012; Urias-Lugo et al., 2015). The addition of 50–75% whole blue maize flour to regular maize flour resulted in an increase in total anthocyanins, phenolic content, and antioxidant capacity, as well as a decrease in starch hydrolysis due to increased SDS and RS (Camelo-Méndez et al., 2018a; Camelo-Méndez et al., 2018b). As a result, blue maize flour can be considered a functional ingredient which could be used to develop functional food products with essential antioxidative potential, substantial levels of ingestible carbohydrates, a reduced pGI, as well as therapeutic benefits for the overall population, such as lowering the risk of nutrition metabolic disorders. Extrusion and subsequent boiling of blue maize-containing pasta resulted in a minor drop in phenolic content and antioxidant properties, revealing that the network formed during manufacturing shielded the phytochemical compounds from heat degradation. In this regard, gluten-free pasta made with blue maize flour may provide a nutraceutical benefit (Camelo-Méndez et al., 2018b). The composite pasta had an adequate gruel loss (9–11%); blue colour maize-based pasta had reduced firmness, gumminess, and dark colour, yet stronger adhesiveness over white maize-based pasta. Following extrusion and heating, the addition of 75% blue maize flour resulted in the maximum overall phenolic content retention of about 80 and 70% respectively.

8.5.5 Yoghurt and milk

Anthocyanins from high-anthocyanin-content colourful maize grains (Arrocillo, Peruano, Purepecha, and Cónico) have a lot of promise for usage as natural colourants in yoghurt (Moreno et al., 2005). The colour of yoghurts was infused with 1 mg of anthocyanin methanol extract from four different cultivars, with the Peruano and Arrocillo extracts showing a much more intense dark red tone than that of the Cónico and Purepecha extracts, and then after 5 to 10 days of refrigerated conditions, the colour of all yoghurt samples changed to a slightly pale yellow tone. The inclusion of 0.3% purple maize pigment comprising pelargonidin, peonidin, cyanidin, malvidin, petunidin, and delphinidin to milk samples resulted in a seven-day delay in the oxidation of unsaturated free fatty acids. Pelargonidin and petunidin severely impacted C14: 1n-5, C17: 1n-7, C18: 2n-6, C20: 2n-6, C20: 3n-3, and C20: 4n-6 degradation, whereas cyanidin and total anthocyanins directly impacted C14: 1n-5, C16: 1n-7, C17: 1n-7, and C20: 4n-6 degradation (Tian et al., 2020).

8.5.6 Beverages

Flores-Calderón et al. (2017) proposed a modified "Sendechó" beverage, which added hops and brewer's yeast to guajillo chilli and blue corn malt (the conventional flavours), resulting in a beer with Sendechó-like character traits as well as corn beer-like character traits due to an ale fermentation process. Eight blue maize malt beers were evaluated for titratable acidity, bitterness units, total reducing sugars, alcohol, cis- and trans-iso-acid concentration, total anthocyanins and phenolics, and radical scavenging activity using varying levels of hop varieties, guajillo, and also the inclusion of caramel malt to blue maize malt. All of these characteristics were measured at three separate stages of the process: boiling wort, fermentation (green beer), and maturation (mature beer), with the exception of pH, total acidity, and alcohol level, which were all influenced particularly in mature beer. In fact, there was a significant drop in polyphenolic concentration and antioxidant activity. Low-alcohol beers were produced with pH, total acidity, bitterness units, iso-acid level, and total reducing sugars comparable to barley beers. The maximum anthocyanin concentration and antioxidant activities were found in beers made with 85% maize malt and 15% caramel malt.

Chicha morada, a beverage comprising an extract of the purple maize pigments, fruits such as pineapple, quince, apple, and peach, which conferred tastes, and a starch source, is made with a Peruvian Andean maize variety, typical of a limited Peruvian zone. Mazamorra morada, a dessert, is made by using a cooking technique and adding dried or fresh fruits. This cultivar had significant nutritional and health benefits, and its flour may be utilized in a range of dishes (pasta, bread, and cakes) to provide colour and taste to processed goods (Salvador-Reyes et al., 2020).

8.5.7 Extruded and expanded snacks

Extrusion is a common method for producing a variety of maize goods, such as expanded snacks. Conventional nixtamalization preparation involves adding $Ca(OH)_2$ to maize and water at 100 °C to generate a solution that is boiled for 20–40 minutes then steeped for at least 16 hours. Lime-cooking extrusion (LCE) is a quicker and more accurate option. Extrusion is a method of making nixtamalized items and is not as common as other methods. Maize exhibits good expansion characteristics for expanded snacks, and coloured maize may also therefore have good physical and technological characteristics for second- and third-generation extruded snack production when used with added calcium hydroxide (Zazueta-Morales et al., 2002). Coloured maize such as blue and purple were widely used for the production of extruded third-generation snacks from whole blue maize and corn starch mixture (Navarro-Cortez et al., 2014). Because anthocyanins seem to be unstable compounds at elevated pH levels and temperatures, extrusion reduces anthocyanin losses in final products by up to 50%–60%. Changing specific processing parameters can result in the creation of a bluish quinoidal base or a colourless carbinol pseudo-base that also affects the colour of goods manufactured with pigmented maize and changes their visual effect. Calcium hydroxide is employed during extrusion to maintain the maximum possible total anthocyanin concentration, strong blue and purple colouring, and maximum expansion of the snacks (Escalante-Aburto et al., 2014). Nixtamalized blue maize can also be used for the development of expanded extruded snacks at 16% of feed moisture, barrel temperature of 130 °C, and particle size index of 83.97, and showed higher total anthocyanin (211.1 mg/kg) and cyanidin 3-glucoside (11.3%) in comparison to raw maize (Escalante-Aburto et al. 2013b). Escalante-Aburto et al. (2013a) also used the blue maize for the development of second generation snacks using extrusion lime cooking instead of the traditional nixtamalization process. Owing to the anthocyanic chemicals found within these grains, extruded lime cooking is able to produce products having similar sensory attributes to conventional nixtamalization products and with extra functional benefits.

Snacks are food products that are ready to eat, lightweight, small, and easily accessible, and most importantly, they are food products that may temporarily fulfil the appetite sensation (Hurtado et al., 2001). Snack foods are consumed worldwide by people from different social classes, ages, and genders, and the snack industry is flourishing, offering a broad variety of snacks, primarily potato and maize derivatives (Camacho-Hernández et al., 2014). The third-generation (3G) snacks, commonly referred as intermediate snack foods or pellets, are made by extrusion, produced at low pressure to prevent expansion, and then dried to make a vitreous pellet (Hollingsworth, 2001). They still necessitate yet another expansion step that can be accomplished using cooking oil, hot air, or electromagnetic treatment, after which the water in the pellet begins to boils and vapour bubble develop, expanding the pellet and giving the snack a porous structure (Boischot et al., 2003). Coloured maize has similar expansion and nutritional properties to regular corn, with the added benefit of anthocyanin and polyphenol chemicals (Pedreschi & Cisneros-Zevallos, 2007; Yan & Zhai, 2010).

Camacho-Hernández et al. (2014) prepared third generation extruded snacks from blue-coloured maize that could be further enhanced by microwave processing to produce 3G snacks with elevated physicochemical parameters and therapeutic potential inferred from nutritional and functional characteristics (dietary fibre and anthocyanins) of the whole blue-coloured maize. Single screw extrusion of blue maize at 120 °C to 126 °C for barrel temperature and 23.8% to 25.2% feed moisture yields a product with good expansion, optimum hardness, and high total anthocyanin content of 71.09 mg/kg.

Roasting is a well-known dry thermal technique for crispy snack production. However, the extreme heat can destroy chemical components like phenolic and anthocyanins, reducing their antioxidative action considerably. Intensification of vapourization by decompression to the vacuum (IVDV) is a new texturizing method suggested as a pre-treatment for roasting purple maize grains that includes subjecting wet grains to a saturated steam pressure accompanied by a decompression to the vacuum preceded by roasting in relatively mild conditions, which also works to help to texturize purple maize grains whilst also conserving as much of their own antioxidant content as possible (Mrad et al., 2014).

8.6 Effect of processing on bioactive compounds

A study on the effect of milling and grinding on the amount of pro-anthocyanidins (PA) in the co-products of blue, red, and purple maize revealed that the PA content was higher in purple-coloured maize co-products. Total PA recovered in wet milling was higher (204.3 ± 6.2 g catechineq/kg maize db) than dry milling co-products (91.0 ± 8.5 g catechineq/kg maize db). In a wet milling process, the maximum yield of PA was reported in steepwater, while the maximum PA content was present in pericarp followed by small and large grits (Chen et al., 2017).

A considerable difference was observed in total (free + bound) phenolics, flavonoids, anthocyanins, β-carotene, and lutein content in ten coloured corn varieties. The amount of β-carotene and lutein concentrations ranged from 0 to 2.42 mg/kg d.m. and 0 to 13.89 mg/kg d.m., respectively. The concentrations of total anthocyanin ranged from 2.50 to 696.07 mg CGE/kg d.m (cyanidin 3-glucoside equivalent). The primary anthocyanin found was cyanidin 3-glucoside (Cy-3-Glu). (Zilić et al., 2012).

Hydroxycinnamic acid followed by hydroxybenzoic and chlorogenic acids are the main phenolics reported in the coloured maize. Ferulic acid is also present in dimeric, trimeric, and tetrameric forms in white and yellow corn (Bento-Silva et al., 2018). The amount of free ferulic acid in Mexican,

Peruvian, and khao niew varieties of pigmented maize ranged from 1.94 mg to.0 mg/100 g, where Peruvian contained the most (Urias-Lugo et al., 2015; López-Martínez et al., 2009; Ramos-Escudero et al., 2012). Apart from ferulic acid, other phenolic compounds present in these coloured varieties were p-coumaric, caffeic, vanillic, chlorogenic and hydroxybenzoic acids. Syringic acid and ferulic acid were the major compounds presents, while vanillic and hydroxybenzoic acid were the minor phenolic acids identified in corn cob (Kapcum et al., 2016). The silk of coloured maize is major source of chlorogenic acids (Muangrat et al., 2018).

Flavonoids were identified in different parts of pigmented corn. High flavonoid content 202–224 mg/100 g was reported in Peruvian purple corn kernels whereas flavonoid content of Serbian pigmented corn phenotypes was 19.90–33.75 mg/100 g. Kaempferol and morin mainly contributed to flavonoid concentration (Ramos-Escudero et al., 2012). The concentration of flavonoids in coloured maize silk reported in Serbian purple corn and Mexican pigmented corn was 3644.9 mg and 2602.4 mg/100 g, respectively (Mendoza-lópez et al., 2018).

About 20 cyanidin, peonidin, and pelargonidin glucosides along with their isomeric forms were detected and estimated in purple and reddish-purple sweetcorn accession developed from purple Peruvian maize. The prominent anthocyanin identified in purple and reddish-purple accession was cyanidin and pelargonidin glucoside derivatives, respectively. Total anthocyanin content showed a constant increase during the optimum sweetcorn-eating period until the maturation of corn seeds (Hong et al., 2020).

8.6.1 Tortilla chips

The red/blue- and blue-coloured corn varieties are mostly enriched in acylated anthocyanins, mostly cyanidin-3-(6″-malonylglucoside) when compared to red and purplish genotypes. Amides derivatives of phenolics were easily extracted from all varieties. Red and blue varieties retained maximum amount of anthocyanins during processing (Collison et al., 2015). In a similar study, variability in colour and phenolics in blue tortillas made from 18 landraces belonging to three blue/purple grained Mexican maize races – Chalqueño (Chal), Elotes Cónicos (Ec) and Bolita (Bol) – revealed that the total anthocyanin content ranged from 1.11 to 3.67 mg equivalent of cyanidin 3-glucoside/kg dry weight. Total soluble phenolics varied between 9.42 and 5.16 mg equivalent of gallic acid/kg DW. These parameters help in appropriate grouping of blue tortillas. The minimum loss of bound phenolic compounds was reported in lime cooking of four different types of whole pigmented Mexican maize – white (WM), yellow (YM), red (RM), blue maize (BM) – extrusion into tortillas. The blue maize (27.52 mg cyaniding 3-glucoside/100 g DW) retained the highest amount of antho-cyanins followed by white-, yellow-, and red-coloured maize (Aguayo-Rojas et al., 2012). Carotenoids in tortillas made from Mexican coloured maize germplasm grains varied from 3.66 to 5.56 mg LE/kg DW for yellow maize and 1.49 to 3.49 mg LE/kg DW for red maize. Around 85% of all carotenoids were lutein and zeaxanthin. In comparison to uncooked grains, the conventional nixtamalization and lime-cooking extrusion technique considerably reduced total carotenoids. The carotenoid-retaining capacity of yellow maize was higher (Corrales-Banuelos et al., 2016).

The study on the effect of cooking on the antioxidant content and antioxidant activity of fresh purple waxy corn revealed that boiling causes significant loss in the antioxidant components in the cooking water as compared to steam cooking. Degradation of components was observed more than leaching into cooking water. In addition, if the kernels were removed before cooking, more loss was observed in antioxidant components (Harakotr et al., 2014).

Nitric oxide and peroxynitrite scavenging activity of coloured maize was significantly affected by the nixtamalization, whereas processing masa into tortilla and tortilla chips showed no effect. Among different varieties tested, scavenging potential of yellow corn and its products was at maximum (López-Martínez et al., 2012).

Beneficial lactobacilli like as Lactobacillus helveticus, Bifidobacterium longum, and Helicobacter pylori were not suppressed by free and bound polyphenol components of Peruvian purple maize at 10 to 50 mg/mL sample concentrations. Free phenolic acids contained anthocyanins as well as hydroxycinnamic acids like p-coumaric acid derivatives, accompanied by caffeic and ferulic acid derivative products, whereas the bound phenolic component contained only hydroxycinnamic acids like ferulic acid, p-coumaric acid, and a ferulic acid derivative with ferulic acid (Ranilla et al., 2017).

The addition of citric acid resulted in an appreciable increase of 60 and 70%, respectively, in the total flavonoid and anthocyanin content in cookies made from dark-red popping corn, blue popping corn, and blue-standard corn flour (Zilic et al., 2016).

On TA98 and TA100, the anthocyanin content was associated with antiradical activity and prevention of 2-aminoanthracene-induced mutagenicity. In white maize, nixtamalization lowered carotenoids by 53 to 56%, but not antioxidant properties or 2-Aa-induced mutagenicity. All the extracts demonstrated antimutagenic action against 2-aminoanthracene-induced mutagenicity (23 to 90%) throughout the nixtamalization processes, indicating a stronger ability to block base changes mutations than frame-shift mutations in the genomic of the tested organism. The findings indicate that, despite pigment degradation, colourful maize has antioxidant and antimutagenic properties after the nixtamalization process (Mendoza-Díaz et al., 2012).

8.7 Health benefits

High variability was observed between the accessions for bioactive compounds and antioxidant activities, and significant interactions were observed between locations and accessions of pigmented grains. Some accessions with remarkable composition and antioxidant activity were RJ02, RJ03, RJ04, RJ06, AZ10, AZ13, AZ16, and AZ18 (Martínez-Martínez et al., 2019). A study revealed that the 2,2-diphenyl-1-picrylhydrazyl (DPPH) radical scavenging activity of kernels from 12 genotypes of waxy corn at two maturation stages (milk and mature) increased, whereas ferric-reducing antioxidant power (FRAP) and Trolox equivalent antioxidant capacity (TEAC) decreased with increases with ripening. Maximum antioxidant potential was shown by purplish black waxy corn genotype (KKU-WX111031) (Harakotr et al., 2014). The four creole corn flours obtained via high-energy milling (HEM) showed high antioxidant activity (ABTS and DPPH assay) as well as improved nutraceutical properties as compared to traditional nixtamalization (TN) flour (Amador-Rodríguez et al., 2019).

Dark red corn grains showed good antioxidant capacity as revealed by electrochemical studies using cyclic voltammetry (CV) on glassy carbon electrode (GC) (Stevanovic et al., 2020). The antioxidant capacity (free+bound) of fifteen native maize genotypes ranged from 1127.70 to 1875.70, 2826.90 to 4263.90, 730.94 to 1340.65, and 3484.80 to 5592.60 μmolTE/100 g in DPPH, ABTS, FRAP, and ORAC, respectively, and IC50 to DPPH and ABTS were expressed in mg/mL (Rodriguez-Salinas et al., 2020). Maximum ABTS radical scavenging activity was reported in the light blue ZPP-2 maize genotype during evaluation of antioxidant activity of whole kernels of ten different coloured maize genotypes (Žilić et al., 2012).

Antioxidant activity of creole maize races decreased via the nixtamalization process (Mendoza-Díaz et al., 2012). Tortillas prepared through lime-cooking extrusion of four different types of whole pigmented Mexican maize – white (WM), yellow (YM), red (RM), blue maize (BM) – retained in the range of 87.2 to 90.7% of total hydrophilic antioxidant activity when compared to raw kernels (Aguayo-Rojas et al., 2012). DPPH and ABTS inhibitory antioxidant activity of white, yellow, and purple corn accessions ranged between 0.73–0.89 and 3.81–4.92 lMtrolox/mg, respectively. Antioxidant activity was correlated to phenolic content of the muffins (Trehan et al., 2018).

Pigmented maize varieties have been reported to contain a high amount of antioxidants, mainly anthocyanins and carotenoids (Mendoza-Díaz et al., 2012). The antioxidant and antimutagenic activities of anthocyanin- and carotenoid-rich extracts from creole maize races revealed higher antioxidant capacity and anthocyanin content in blue and red genotypes than yellow- and white-coloured grains. The highest carotenoid levels were shown by red colour grains. Anthocyanin losses were observed after processing of blue grains into masa and tortillas (64%). Carotenoid content was also found to be reduced (53 to 56%) after nixtamalization. Antimutagenic activity was identified in all extracts against 2-aminoanthracene-induced mutagenicity (23 to 90%), indicating a higher potential. The findings demonstrate that, despite pigment losses, creole maize pigments have antioxidant and antimutagenic properties following the nixtamalization process.

The effects of nixtamal consumption from raw and alkaline-treated blue hybrid maize (*Zea mays* L. var. E3E4) on liver oxidative stress and inflammation on rats fed a high-fat diet were investigated (Magaña-Cerino et al., 2020). The nixtamalization process modified the chemical profile with cyanidin-3-O-glucoside being the main anthocyanin compound (92.19%). Antioxidant capacity by 2,2-azino-bis (3-ethylbenzothia zoline-6-sulphonic acid) (ABTS) was found to increase (IC$_{50}$ 668.70 µg/mL). Sod1gene expression and total liver antioxidant capacity were augmented after nixtamal treatment while malondialdehyde levels were found to decrease.

Dietary purple corn colour (PCC) significantly suppressed the high-fat-diet-induced increase in body weight gain, and white and brown adipose tissue weights (Tsuda et al., 2003). Feeding the high-fat diet markedly induced hypertrophy of the adipocytes in the epididymal white adipose tissue. The high-fat-diet-induced hyperinsulinemia, hyperglycaemia, and hyperleptinemia. These effects were totally normalized in rats fed high fat plus PCC. It was suggested that dietary PCC may improve high-fat-diet-induced insulin resistance in mice and can be used as functional food factor to prevent diabetes and obesity.

By hypothetical anthocyanin-starch interactions, blue maize extract has been claimed to operate as a spontaneous starch-modifier. Anthocyanin extract can be used to make food products with a higher resistant starch, which may have beneficial effects on human health (Camelo-Méndez et al., 2016). Antioxidant capacity of blue maize flour and the effects of polyphenol-containing extracts and blue maize wholegrain flour on starch digestion under uncooked and cooked conditions were investigated (Camelo-Méndez et al., 2017). Extracts of blue maize flour reduced amylase activity (>90% of inhibition). Blue maize flour showed higher slowly digestible and resistant starch contents, thus exhibiting lower predicted glycaemic index than white maize. Blue maize flour is a nutraceutical component which can be utilized to create functional foods with strong antioxidant capabilities, ingestible carbohydrate levels, and possible health advantages for the general public.

Coloured maize anthocyanins have been reported to regulate activation of inflammatory pathways and production of pro-inflammatory mediators (Zhang et al., 2019). Also, coloured maize anthocyanins are known to combat obesity-associated inflammation in relation to adipocytes and adipose tissue macrophages interactions. Antiproliferative effects of anthocyanins from blue corn and tortilla extracts have been suggested to serve as complementary strategies to treat the proliferation of prostate and breast cancer cells (Herrera-Sotero et al., 2019).

The effects of anthocyanins as well as other polyphenols found in free and bound phenolic sections of Peruvian purple maize on the development of probiotic lactic acid bacteria have been studied (Ranilla et al., 2017). The multiplication of good probiotic lactic acid bacteria such as L. helveticus and B. longum was not inhibited by the free or bound phenolic fractions. When targeted for different health uses, polyphenols from Peruvian purple maize can potentially be used as functional food additives, and they will be appropriate without exhibiting any detrimental side effects on intestinal health-associated healthy microbes.

The strongest *in vitro* antidiabetic effect has been reported in fat fractions of blue corn extracts of the blue corn variety (Smorowska et al., 2021) which was attributed to high content of polyphenolic compounds. In general, eight anthocyanin compounds were isolated from the extracts with pelargonidin, cyanidin-3-glucoside as major compounds. Unsaturated fatty acids, such as linoleic acid and oleic acid, were identified in abundance while saturated fatty acids like stearic and palmitic acid were found in lesser quantities in the lipid fraction of blue- and yellow-coloured maize. The blue maize had the highest quantities of sitosterol and campesterol, with tiny portions of stigmastanol and stigmasterol. Caffeic acid, procyanidin B2, and gallic acid are among the phenolic acid compounds found in the highest levels in both blue- and yellow-coloured maize.

The crude extracts derived from various varieties of maize have a number of possible health benefits, including suppression of -glucosidase, ACE, and AR. Total phenolics, total anthocyanins, and suppression of peroxynitrite production, -glucosidase, and AR inhibition are all significantly different. Purple maize had the maximum total polyphenolic content and AR inhibiting action amongst uncooked maize, while yellow had higher inhibitory -glucosidase action and modest ACE suppression, implying that non-phenolic chemicals are involved. The antioxidant properties of maize, combined with its antidiabetic properties, imply that it can be utilized to control and prevent type 2 diabetes.

Purple waxy corn and ginger (PWCG) is a potential candidate to serve as functional food to improve peripheral neuropathy in diabetic patients. PWCG was observed to increase sciatic function index (SFI), paw withdrawal threshold intensity (PWTI), paw withdrawal latency (PWL), and nerve conduction velocity (NCV) in rats, as well as oxidative stress status and axon density (Wattanathorn et al., 2015). PWCG exerts these effects through improved oxidative stress. The effects of chokeberry and purple maize in a diet-based rat model of human metabolic syndrome are consistent with the reported effects of anthocyanins, as they are the only polyphenolic compounds present in sufficient doses in both the treatments. These findings suggest that anthocyanin interventions using chokeberry or purple maize might be beneficial in attenuating obesity and metabolic syndrome in humans (Bhaswant et al., 2017).

Purple maize colour is a common food colourant that has been linked to lower blood pressure, obesity, and anti-cancer properties in colon cancers (Long et al., 2013). Purple maize colour delayed the development of hepatocellular carcinoma and reduced the prevalence of adenocarcinoma in the lateral prostate. It was observed that cyanidin-3-glucoside and pelargonidin-3-glucoside are the active compounds involved in these types of effects. Purple maize has also been found to contain the highest concentration of pro-anthocyanidins, followed by red and blue maize (Chen et al., 2017). These pro-anthocyanidins showed good anti-inflammatory effects, inhibiting nitric oxide synthase (66%) and cyclooxygenase-2 activities (89%), showing them to be active ingredients for potential pharmaceutical use.

References

Abdel-Aal, E. S. M., Young, J. C., & Rabalski, I. (2006). Anthocyanin composition in black, blue, pink, purple, and red cereal grains. *Journal of Agricultural and Food Chemistry, 54*(13), 4696–4704.

Aguayo-Rojas, J., Mora-Rochin, S., Cuevas-Rodriguez, E. O., Serna-Saldivar, S. O., Gutierrez-Uribe, J. A., Reyes-Moreno, C., & Milan-Carrillo, J. (2012). Phytochemicals and antioxidant capacity of tortillas obtained after lime-cooking extrusion process of whole pigmented Mexican maize. *Plant Foods for Human Nutrition, 67*(2), 178–185.

Amador-Rodríguez, K. Y., Martínez-Bustos, F., & Silos-Espino, H. (2019). Effect of high-energy milling on bioactive compounds and antioxidant capacity in nixtamalized creole corn flours. *Plant Foods for Human Nutrition, 74*(2), 241–246.

Bakowska-Barczak, A. (2005). Acylated anthocyanins as stable, natural food colourants—A review. *Polish Journal of Food and Nutrition Sciences, 14/55*(2), 107–116.

Barrientos-Ramírez, L., Ramírez-Salcedo, H. E., Fernández-Aulis, M. F., Ruíz-López, M. A., Navarro-Ocaña, A., & Vargas-Radillo, J. J. (2018). Anthocyanins from rose maize (*Zea mays* L.) grains. *Interciencia, 43*(3), 188–192.

Bello-Pérez, L. A., Flores-Silva, P. C., Camelo-Méndez, G. A., Paredes-López, O., & Figueroa-Cárdenas, J. D. D. (2015). Effect of the nixtamalization process on the dietary fiber content, starch digestibility, and antioxidant capacity of blue maize tortilla. *Cereal Chemistry, 92*(3), 265–270.

Bento-Silva, A., VazPatto, M. C., & do Rosário Bronze, M. (2018). Relevance, structure and analysis offerulic acid in maize cell walls. *Food Chemistry, 246*, 360–378.

Betrán, F. J., Bockholt, A. J., & Rooney, L. W. (2000). Blue corn. In A. R. Hallauer (Ed.), *Specialty corns* (2nd ed., pp. 305–314). Boca Raton, FL: CRC Press.

Bhaswant, M., Shafie, S. R., Mathai, M. L., Mouatt, P., & Brown, L. (2017). Anthocyanins in chokeberry and purple maize attenuate diet-induced metabolic syndrome in rats. *Nutrition, 41*, 24–31.

Boischot, C., Moraru, C. I., & Kokini, J. L. (2003). Factors that influence the microwave expansion of glassy amylopectin extrudates. *Cereal Chemistry, 80*(1), 56–61.

Bugusu, B. A., Campanella, O., & Hamaker, B. R. (2001). Improvement of sorghum-wheat composite dough rheological properties and bread-making quality through zein addition. *Cereal Chemistry, 78*(1), 31–35.

Camacho-Hernández, I. L., Zazueta-Morales, J. J., Gallegos-Infante, J. A., Aguilar-Palazuelos, E., Rocha-Guzmán, N. E., Navarro-Cortez, R. O., & Gómez-Aldapa, C. A. (2014). Effect of extrusion conditions on physicochemical characteristics and anthocyanin content of blue corn third-generation snacks. *CyTA: Journal of Food, 12*(4), 320–330.

Camelo-Méndez, G. A., Agama-Acevedo, E., Sanchez-Rivera, M. M., & Bello-Pérez, L. A. (2016). Effect on *in vitro* starch digestibility of Mexican blue maize anthocyanins. *Food Chemistry, 211*, 281–284.

Camelo-Méndez, G. A., Agama-Acevedo, E., Tovar, J., & Bello-Pérez, L. A. (2017). Functional study of raw and cooked blue maize flour: Starch digestibility, total phenolic content and antioxidant activity. *Journal of Cereal Science, 76*, 179–185.

Camelo-Méndez, G. A., Ferruzzi, M. G., González-Aguilar, G. A., & Bello-Pérez, L. A. (2016). Carbohydrate and phytochemical digestibility in pasta. *Food Engineering Reviews, 8*(1), 76–89.

Camelo-Méndez, G. A., Flores-Silva, P. C., Agama-Acevedo, E., Tovar, J., & Bello-Pérez, L. A. (2018a). Incorporation of whole blue maize flour increases antioxidant capacity and reduces in vitro starch digestibility of gluten-free pasta. *Starch/Stärke, 70*(1–2), 1700126.

Camelo-Méndez, G. A., Tovar, J., & Bello-Pérez, L. A. (2018b). Influence of blue maize flour on gluten-free pasta quality and antioxidant retention characteristics. *Journal of Food Science and Technology, 55*(7), 2739–2748.

Campas-Baypoli, O. N., Rosas-Burgos, E. C., Torres-Chavez, P. I., Ramirez-Wong, B., & Serna-Sald'ivar, S. O. (1999). Physicochemical changes of starch during maize tortilla production. *Starch/Stärke, 5*, 173–177.

Capocchi, A., Bottega, S., Spano, C., & Fontanini, D. (2017). Phytochemicals and antioxidant capacity in four Italian traditional maize (*Zea mays* L.) varieties. *International Journal of Food Science and Nutrition, 68*(5), 515–524.

Casas, M. I., Duarte, S., Doseff, A. I., & Grotewold, E. (2014). Flavone-rich maize: An opportunity to improve the nutritional value of an important commodity crop. *Frontiers in Plant Science, 5*, 1–11.

Castañeda-Ovando, A., Galán-Vidal, C. A., Pacheco, L., Rodriguez, J. A., & Paez-Hernandez, M. E. (2010). Characterization of main anthocyanins extracted from pericarp blue corn by MALDI-ToF MS. *Food Analytical Methods, 3*(1), 12–16.

Chalorcharoenying, W., Lomthaisong, K., Suriharn, B., & Lertrat, K. (2017). Germination process increases phytochemicals in corn. *International Food Research Journal, 24*(2), 552–558.

Chen, C., Somavat, P., Singh, V., & de Mejia, E. G. (2017). Chemical characterization of proanthocyanidins in purple, blue, and red maize coproducts from different milling processes and their anti-inflammatory properties. *Industrial Crops and Products, 109*, 464–475.

Colín-Chávez, C., Virgen-Ortiz, J. J., Serrano-Rubio, L. E., Martínez-Téllez, M. A., & Astier, M. (2020). Comparison of nutritional properties and bioactive compounds between industrial and artisan fresh tortillas from maize landraces. *Current Research in Food Science, 3*, 189–194.

Collison, A., Yang, L., Dykes, L., Murrayand, S., & Awika, J. M. (2015). Influence of genetic background on anthocyanin and copigment composition and behavior during thermoalkaline processing of maize. *Journal of Agricultural and Food Chemistry, 63*(22), 5528–5538.

Corrales-Bañuelos, A. B., Cuevas-Rodríguez, E. O., Gutiérrez-Uribe, J. A., Milán-Noris, E. M., Reyes-Moreno, C., Milán-Carrillo, J., & Mora-Rochín, S. (2016). Carotenoid composition and antioxidant activity of tortillas elaborated from pigmented maize landrace by traditional nixtamalization or lime cooking extrusion process. *Journal of Cereal Science, 69*, 64–70.

Cortes-Gomez, A., San, M., Martínez-Bustos, F., & Vázquez-Carrillo, G. M. (2005). Tortillas of blue maize (*Zea mays* L.) prepared by a fractionated process of nixtamalization: Analysis using response surface methodology. *Journal of Food Engineering 66*(3), 273–281.

De La Parra, C., Serna Saldivar, S. O., & Liu, R. H. (2007). Effect of processing on the phytochemical profiles and antioxidant activity of corn for production of masa, tortillas, and tortilla chips. *Journal of Agricultural and Food Chemistry, 55*(10), 4177–4183.

Del Pozo-Insfran, D., Brenes, C. H., Saldivar, S. O. S., & Talcott, S. T. (2006). Polyphenolic and antioxidant content of white and blue corn (*Zea mays* L.) products. *Food Research International, 39*(6), 696–703.

dePascual-Teresa, S., Santos-Buelga, C., & Rivas-Gonzalo, J. (2002). LC-MS analysis of anthocyanins extracts from purple corn cob. *Journal of the Science of Food and Agriculture, 82*(9), 1003–1006.

El-Megeid, A. A. A., AbdAllah, I. Z. A., Elsadek, M. F., & El-Moneim, Y. F. A. (2009). The protective effect of the fortified bread with green tea against chronic renal failure induced by excessive dietary arginine in male albino rats. *World Journal of Dairy and Food Science, 4*, 107–117.

Erickson, D. P., Campanella, O. H., & Hamaker, B. R. (2012). Functionalizing maize zein in viscoelastic dough systems through fibrous, β sheet-rich protein networks: An alternative, physicochemical approach to gluten-free breadmaking. *Trends in Food Science and Technology, 24*(2), 74–81.

Escalante-Aburto, A., Ramirez-Wong, B., Torres-Chavez, P. I., Barron-Hoyos, J. M., Figueroa-Cardenas, J. D., & Lopez-Cervantes, J. (2013a). The nixtamalization process and its effect on anthocyanin content of pigmented maize: A review. *Revista Fitotecnia Mexicana*, *36*(4), 429–437.

Escalante-Aburto, A., Ramirez-Wong, B., Torres-Chavez, P. I., Figueroa-Cardenas, J. D., Lopez-Cervantes, J., Barron-Hoyos, J. M., & Morales Rosas, I. (2013b). Effect of extrusion processing parameters on anthocyanin content, physicochemical properties of nixtamalized blue corn expanded extrudates. *CyTA: Journal of Food*, *11*(sup1), 29–37.

Escalante-Aburto, A., Ramírez-Wong, B., Torres-Chávez, P. I., López-Cervantes, J., Figueroa-Cárdenas, J. D. D., Barrón-Hoyos, J. M., & Gutiérrez-Dorado, R. (2014). Obtaining ready-to-eat blue corn expanded snacks with anthocyanins using an extrusion process and response surface methodology. *Molecules*, *19*(12), 21066–21084.

Espinosa Trujillo, E., Mendoza Castillo, M., Castillo Gonzalez, F., Ortiz Cereceres, J., Delgado Alvarado, A., & Carrillo Salazar, A. (2009). Anthocyanin accumulation in pericarp and aleurone layer of maize kernel and their genetic effects on native pigmented varieties. *Revista Fitotecnia Mexicana*, *32*(4), 303–309.

Flores-Calderón, A. M. D., Luna, H., Escalona-Buendía, H. B., & Verde-Calvo, J. R. (2017). Chemical characterization and antioxidant capacity in blue corn (*Zea mays* L.) malt beers. *Journal of the Institute of Brewing*, *123*(4), 506–518.

Giusti, M. M., & Wrolstad, R. E. (2003). Acylated anthocyanins from edible sources and their applications in food systems. *Biochemical Engineering Journal*, *14*(3), 217–225.

Guillén-Sánchez, J., Mori-Arismendi, S., & Paucar-Menacho, L. M. (2014). Characteristics and functional properties of purple corn (*Zea mays* L.) var. subnigro violaceo. *Scientia Agropecuaria*, *5*, 211–217.

Haggard, S., Luna-Vital, D., West, L., Juvik, J. A., Chatham, L., Paulsmeyer, M., & de Mejia, E. G. (2018). Comparison of chemical, colour stability, and phenolic composition from pericarp of nine coloured corn unique varieties in a beverage model. *Food Research International*, *105*, 286–297.

Harakotr, B., Suriharn, B., Scott, M. P., & Lertrat, K. (2015). Genotypic variability in anthocyanins, total phenolics, and antioxidant activity among diverse waxy corn germplasm. *Euphytica*, *203*(2), 237–248.

Harakotr, B., Suriharn, B., Tangwongchai, R., Scott, M. P., & Lertrat, K. (2014a). Anthocyanin, phenolics and antioxidant activity changes in purple waxy corn as affected by traditional cooking. *Food Chemistry*, *164*, 510–517.

Harakotr, B., Suriharn, B., Tangwongchai, R., Scott, M. P., & Lertrat, K. (2014b). Anthocyanins and antioxidant activity in coloured waxy corn at different maturation stages. *Journal of Functional Foods*, *9*, 109–118.

Hernández-Martínez, V., Salinas-Moreno, Y., Ramírez-Díaz, J. L., Vázquez-Carrillo, G., Domínguez-López, A., & Ramírez-Romero, A. G. (2016). Colour, phenolic composition and antioxidant activity of blue tortillas from Mexican maize races. *CyTA: Journal of Food, 14*(3), 473–481.

Hernández-Uribe, J. P., Agama-Acevedo, E., Islas-Hernández, J. J., Tovar, J., & Bello-Pérez, L. A. (2007). Chemical composition and *in vitro* starch digestibility of pigmented corn tortilla. *Journal of the Science of Food and Agriculture, 87*(13), 2482–2487.

Herrera-Sotero, M. Y., Cruz-Hernández, C. D., Oliart-Ros, R. M., Chávez-Servia, J. L., Guzmán-Gerónimo, R. I., González-Covarrubias, V., ... Rodríguez-Dorantes, M. (2019). Anthocyanins of blue corn and tortilla arrest cell cycle and induce apoptosis on breast and prostate cancer cells. *Nutrition and Cancer, 72*(5), 768–777.

Hollingsworth, P. (2001). Third-generation snacks take aim at popcorn market. *Food Technology, 55*(6), 20.

Hong, H. T., Netzel, M. E., & O'Hare, T. J. (2020a). Anthocyanin composition and changes during kernel development in purple-pericarp super-sweet sweetcorn. *Food Chemistry, 315*, 126284.

Hong, H. T., Netzel, M. E., & O'Hare, T. J. (2020b). Optimization of extraction procedure and development of LC-DAD-MS methodology for anthocyanin analysis in anthocyanin-pigmented corn kernels. *Food Chemistry, 319*, 126515.

Hu, C. Y., Li, Q. L., Shen, X. F., Quan, S., Lin, H., Duan, L., ... Shi, J. X. (2016). Characterization of factors underlying the metabolic shifts in developing kernels of coloured maize. *Scientific Reports, 6*(1), 1–12.

Hurtado, M., Escobar, B., & Estévez, A. M. (2001). Mezclas legumbre/cereal por fritura profunda de maíz amarillo y de tres cultivares de frejol para consumo "snacks". *Archivos Latinoamericanos De Nutrición, 5*, 303–308.

Jing, P., Noriega, V., Schwartz, S. J., & Giusti, M. M. (2007). Effects of growing conditions on purple corncob (*Zea mays* L.) anthocyanins. *Journal of Agricultural and Food Chemistry, 55*(21), 8625–8629.

Kahriman, F., Akta, F., Pinar, G., Songur, U., & Egesel, C. O. (2021). Assessment of genetic diversity of Turkish maize landraces for phytic acid and total phenolic contents. *Acta Fytotech n Zoo Techn, 24*(1), 16–24.

Kapcum, N., Uriyapongson, J., Alli, I., & Phimphilai, S. (2016). Anthocyanins, phenolic compounds and antioxidant activities in coloured corn cob and coloured rice bran. *International Food Research Journal, 23*(6), 2347–2356.

Kim, J.-T., Yi, G., Chung, I.-M., Son, B.-Y., Bae, H.-H., Go, Y. S., ... Kim, S.-L. (2020). Timing and pattern of anthocyanin accumulation during grain filling in purple waxy corn (*Zea mays* L.) Suggest Optimal Harvest Dates. *ACS Omega, 5*(25), 15702–15708.

Kocadağlı, T., Žilić, S., Taş, N. G., Vančetović, J., Dodig, D., & Gökmen, V. (2016). Formation of α-dicarbonyl compounds in cookies made from wheat, hull-less barley and coloured corn and its relation with phenolic compounds, free amino acids and sugars. *European Food Research and Technology, 242*(1), 51–60.

Kokkaew, H., & Pitirit, T. (2016). Optimization for anthocyanin and antioxidant contents and effects of acidulants on purple corn cake containing corn silk powder qualities. *International Food Research Journal, 23*, 2390–2398.

Lago, C., Cassani, E., Zanzi, C., Landoni, M., Trovato, R., & Pilu, R. (2014). Development and study of a maize cultivar rich in anthocyanins: Coloured polenta, a new functional food. *Plant Breeding, 133*(2), 210–217.

Lago, C., Landoni, M., Cassani, E., Atanassiu, S., Cantaluppi, E., & Pilu, R. (2020). Development and characterization of a coloured sweet corn line as a new functional food. *Maydica, 59*, 2014.

Lao, F., & Giusti, M. M. (2016). Quantification of purple corn (*Zea mays* L.) anthocyanins using spectrophotometric and HPLC approaches: Method comparison and correlation. *Food Analytical Methods, 9*(5), 1367–1380.

Li, Q., Somavat, P., Singh, V., Chatham, L., & de Mejia, E. G. (2017). A comparative study of anthocyanin distribution in purple and blue corn coproducts from three conventional fractionation processes. *Food Chemistry, 231*, 332–339.

Li, T. C., Zhang, W., Yang, H. Y., Dong, Q., Ren, J., Fan, H. H., ... Zhou, Y. B. (2019). Comparative transcriptome analysis reveals differentially expressed genes related to the tissue specific accumulation of anthocyanins in pericarp and aleurone layer for maize. *Scientific Reports, 9*(1), Art. No. 2485. https://doi.org/10.1038/s41598-018-37697-y.

Long, N., Suzuki, S., Sato, S., Naiki-Ito, A., Sakatani, K., Shirai, T., & Takahashi, S. (2013). Purple corn colour inhibition of prostate carcinogenesis by targeting cell growth pathways. *Cancer Science, 104*(3), 298–303.

Lopez-Martinez, L. X., Oliart-Ros, R. M., Valerio-Alfaro, G., Lee, C. H., Parkin, K. L., & Garcia, H. S. (2009). Antioxidant activity, phenolic compounds and anthocyanins content of eighteen strains of Mexican maize. *LWT - Food Science and Technology, 42*(6), 1187–1192.

Lopez-Martınez, L. X., Parkin, K. L., & Garcia, H. S. (2012). Effect of processing of corn for production of masa, tortillas and tortilla chips on the scavenging capacity of reactive nitrogen species. *International Journal of Food Science and Technology, 47*(6), 1321–1327.

Magaña-Cerino, J. M., Peniche-Pavía, H. A., Tiessen, A., & Gurrola-Díaz, C. M. (2020). Pigmented maize (*Zea mays* L.) contains anthocyanins with potential therapeutic action against oxidative stress - A review. *Polish Journal of Food and Nutrition Science, 70*(2), 85–89.

Mariscal Moreno, R. M., Figuero, J. D. C., Santiago-Ramos, D., Villa, G. A., Sandoval, S. J., Rayas-Duarte, P., ... Martínez Flores, H.E. (2015). The effect of different nixtamalization processes on some physicochemical properties, nutritional composition and glycemic index. *Journal of Cereal Science, 65*, 140–146.

Martínez-Martínez, R., Vera-Guzmán, A. M., Chávez-Servia, J. L., Bolaños, E. N. A., Carrillo-Rodríguez, J. C., & Pérez-Herrera, A. (2019). Bioactive compounds and antioxidant activities in pigmented maize landraces. *Interciencia, 44*(9), 549–556.

Matthias, T., Neidhöfer, S., Pfeiffer, S., Prager, K., Reuter, S., & Gershwin, M. E. (2011). Novel trends in celiac disease. *Cellular and Molecular Immunology, 8*(2), 121–125.

Méndez-Lagunas, L. L., Cruz-Gracida, M., Barriada-Bernal, L. G., & Rodríguez-Méndez, L. I. (2020). Profile of phenolic acids, antioxidant activity and total phenolic compounds during blue corn tortilla processing and its bioaccessibility. *Journal of Food Science and Technology, 57*(12), 4688–4696.

Mendoza-Díaz, S., Ortíz-Valerio, M. A., Castaño-Tostado, E., Figueroa-Cárdenas, J. D., Reynoso Camacho, R., ... Loarca-Piña, G. (2012). Antioxidant capacity and antimutagenic activity of anthocyanin and carotenoid extracts from nixtamalized pigmented creole maize races (*Zea mays* L.). *Plant Foods for Human Nutrition, 67*(4), 442–449.

Mendoza-López, M. L., Alvarado-Díaz, C. S., Pérez-Vega, S. B., Leal-Ramos, M. Y., Gutiérrez-Méndez, N., Alvarado-Díaz, C. S., Pérez-Vega, S. B., & Gutiérrez-Méndez, N. (2018). Compositional and free radical scavenging properties of *Zea mays* female inflorescences (maize silks) from Mexican maize landraces inflorescences (maize silks) from Mexican maize landraces. *CyTA: Journal of Food, 16*(1), 96–104.

Mendoza-Mendoza, C. G., Mendoza-Castillo, M. C., Delgado-Alvarado, A., Sánchez-Ramírez, F. J., & Kato-Yamakake, T. Á. (2020). Anthocyanins content in the kernel and corncob of Mexican purple corn populations. Maydica electronic publication.

Milan-Carrillo, J., Gutierrez-Dorado, R., Cuevas-Rodríguez, E. O., Garzon-Tiznado, J. A., & Reyes-Moreno, C. (2004). Nixtamalized flour from quality protein maize (*Zea mays* L). Optimization of alkaline processing. *Plant Foods for Human Nutrition, 59*(1), 35–44.

Mildner-Szkudlarz, S., Zawirska-Woitasiak, R., Obuchowski, W., & Goslinski, M. (2009). Evaluation of antioxidant activity of green tea extract and its effect on the biscuits lipid fraction oxidative stability. *Journal of Food Science, 74*(8), S362–S370.

Monroy, Y. M., Rodrigues, R. A., Sartoratto, A., & Cabral, F. A. (2016). Extraction of bioactive compounds from cob and pericarp of purple corn (*Zea mays* L.) by sequential extraction in fixed bed extractor using supercritical CO_2, ethanol, and water as solvents. *Journal of Supercritical Fluids, 107*, 250–259.

Monroy, Y. M., Rodrigues, R. A., Sartoratto, A., & Cabral, F. A. (2020). Purple corn (*Zea mays* L.) pericarp hydroalcoholic extracts obtained by conventional processes at atmospheric pressure and by processes at high pressure. *Brazilian Journal of Chemical Engineering, 37*(1), 237–248.

Montilla, E. C., Hillebrand, S., Antezana, A., & Winterhalter, P. (2011). Soluble and bound phenolic compounds in different Bolivian purple corn (*Zea mays* L.) Cultivars. *Journal of Agricultural and Food Chemistry, 59*(13), 7068–7074.

Mora-Rochín, S., Gaxiola-Cuevas, N., Gutiérrez-Uribe, J. A., Milán-Carrillo, J., Milán-Noris, E. M., Reyes-Moreno, C., … Cuevas-Rodríguez, E. O. (2016). Effect of traditional Nixtamalization on anthocyanin content and 1 profile in Mexican blue maize (*Zea mays* L.) landraces. *LWT - Food Science and Technology.* https://doi.org/10.1016/j.lwt.2016.01.009

Moreno, Y. S., Hernández, D. R., & Velázquez, A. D. (2005). Extraction and use of pigments from maize grains (*Zea mays* L.) as colourants in yogurt. *Archivos Latinoamericanos de Nutricion, 55*(3), 293–298.

Moreno, Y. S., Sánchez, G. S., Hernández, D. R., & Lobato, N. R. (2005). Characterization of anthocyanin extracts from maize kernels. *Journal of Chromatographic Science, 43*(9), 486–487.

Morohashi, K., Casas, M. I., Falcone Ferreyra, M. L., Mejía-Guerra, M. K., Pourcel, L., Yilmaz, A., … Grotewold, E. (2012). A genome-wide regulatory framework identifies maize pericarp colour1 controlled genes. *Plant Cell, 24*(7), 2745–2764.

Mrad, R., Debs, E., Saliba, R., Maroun, R. G., & Louka, N. (2014). Multiple optimization of chemical and textural properties of roasted expanded purple maize using response surface methodology. *Journal of Cereal Science, 60*(2), 397–405.

Muangrat, R., Pongsirikul, I., & Blanco, P. H. (2018). Ultrasound assisted extraction of anthocyanins and total phenolic compounds from dried cob of purple waxy corn using response surface methodology. *Journal of Food Processing and Preservation, 42*(2), 1–11.

Nankar, A. N., Dungan, B., Paz, N., Sudasinghe, N., Schaub, T., Holguin, F. O., & Pratt, R. C. (2016). Quantitative and qualitative evaluation of kernel anthocyanins from southwestern United States blue corn. *Journal of the Science of Food and Agriculture, 96*(13), 4542–4552.

Navarro-Cortez, R. O., Palazuelos, E. A., Zazueta-Morales, J. Z., Rosas, J. C., Gãmez-Aldapa, C. A., & Aguirre-Tostado, F. S. (2014). Microstructure of an extruded third-generation snack made from a whole blue corn and corn starch mixture. *International Journal of Food Processing and Technology, 1*(1):10–17.

Pasqualone, A., Bianco, A. M., Paradiso, V. M., Summo, C., Gambacorta, G., & Caponio, F. (2014). Physico-chemical, sensory and volatile profiles of biscuits enriched with grape marc extract. *Food Research International, 65*, 385–393.

Pedreschi, R., & Cisneros-Zevallos, L. (2007). Phenolic profiles of Andean purple corn (*Zea mays* L.). *Food Chemistry, 100*(3), 956–963.

Petroni, K., Pilu, R., & Tonelli, C. (2014). Anthocyanins in corn: A wealth of genes for human health. *Planta, 240*(5), 901–911.

Psolova, A. O., Derkach, K. V., & Satarova, T. M. (2019). Content of anthocyanin sweet corn with different grain colouring. *Agrology, 2*(2), 128–133.

Ramírez, L. B., Salcedo, H. E. R., Fernández Aulis, M. F., RuízLópez, M. A., Ocaña, A. N., & Vargas Radillo, J. J. (2018). Anthocyanins from rose maize (*Zea mays* L.) Grains. *Interciencia, 43*, 188–192.

Ramos-Escudero, F., Muñoz, A. M., Alvarado-Ortíz, C., Alvarado, Á., & Yáñez, J. A. (2012). Purple corn (*Zea mays* L.) phenolic compounds profile and its assessment as an agent against oxidative stress in isolated mouse organs. *Journal of Medicinal Food, 15*(2), 206–215.

Ranilla, L. G., Christopher, A., Sarkar, D., Shetty, K., Chirinos, R., & Campos, D. (2017). Phenolic composition and evaluation of the antimicrobial activity of free and bound phenolic fractions from a Peruvian purple corn (*Zea mays* L.) accession. *Journal of Food Science.* https://doi.org/10.1111/1750-3841.13973

Rodriguez-Salinas, P. A., Zavala-Garcia, F., Urias-Orona, V., Muy-Rangel, D., Heredia, J. B., & Nino-Medina, G. (2020). Chromatic, nutritional and nutraceutical properties of pigmented native maize (*Zea mays* L.) genotypes from the northeast of Mexico. *Arabian Journal for Science and Engineering.* https://doi.org/10.1007/s13369-019-04086-0

Sales, Z. G., Juanico, C. B., Dizon, E. I., & Hurtada, W. A. (2018). Indigenous pigmented corn (*Zea mays* L.) flour as substitute for all-purpose flour to improve the sensory characteristics and nutrient content of crackers. *Nutritional Status, Dietary Intake and Body Composition, 24*(4), 617.

Salinas-Moreno, Y., Cruz-Chavez, F., Diaz-Ortiz, S., & Castillo-Gonzalez, F. (2012). Pigmented maize grains from Chiapas, physical characteristics, anthocyanin content and nutraceutical value. *Revista de Fitotecnia Mexicana, 35*(1), 33–41 (in Spanish; English abstract).

Salinas-Moreno, Y., Salas-Sanchez, G., Rubio-Hernandez, D., & Ramos-Lobato, N. (2005). Characterization of anthocyanins extracts from maize kernels. *Journal of Chromatographic Science, 43*(9), 483–487.

Salvador-Reyes, R., Silva, P., & Clerici, M. T. (2020). Peruvian Andean maize: General characteristics, nutritional properties, bioactive compounds, and culinary uses. *Food Research International, 130,* 108934.

Sharma, M., Cortes-Cruz, M., Ahern, K. R., McMullen, M., Brutnell, T. P., & Chopra, S. (2011). Identification of the *Pr1* gene product completes the anthocyanin biosynthesis pathway of maize. *Genetics, 188*(1), 69–79.

Shen, L. Y., & Petolino, J. F. (2006). Pigmented maize seed via tissue-specific expression of anthocyanin pathway gene transcription factors. *Molecular Breeding, 18*(1), 57–67.

Simic, M., Zilic, S., Simuruna, O., Filipcev, B., Skrobot, D., & Vancetovic, J. (2018). Effects of anthocyanin-rich popping maize flour on the phenolic profile and the antioxidant capacity of mix-bread and its physical and sensory properties. *Polish Journal of Food and Nutrition Sciences, 68*(4), 299–308.

Simla, S., Boontang, S., & Harakotr, B. (2016). Anthocyanin content, total phenolic content, and antiradical capacity in different ear components of purple waxy corn at two maturation stages. *Australian Journal of Crop Science, 10*(5), 675–682.

Smorowska, A. J., Żołnierczyk, A. K., Nawirska-Olszańska, A., Sowiński, J., & Szumny, A. (2021). Nutritional properties and in vitro antidiabetic activities of blue and yellow corn extracts: A comparative study. *Journal of Food Quality.* Article ID 8813613, 10 pages.

Stevanovic, M., Stevanovic, S., Mihailovic, M., Kiprovski, B., Bekavac, G., Mikulic-Petkovsek, M., & Lovic, J. (2020). Antioxidant capacity of dark red corn–biochemical properties coupled with electrochemical evaluation. *Revista de Chimie, 71*(6), 31–41.

Suriano, S., Balconi, C., Valoti, P., & Redaelli, R. (2021). Comparison of total polyphenols, profile anthocyanins, colour analysis, carotenoids and tocols in pigmented maize. *LWT, 144*, 111257.

Tanaka, Y., Sasaki, N., & Ohmiya, A. (2008). Biosynthesis of plant pigments: Anthocyanins, betalains and carotenoids. *Plant Journal, 54*(4), 733–749.

Tian, X. Z., Lu, Q., Paengkoum, P., & Paengkoum, S. (2020). Short communication: Effect of purple corn pigment on change of anthocyanin composition and unsaturated fatty acids during milk storage. *Journal of Dairy Science, 103*(9), 7808–7812.

Trehan, S., Singh, N., & Kaur, A. (2018). Characteristics of white, yellow, purple corn accessions: Phenolic profile, textural, rheological properties and muffin making potential. *Journal of Food Science and Technology, 55*(6), 2334–2343.

Tsuda, T., Horio, F., Uchida, K., Aoki, H., & Osawa, T. (2003). Dietary cyanidin 3-O-β-D-glucoside-rich purple corn colour prevents obesity and ameliorates hyperglycemia in mice. *Journal of Nutrition, 133*(7), 2125–2130.

Urias-Lugo, D. A., Heredia, J. B., Muy-Rangel, M. D., Valdez-Torres, J. B., Serna-Saldívar, S. O., & Gutiérrez-Uribe, J. A. (2015). Anthocyanins and phenolic acids of hybrid and native blue maize (*Zea mays* L.) extracts and their antiproliferative activity in mammary (MCF7), liver (HepG2), colon (Caco2 and HT29) and prostate (PC3) cancer cells. *Plant Foods for Human Nutrition, 70*(2), 193–199.

Urias-Peraldi, M., Gutierrez-Uribe, J. A., Preciado-Ortiz, R. E., Cruz-Morales, A. S., Serna-Saldivar, S. O., & Garcia-Lara, S. (2013). Nutraceutical profiles of improved blue maize (*Zea mays*) hybrids for subtropical regions. *Field Crops Research, 141*, 69–76.

Vázquez-Carrillo, M. G., Aparicio-Eusebio, L. A., Salinas-Moreno, Y., Buendía-Gonzalez, M. O., & Santiago-Ramos, D. (2018). Nutraceutical, physicochemical, and sensory properties of blue corn polvorones, a traditional flour-based confectionery. *Plant Foods for Human Nutrition*, *73*(4), 321–327.

Wang, H., Cao, G. H., & Prior, R. L. (1997). Oxygen radical absorbing capacity of anthocyanins. *Journal of Agricultural and Food Chemistry*, *45*(2), 304–309.

Wattanathorn, J., Thiraphatthanavong, P., Muchimapura, S., Thukhammee, W., Lertrat, K., & Suriharn, B. (2015). The combined extract of *Zingiber officinale* and *Zea mays* (purple colour) improves neuropathy, oxidative stress, and axon density in streptozotocin induced diabetic rats. *Evidence-Based Complementary and Alternative Medicine*. http:// doi.org/10.1155/2015/301029

Yan, Z., & Zhai, W. (2010). Identification and antioxidant activity of anthocyanins extracted from the seed and cob of purple corn (*Zea mays* L). *Innovative Food Science and Emerging Technologies*, *11*(1), 169–176.

Yang, Z., Chen, Z., Yuan, S., Zhai, W., Piao, X., & Piao, X. (2009). Extraction and identification of anthocyanin from purple corn (*Zea mays* L.). *International Journal of Food Science and Technology*, *44*(12), 2485–2492.

Yang, Z., & Zhai, W. (2010). Identification and antioxidant activity of anthocyanins extracted from the seed and cob of purple corn (*Zea mays* L.). *Innovative Food Science and Emerging Technologies*, *11*(1), 169–176.

Zazueta-Morales, J., Martínez-Bustos, F., Jacobo-Valenzuela, N., Ordorica-Falomir, C., & Paredes-Lopez, O. (2002). Effects of calcium hydroxide and screw speed on physicochemical characteristics of extruded blue maize. *Journal of Food Science*, *67*(9), 3350–3358.

Zhang, Q., Luna-Vitala, D., & de Mejia, E. G. (2019). Anthocyanins from coloured maize ameliorated the inflammatory paracrine interplay between macrophages and adipocytes through regulation of NF-κB and JNK-dependent MAPK pathways. *Journal of Functional Foods*, *54*, 175–186.

Zhao, X., Zhang, C., Guigas, C., Ma, Y., Corrales, M., Tauscher, B., & Hu, X. (2009). Composition, antimicrobial activity, and antiproliferative capacity of anthocyanin extracts of purple corn (*Zea mays* L.) from China. *European Food Research and Technology*, *228*(5), 759–765.

Zilic, S., Kocadağlı, T., Vančetović, J., & Gökmen, V. (2016). Effects of baking conditions and dough formulations on phenolic compound stability, antioxidant capacity and colour of cookies made from anthocyanin-rich corn flour. *LWT - Food Science and Technology*, *65*, 597–603.

Zilic, S., Serpen, A., Akıllıoğ, L. G., Gökmen, V., & Vančetović, J. (2012). Phenolic compounds, carotenoids, anthocyanins, and antioxidant capacity of coloured maize (*Zea mays* L.) kernels. *Journal of Agricultural and Food Chemistry*, *60*(5), 1224–1231.

Chapter 9

Quality improvement through bio-fortification

Maninder Kaur, Pooja Manchanda,
and Arashdeep Singh

DOI: 10.1201/9781003245230-9

9.1 Introduction

Maize (*Zea mays* L.) is one of the dominant subsistence crops in the world to provide proper nutrition and health care to the livelihood of growing populations. During 2020 and 2021, it was estimated that the world's maize constitutes 1145.90 million metric tonnes of production, and India comprises of 30,250 thousand metric tonnes maize production (maize oulook-2021, International Grains Council, accessed 26 November 2020, https://ipad.fas.usda.gov/). Having a healthy diet is a fundamental right of all human beings, but micronutrient deficiencies serve as a major bottleneck in the development of human social and economic development, affecting approximately three billion people worldwide (Khush et al., 2012). Mineral malnutrition is a serious global challenge affecting mankind (www.copenhagenconsensus.com). In this scenario, the bio-fortification approach serves as a virtuous platform to inculcate the minerals and vitamins in food crops to feed a high proportion of the world's population, which is expected to increase to eight billion by 2030 (Stein 2010; Khush et al., 2012). It is a promising method to enhance the micronutrient content through dietary diversification, mineral supplementation, food fortification, and increasing the bioavailability of mineral elements in crops. Bio-fortification differs from fortification, as it results in more nutritive plant foods naturally as well as the addition of nutrient supplements to the foods during food processing (Malik and Maqbool 2020). Moreover, bio-fortification is more economically sustainable than fortification methods, as the developed crop does not require fortificants to be bought and added to foods. Generally food is equipped with bulk elements like carbohydrates, lipids, essential amino acids, and fibre, but its chronic lack of micronutrients create hidden hunger in the human diet. Exogenous fortification includes the addition of multivitamin premixes to maize flour, while endogenous fortification, referred to as "bio-fortification", enables a sustainable and practical solution to combat the micronutrient deficiency (Nuss and Tanumihardjo, 2010). Its major objective is to enhance the micronutrient content of the edible part of staple crops and reduce the anti-nutrient substances inhibiting micronutrient bioavailability to increase the micronutrient content in the crops. The bioavailability refers to the innate availability of nutrients in the crops essential for metabolic processes in the human body. Different interventions, including dietary diversification, industrial fortification, pharmaceutical supplementation, and crop bio-fortification, serve as the best methods to impart nutrients in the human diet. Among all these interventions, bio-fortification is one of the most acceptable and preferred method among researchers, growers, and consumers based on its acceptability, availability, affordability, sustainability, and accessibility (Maqbool et al., 2018).

It has been estimated that the basic nutritional requirement of humans is the consumption of at least 22 mineral elements for proper nutrition (Welch and Graham, 2004; White and Broadley, 2005a; Graham et al., 2007). Food crops generally have a scarcity of the mineral elements viz. iron (Fe), zinc (Zn), copper (Cu), calcium (Ca), magnesium (Mg), iodine (I), and selenium (Se). Moreover, it has been estimated that over 60% of the world's six billion people are Fe deficient, over 30% are Zn deficient, 30% are iodine deficient and 15% are Se deficient (White and Broadley 2009). Maize staple crop majorly accounts for 15 to 56% of the total daily calories of humans in developing countries (Food and Agriculture Organization, 2008). The imbalances in the mineral content in soil occur due to the lack of nutrients viz. Fe, Zn, Ca, Mg, Cu, I, and Se, which are required for a balanced human diet. Cereal crops are staple foods that provide a large portion of energy and macronutrients to people. Among food crops, maize is a model cereal crop to alleviate the global micronutrient deficiencies through bio-fortification programmes. The second most important maize growing country in Asia is India (Prasanna, 2014). It has been estimated that 23% of maize production is utilized for human food and around 63% intended for poultry and animal feed (Yadav et al., 2015). Maize is capable for providing at least 30% of the food calories to more than 4.5 billion people in 94 developing countries and constitutes about 15% of all food-crop protein (Shiferaw et al., 2011). Although maize is equipped with nutritious minerals and is an extensive energy provider, it is deficient in the essential amino acids viz. lysine and tryptophan (Olson and Frey 1987). Maize serves as a valuable model due to the presence of historical collections of carotenoid mutants, genome sequence, and other molecular resources. The ability of natural accumulation of carotenoids in the edible seed endosperm makes maize crop an obvious target for bio-fortification projects. Researchers exploited the use of early maize mutant analyses that enable understanding of the biosynthesis of carotenoids. Maize kernels can be consumed in several ways, such as being parched, fried, boiled, roasted, fermented, and ground for use in the making of breads, tortillas, gruel, porridges, cakes, and alcoholic beverages (Gardner and Inglett 1971), so bio-fortification of the maize crop is a current requirement to meet dietary requirements. It has been reported that vitamin A-deficit countries consume maize as a staple food crop, so the development of provitamin A-bio-fortified maize cultivars is a must to alleviate vitamin A deficiency in the masses (Ortiz-Monasterio et al., 2007). Maize crop is preferred for bio-fortification due to its reliance on higher bioavailability and the bioconversion of provitamin A carotenoids into vitamin A to alleviate human malnutrition. Maize hybrids viz. HP1097-18, HP1097-11, and HP1097-2 are high-yielding hybrids along with high provitamin A content developed through bio-fortification programmes (Maqbool 2017).

Specialized agencies viz. World Health Organization (WHO) and the Consultative Group on International Agricultural Research (CGIAR) are keen to combat hidden hunger due to micronutrient deficiencies, as the symptoms and effects are not easily or visibly identified. Developing countries with high birth rates and adult mortality rates have has numerous people deficient in Zn, Fe, and vitamin A, which are among the top ten risk factors contributing to the possibility of disease (WHO 2002). The CGIAR support bio-fortification through the Harvest-Plus program to improve micronutrient in crops like rice, wheat, maize, beans, cassava, pearl millet, and sweet potato through breeding and biotechnological approaches. The implementation of micronutrient supplementation through cost effective bio-fortification has the potential to alleviate the malnutrition limitation in an efficient manner (Gómez-Galera et al., 2010). The maize bio-fortification efforts are proceeding in Zambia, Ghana, Ethiopia, and Guatemala, while India and Pakistan are producing bio-fortified wheat. Alternatively, the bio-fortification of basic processed foods with essential minerals resulted in the reduction of micronutrient malnutrition (Lyddon, 2004), but it does not reach poor populations so it is not a feasible strategy for people of low income groups. It has been reported that bio-fortified crops are consumed by more than 20 million people (Bouis and Saltzman, 2017). The need of essential amino acids such as lysine and tryptophan in the bio-fortification program lies in the fact that these are not synthesized in human body (Gupta et al., 2015). The supplementation of these micronutrients is a must for sustained life. Simultaneously, the bio-fortification requires policy guidelines to streamline the bio-fortified programmes. The policy regulations in India will facilitate nutri-farms for adoption of micronutrient-enriched maize, Fe-enriched pearl millet, and Zn-enriched rice in 100 districts of nine different states (Economic Survey of India 2014). The aim of this chapter is to summarize the development of bio-fortification-based quality improvement in maize through plant breeding and biotechnological approaches along with combinatorial approach of plant breeding strategies and biotechnological approaches capable of enhancing provitamin A for accelerating genetic improvements in the plant material.

9.2 Micronutrients contribution

The micronutrients such as iron, zinc, and vitamin A are vital for the world's population growth. Bio-fortification has the potential to combat widespread micronutrient deficiencies in humans through micronutrient implementation in food crops. Deficiencies in micronutrients viz. iron, zinc, and vitamin A deteriorating human health are mentioned here briefly. Iron deficiency ranked among the most widespread nutrient deficiencies affecting people throughout the world (Stoltzfus and Dreyfuss 1998). Its deficiency causes fatigue, weakness, irritability, and impaired mental development. Moreover, it resulted in maternal and perinatal mortality, and impairment of cognitive skills and physical abilities (Stein et al., 2005; Stoltzfus et al., 2004). Moreover, it led to high mortality and morbidity rates, diminished cognitive abilities, and reduced labour productivity hampering national development.

Zinc micronutrient is an essential trace mineral affecting gene expression, cell development, and replication (Hambridge, 2000). Zinc deficiency leads to infectious diseases causing immune and nervous systems impairment. Deficiency in zinc increases the risk of diarrhoea, pneumonia, malaria, morbidity, and mortality. The major cause of zinc deficiency is the lack of zinc micronutrients in the soil, so the combination of plant breeding with genetic approaches serves as a prerequisite to make Zn in soils available. Techniques like seed priming, and foliar and soil applications are capable of enhancing Zn enrichment in the soil surface (Maqbool and Beshir, 2018). Seed priming basically involves soaking the seed in a micronutrient solution with a specific concentration for a specific duration. It has been reported that in maize, seed priming with 1% zinc sulphate ($ZnSO_4$) solution for a 16 hour treatment led to an enhancement in the crop growth, grain yield, and grain Zn content (Harris et al., 2007). Fe and Zn chelate foliar spraying also significantly increased the maize crop yield (Hagh et al., 2016). Bio-fortification serves to improve the Zn uptake from soil, xylem loading, and remobilization in grains and sequestration in endosperm to improve the maize Zn concentration. It has been reported that the intake of bio-fortified maize genotypes comprising of 34 µg zinc/g grain zinc results in the reduction of phytate levels as compared to normal maize content with 21 µg zinc/g among young children (Chomba et al., 2015). The Zn bio-fortified maize genotype BIO-MZN01 led up to 8.0 t/h yield with both high-yielding and Zn-enriched potential (Johnson 2018b). The plants have the capability to acquire Fe and Zn from the surrounding rhizosphere and environment because they are unable to synthesize these nutrients in the plant body.

The inculcation of vitamin A in human diet is a must because humans are not capable of synthesizing vitamin A de novo in their bodies. The vitamin A requirement can be fulfilled directly as retinol from animal sources and its utilization as provitamin A carotenoids from plants. The major carotenoid is β-carotene, resulting in conversion into two molecules of retinol. The other carotenoids, including α-carotene, γ-carotene, and β-cryptoxanthin, are capable of generating one molecule of retinol each (Bai et al., 2011). All these compounds are members of a large group of isoprenoid compounds synthesized in plants (Owens et al., 2019). The reduced form of retinal is vital for rhodopsin biosynthesis, normal vision, and healthy epithelial and immune system cells. The acidic form of vitamin A, i.e. retinoic acid, is a key developmental morphogen. Thus, vitamin A is vital for normal metabolic activities in the human body. Its deficiency leads to xerophthalmia (progressive blindness), infant morbidity and mortality, altered immunological responses, and blindness in children. External sources viz. meat and milk products can be converted into retinol followed by conversion into retinal or retinoic acid. The carotenoids exist as provitamin A carotenoids in plants. Humans are capable of acquiring vitamin A due to the presence of β-carotene 15, 15′-monooxygenase enzyme in the human body, resulting in the conversion of β-carotene into retinal. The bio-fortification strategies to enhance the vitamin A or carotenoid content mainly include push, block, and metabolic sink, and post-harvest carotenoid stability strategies. The "push" strategy basically includes the over-expression of biosynthetic enzymes to enhance the metabolic flux upstream of a target compound. On the other hand, the "block" strategy leads to decreases in the metabolic flux downstream of a target compound through gene silencing. Third, metabolic sink results in the accumulation of carotenoid due to over-expression of gene and gene products. Lastly, the post-harvest carotenoid stability leads to an increase in the carotenoid content through metabolic engineering (Giuliano, 2017)

Moreover, it has been estimated that micronutrient deficiency including vitamin A, iron, and zinc is the cause of death of more than 1.5 million children mostly in South Asia and Sub-Saharan Africa (Caulfield et al., 2005). The malnutrition problem is further exacerbated by the increasing world population, which is likely to reach nine billion by 2050 and is a serious issue in the developing world (Khush et al., 2012). The maize endosperm contains low levels of the provitamin A carotenoid β-carotene and also does not accumulate zeaxanthin carotenoids causing vitamin A deficiency disease in populations consuming cereal-based diets (Berman et al., 2017).

9.3 Double bio-fortification

"Double bio-fortification" is the generation of crop varieties enriched with two diverse micronutrients in two manners viz. sequential and simultaneous gene stacking employing higher micronutrient concentration. The process of introgression of different favourable alleles of various genes in a single variety generally leads to the development of double-bio-fortified crops referred as "double fortification". The advent of molecular approaches comprising of molecular markers has led to the reduction of phenotypic screening, and the employment of different gene-targeting micronutrients in a single variety increases the nutrient content in crops thus eliminating the malnutrition bottleneck among humans. It has been reported that marker-assisted stacking of crtRB1, lcyE, and o2 led to the development of varieties enriched with both essential amino acids and provitamin A concentration (Muthusamy et al. 2014; Liu et al. 2015; Zunjare et al. 2018a). The bio-fortification strategy coupled with molecular approaches results in imparting both crtRB1 and lcyE genes contributing towards enhanced provitamin A, while o2 allele enhanced the lysine and tryptophan in the same genetic background. The strategy to introgress different favourable alleles of various genes in a single variety through biotechnological methods saves time and results in the reduction of phenotypic screening of individuals drastically (Hossain et al., 2019). The first double-bio-fortified maize hybrid "PusaVivek QPM 9 Improved" possesses higher provitamin A (8.15 ppm after two months of storage) and high tryptophan (0.74%) and lysine (2.67%) (Muthusamy et al. 2014; Yadava et al. 2017). This hybrid followed the introgression of crtRB1 allele in opaque-2-based hybrid, "Vivek QPM9" with 41% more tryptophan and 30% more lysine over the original hybrid "Vivek QPM9" at ICAR-IARI, New Delhi (Prasanna et al., 2020). Simultaneously, four QPM hybrids viz. HQPM1, HQPM4, HQPM5, and HQPM7 pyramided with crtRB1 and lcyE favourable alleles enhances the provitamin A concentration in the QPM genetic background via single-cross hybrid approach (https://iimr.icar.gov .in/). The pro A improved versions of the HQPM1, HQPM5, and HQPM7 hybrids developed through bio-fortification mechanisms (Jaiwal et al., 2019).

9.4 Bio-fortification approaches

Plant breeding and biotechnological interventions have the potential to enhance sufficient micronutrients in crops to feed the growing world population in the present times. There are basically two major complementary approaches for bio-fortification in crop plants (see Figure 9.1) (White and

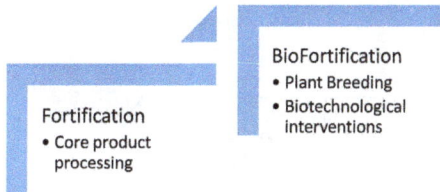

Figure 9.1 Types of fortification

Broadley, 2009). Primarily, agronomic approaches, through plant breeding including optimization of mineral fertilizers and improving the solubilization and mobilization of mineral elements in the soil and development of bio-fortified food, led to an increase in the concentrations of bioavailable mineral elements in food crops. Secondly, the biotechnology-based genetic engineering to acquire mineral elements for accumulation in edible tissues increased concentrations of "promoter" substances such as ascorbate, β-carotene, and cysteine-rich polypeptides, which stimulate the absorption of essential mineral elements by the gut and reduce concentrations of "antinutrients" such as oxalate, poly phenolics, and phytate which interfere with their absorption.

9.4.1 Plant breeding

The development of micronutrient-enriched foods through plant breeding methods has the power to alleviate the health issues among human beings with the development of highly nutritious foods (see Figure 9.2). Plant breeding approaches have immense application in the enhancement of productivity, resistance to biotic and abiotic stresses, and food palatability. The exploitation of genetic variation in micronutrients through conventional plant breeding has the potential to improve the mineral and vitamin levels in crops (Hirschi 2009). Maize bio-fortification programmes led to the development of several commercialized maize cultivars with high levels of provitamin A, lysine, tryptophan, and Zn and Fe content (Goredema-Matongera et al., 2021).

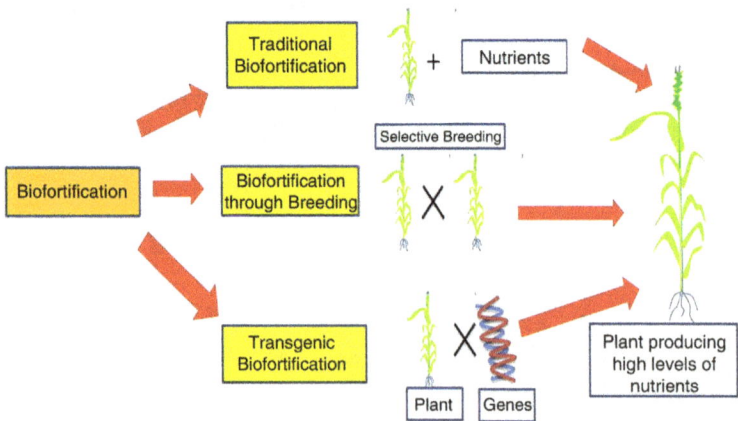

Figure 9.2 Fortification strategies to enhance nutrient levels in plants

9.4.1.1 Mineral enhancement and absorption

Plant breeding programmes have the objective of increasing levels of enhancers, viz. inulins meant to enhance the absorption of iron and zinc, and decreasing the levels of inhibitors, such as phytic acid that reduces the availability of iron and zinc, which can be helpful in combating micronutrient deficiencies. It has been reported that sweet corn varieties bred to have sugary (su) or shrunken (sh) gene mutations led to sugary enhancer (se) sweet corn with increased sweetness content (Gonzales and others 1976). It has also been reported that inclusion of vitamin A increased iron absorption, which positively correlated with the enhanced bioavailability of iron.

9.4.1.2 Provitamins and carotenoids

The quantification of β-carotene–bio-fortified maize based on consumption of a single serving of maize porridge is considered as a good bioavailability plant source of vitamin A content (Li et al., 2010). The presence of genetic diversity is a remarkable feature to create hybridization of lines between different clusters for agronomic traits and carotenoids (Halilu et al., 2016). The widely accepted breeding strategy in maize for increased concentrations of provitamins A include a) selection of maize genotypes with high concentration of provitamins A; b) development of hybrids based on the specific combining ability, and c) identification of alleles responsible for carotenoid biosynthetic pathway like increased activity of phytoene synthase, decreased activity of epsilon cyclase, and decreased activity of hydroxylases that catalyze the conversion of β-cryptoxanthin to zeaxanthin. The reduction of the phytate content through agricultural plant breeding techniques has the potential to improve the bioavailability of zinc from 20% to 40% in maize (Hotz 2009). The presence of additive gene action and medium to high heritability for provitamin A trait in maize represents the recurrent selection for provitamin A bio-fortification (Suwarno et al., 2014).

9.5 Plant breeding strategies

The plant breeding strategies for the enhancement of provitamin A content in maize through intra-population recurrent selection, pedigree selection, and conversion of white lines into yellow coloured prove to be effective for bio-fortification programmes.

9.5.1 Conventional approach

The conventional breeding mineral bio-fortification approaches led to increasing the density of the mineral nutrient of interest, decreasing the density of anti-nutritive compounds (nutrient inhibitors), and increasing the density of compounds to enhance bioavailability of micronutrients (Lung'aho et al., 2011). The backcrossing methodology employed the use of donor maize genotype to transfer the β-carotene gene into desired target line. The evaluation through comparative yield trials and multi-location trials has the potential for the identification of superior hybrid with increased β-carotene concentration (Natesan et al., 2020). The backcrossing programmes serve as an effective way to convert white lines into yellow genotypes aiming to increase the provitamin A concentrations (Shrestha and Karki, 2014). Backcross breeding also serves as a good method for the development of provitamin A based bio-fortified maize through phytic acid reduction (Pixley et al., 2013a). It has been reported that reduced phytic acid cultivar backcrossed with elite maize cultivar to enhance the iron absorption without alteration in the agronomic factors (Hurrell et al., 1992). Three-way crosses and single-cross hybrids are also employed for enrichment of provitamin A in maize endosperm and act as testers for the development of economical and optimum uniformity for hybrid maize plants. The backcrossing approach of mutant maize line imparts high β-carotene content in maize lines (Harjes et al., 2008). The utilization of mutation breeding explored a wide range of natural or induced viz. y1, vp2, vp5, vp7, vp9, w3, and y9 mutants linked with carotenoid content that are present in crop plants. The breeding of low-phytic acid mutations enhance the uptake of iron and zinc in the human diet (Adams et al., 2002).

9.5.2 Combinatorial approach of molecular and plant breeding approaches

The identification of effective molecular markers linked with the location of both endosperm hardness and amino acid levels is an effective way to combat the micronutrient deficiency at molecular level. The marker-assisted backcrossing i.e. foreground selection, recombinant selection, and background selection also has the potential to increase the provitamin A content in maize. The molecular markers linked with *PSY1*, *LCYE*, and *CRTRB1* genes facilitate the selection of favourable alleles in backcrossing or introgression breeding programmes. The marker-assisted backcross breeding employed the introgression of the favourable alleles of *CRTRB1* gene in elite maize genotypes followed by identification of superior hybrid with higher micronutrient content efficient to combat the micronutrient deficiency (Muthusamy et al. 2014). The molecular characterization of 5′ UTR of the lycopene epsilon cyclase (lcyE) gene deciphered four SNPs (SNP1: position 446, SNP2: position 458, SNP3: position 459, and SNP4: position 483) to differentiate low and high provitamin A lines having favourable alleles of lcyE 5′TE (Zunjare et al., 2018b). Recently, Abhijith et al. (2020) gave the first report of the development of breeder-friendly gene-based markers for the lpa1-1 and lpa2-1 mutant allele employed for maize bio-fortification. The use of gene-specific marker and simple sequence repeat markers for foreground and background selection led to the development of improved lines with recurrent parent genome recovery with good agronomic performance and high β-carotene concentration. All the above-mentioned investigations tend to reflect the potential of plant breeding approaches and molecular approaches in maize bio-fortification programmes.

9.5.2.1 Screening and characterization

The simplest plant breeding approach is the selection of high-carotenoid maize from the subset of maize on colour basis in the range from the darkest orange to yellow colour. The screening and characterization of varieties with high and low micronutrients are a prerequisite for the assessment of micronutrients. It has been observed that there is less correlation between carotenoid, provitamin A content, and colour. The most abundant provitamins A in maize include β-carotene, β-cryptoxanthin, and α-carotene, which are present in much smaller amounts. There is considerable variation in the ratios of total provitamins A to total carotenoid concentrations, as well as in the ratio of β-carotene to β-cryptoxanthin. Likewise, the selection for crossing of desirable parents with favourable specific combining ability has the potential to improve the high concentrations of provitamin A. Yellow genotypes of maize equipped with carotenoids, provitamin A activity, zeaxanthin, and lutein have been observed (Brenna and Berardo, 2004; Howe and Tanumihardjo 2006). The provitamin A activity resulted in the conversion of β-cryptoxanthin, and α- and β-carotene into vitamin A. (Monasterio et al., 2007). High performance liquid chromatography (HPLC) is an efficient approach for carotenoid determinations (Gama et al., 2005). Another methodology, namely near-infrared reflectance spectroscopy, also makes the screening of maize samples with provitamin A, Fe, and Zn easier (Monasterio et al., 2007). The adapted spectrophotometer methodologies are capable of screening maize germplasm employed for bio-fortification breeding programmes (Monasterio et al., 2007). Oikeh et al. (2003) observed significant (P<0.001) varietal differences in the mean of kernel-Fe and -Zn levels in tropical maize varieties. The selection criteria led to changes in the colour, taste, and aroma, affecting the consumer acceptance. African countries prefer white maize over yellow maize, and changes in dietary preferences for provitamin A bio-fortified orange-coloured maize is a challenging task (Pillary et al., 2011). The presence of proportional differences among provitamin A and non-provitamin A carotenoids in the maize kernel serve as an indicator of variability and require screening and characterization of maize germplasm for the selection of high provitamin A maize genotypes and for the development of bio-fortified commercial maize cultivars (Maqbool et al., 2018). The purpose of intra-population recurrent selection employed the use of full sib recurrent selection and S1 recurrent selection to enrich provitamin A in maize kernels. The plant breeding based pedigree selection results in the selection of superior individuals through selection and characterization of high nutritious maize hybrids (Shrestha and Karki, 2014).

9.5.3 Genetic engineering

The advent of genetic engineering-based transgenic techniques serve as an alternative to improve micronutrients and lead to the development of highly nutritious sustainable crops (Zhao and Shewry 2011). This has the potential to solve the challenging endeavour of bio-fortification through biotechnology-based techniques. Genetic engineering to enhance iron and zinc through introduction of genes of siderophores, chelating, reducing agents, and transporter proteins led to an increase in the uptake of nutrients from the soil surface. The genetic modification approaches resulted in an increased ascorbate concentration in maize kernels (Chen et al., 2003). Ortiz-Monasterio et al. (2007) employed biotechnological interventions and identified maize varieties with 15 µg β-carotene/g as compared to 0.1 µg/g standard varieties. The maize bio-fortification has been done with vitamin C, vitamin A, and vitamin B_9 content. Transgenic maize has been developed through the introduction of rice dehydroascorbatereductase (DHAR) gene under the control of the barley D-hordein promoter to enhance vitamin C, β-carotene, ascorbate, and folate in maize endosperm through transgenical engineering of carotenoid metabolic pathways (Naqvi et al., 2009). The introduction of bacterial genes for dihydrodipicolinic acid (DHPHS) and aspartokinase (AK) enzymes encoded by dapA gene from *Corynebacterium* and *lysC* gene led to a five-fold increase in the lysine content of rapeseed, corn, and soybean (Falco et al., 1995).

The exploration of carotenoid genes, including phytoene synthase 1 (PSY1), phytoenedesaturase (PDS), ζ-carotene desaturase (ZDS), lycopene β-cyclase (LCYB), lycopene ε-cyclase (LCYE), and β-carotene hydroxylase 1 (CRTRB1), has the potential to enhance the production of provitamin A carotenoids in maize (Maqbool et al., 2018). Savage (1960) observed that the phytate content causes an inhibition to zinc absorption. Generally, maize is equipped with a high phytate content which is undesirable. The identification of low-phytate alleles has the power to increase the bioavailability of zinc (Gibson, 1994). The presence of low-phytate content in maize grains contributing to homozygosity for allelic variants of a single gene has the potential to exploit the dietary phytate reduction on human mineral bioavailability and homeostasis (Hambidge et al., 2004). Drakakaki et al. (2005) demonstrated that the *phytase* gene from

Aspergillus niger resulted in reduction of phytate by up to 95% while soybean ferritin gene expression increased Fe content by 20–70%. The transgenic technology led to the development of transgenic corn plants with increased levels of three vitamins in the endosperm through the simultaneous modification of three separate metabolic pathways. It resulted in a 169-fold increase in β-carotene, six-fold increase in ascorbate, and double the folate content than normal maize (Naqvi et al., 2009). Harjes et al. (2008) worked on lycopene epsilon cyclase tapped for maize bio-fortification. It has been observed that psy1 in the carotenoid pathway resulted in the selection for endosperm accumulating carotenoids and conversion from white to yellow kernels (Palaisa et al., 2004).

The allelic variation at *crtRB1 3'TE* gene resulted in wide variation of kernel β-carotene and can be exploited for bio-fortification for crop improvement in maize (Vignesh et al., 2012). Transgenic strategies are not exploited in bio-fortified maize because of the presence of high natural genetic variation in attainable 15 µg/g of provitamin A carotenoids (Pixley et al., 2013b). The recessive o2 reduced the transcription of lysine ketoreductase (LKR), which reduces lysine in maize endosperm with the aim of enhancing the concentration of lysine (Kemper et al., 1999). The nas5 gene has an association with the Fe concentration and eases the multi-trait selection for the development of bio-fortified maize varieties through bio-fortification (Jia & Jannink, 2012). The purpose of transgenes is to specifically target the redistribution of micronutrients between tissues, increasing the biochemical pathways efficiency in edible tissues, reconstructing system biology pathways, increasing bioavailability of micronutrients through removal of antinutrients, and allowing multigene transfer. The transgenic approach has led to the development of "multivitamin corn" with high levels of β-carotene, ascorbate (vitamin C), and folate (vitamin B₉) in a single go (Carvalho & Vasconcelos, 2013). The main purpose of transgenic approaches is to modulate the transporters' expression through increasing the uptake and utilization efficiency of plants to increase the Fe and Zn content in the crop plants (Kerkeb et al., 2008). The reduction of the concentration of anti-nutritional factors like phytic acid also led to improving the bioavailability of micronutrients viz. Fe and Zn in the plant.

The over-expression of the bacterial crtB and crtI genes based on metabolic sink strategy under the control of a "super γ-zein promoter" led to a 34-fold increase of β-carotene content in the maize endosperm utilizing particle bombardment genetic transformation technology (Aluru et al., 2008). It has also been observed that over-expressed Zmpsy1 led to increased carotenoid levels up to 53-fold with triple carotenoid levels (Zhu et al., 2008). Researchers transformed the five carotenoid genes including three from plant sources PSY1, LCY-B, and CHY and two from bacterial sources CRTI and CRTW under the control of an endosperm-specific promoter with the aim of increasing the carotenoid content in maize. The diverse activities of carotenoid β-hydroxylases and the functional analysis of the corresponding genes through biotechnology programmes play a vital role in increasing carotenoid content in maize due to the low ratio of pro-vitamin A to non-provitamin A carotenoids in the endosperm and varying carotenoid profiles of different maize varieties (Berman et al., 2017). The extent of provitamin A bio-fortification of maize requires the understanding of carotenoid biosynthesis pathway, key genes involved in provitamin A bio-synthesis in maize through biotechnological interventions. Yan et al. (2010) recommend the use of user-friendly PCR-based markers for introgression of favourable CRTRB1 and LCYE alleles in tropical maize germplasm to enhance the provitamin A content in maize endosperm. The prevalence of carotenoid content in maize endosperm linked with expression of dif-ferent genes like, *PSY* (Vallabhaneni and Wurtzel, 2009), *LCYE* (Harjes et al., 2008), *LCYB* (Bai et al., 2009), and *HYD3/CRTRB1* (Vallabhaneni et al., 2009) is specifically involved in carotenoid biosynthesis pathway. It has been observed that *CRTRB1* results in the conversion of β-carotene to β-cryptoxanthin in maize endosperm with respect to carotenoid pathway and low hydroxylation level and has significant correlation with decreased β-cryptoxanthin and increased β-carotene concentration in maize (Babu et al., 2013). Multigene engineering through DNA transformation methods has the potential to develop the nutritionally improved genotypes utilizing synthetic promoters with synthetic transcription factors in coordination with transcriptional control of multiple genes and its implementation in the plant system (Altpeter et al., 2005). High-carotenoid corn (Carolight®) was developed through the introduction of PSY1 gene from corn and CRT1 gene from *Pantoeaannatis* in the genetic background of white endosperm M37W variety (Zanga et al., 2016). The genetic manipulation serves as an effective way to explore the wide range of variation to enhance the low level of mineral content in maize. The allelic variation in 5′ untranslated region (UTR) of lycopene epsilon cyclase (lcyE) gene is efficient in increasing the provitamin A concentration in maize. The identified SNPs hold promise in enrichment of provitamin A in maize for marker development and marker-assisted selection.

9.5.4 Marker-assisted selection

The molecular markers target many different traits employed for genetic improvement of numerous crop plants. The application of DNA molecu-lar markers for plant breeding referred to as marker-assisted selection improves the precision and efficacy of plant breeding. MAS has several advantages including selection of targeted gene, shortening the duration of varietal development, introgression of desirable trait into agronomi-cally superior variety, and improvement of nutritional quality in maize.

Although HPLC- and Ultra-performance liquid chromatography (UPLC)-based quantification eases the screening of maize germplasm with high carotenoid, it is time-consuming and highly expensive. Moreover, colour quantification is not a clear indication of β-carotene quantification, so, MAS serves as the best approach for bio-fortification. The introgression of high β-carotene content through marker-assisted selection enables the development of bio-fortified tropical maize (Yan et al., 2010). The low and poor correlations between colour and carotenoid revealed that marker-assisted selection is a better method than colour basis selection method (Harjes et al., 2008). The introgression of rare alleles namely *lcyE* and *crtRB1* into elite maize germplasm through molecular marker-assisted breeding in maize is efficient in increasing levels of provitamin A in Africa, leading to the reduction of vitamin A deficiency (Fiedler et al., 2013). Moreover, the introgression of alleles of crtRB1 and lcyE gene(s) in commercial hybrids using MAS lycopene ε-cyclase (lcyE) and β-carotene hydroxylase1 (crtRB1) for high provitamin A content resulted in four and a half times more provitamin A as compared to original hybrids (Hossain et al., 2019). The marker-assisted stacking of crtRB1, lcyE, and o2 in the four maize hybrids (HQPM1, HQPM4, HQPM5, and HQPM7) popularly grown in India resulted in enriched pro A, lysine, and tryptophan (Zunjare et al., 2018a). The stacking of shrunken2 (sh2), opaque2 (o2), lycopene epsilon cyclase (lcyE), and β-carotene hydroxylase (crtRB1) genes in the parents of four hybrids viz. APQH1, APHQ4, APHQ5, and APHQ7 employed for marker-assisted backcross breeding enabled the enrichment of sweet corn hybrids with multiple essential nutrients (Baveja et al., 2021). The International Maize and Wheat Improvement Center (CIMMYT), in collaboration with the International Institute of Tropical Agriculture (IITA) and the National Agricultural Research Systems (NARS), developed in 17 countries of Sub-Saharan Africa a broad range of QPM cultivars with twice the content of limiting amino acids lysine and tryptophan (Krivanek et al., 2007). Marker-assisted breeding is the best approach for the introgression of the crtRB1 3'TE favourable allele using high β-carotene CIMMYT in-breeds as donors for the development of provitamin A-rich maize cultivars in India (Vignesh et al., 2012). The SSR markers for the opaque-2 locus phi057 and phi112 were reported to be highly polymorphic and extensively used for the introgression of the opaque-2 gene in non-QPM background for the development of QPM maize. The exploitation of favourable alleles like PSY1, LCYE, CRTRB1, and HYD3 in maize through MAS increased the enrichment of provitamin A in endosperms (Maqbool et al., 2018). QPM is genotype developed through the incorporation of opaque-2 gene linked with modifiers and results in the increment of twice the amount of lysine and tryptophan as compared to normal maize endosperm. The simultaneous use of colour marker with MAS strategy to enhance the LCYE content serves as an economical option for provitamin A bio-fortification (Maqbool et al., 2018). The rapid advances in genome research and molecular technology have led to the use of DNA marker-assisted selection (MAS), which holds promise in enhancing selection efficiency and expediting the process of development of new varieties/hybrids with higher yield potential (Ribaut and Hoisington, 1998). The association of DNA markers with favourable alleles is linked with high provitamin A content in maize germplasm (Babu et al., 2013).

9.5.5 Mapping strategies

The association and linkage mapping populations in maize revealed that the gene encoding β-carotene hydroxylase 1 (crtRB1) is a principal quantitative trait locus associated with the provitamin A β-carotene concentration in maize kernels. The gene exploitation of crtRB1 has significant correlation with higher β-carotene concentrations. Another association-mapping approach also eases the selection of the most favourable allele within a diverse genetic background with resolution within 2000 base pairs (Flint-Garcia et al., 2005). Yan et al. (2010) confirmed that crtRB1 (β-carotene hydroxylase) gene is linked with β-carotene accumulation in maize kernels, confirmed through association-mapping approach following linkage disequilibrium. The existence of quantitative trait locus (QTL) for carotenoid content was deciphered by Wong et al. (2004) and Islam (2004) in maize populations. Wong et al. (2004) performed the first QTL mapping study of maize grain carotenoids and showed that the identified QTL has an association with carotenoid biosynthetic genes y1 and zds1. The major QTLs were present on chromosomes 6 and 7 linked with phytoene synthase and ζ-carotene desaturase vital for the carotenoid content pathway. For QTL detection, the existence of natural genetic variation is vital but the micronutrient Fe content is present in low amounts in maize, so here transgenic approaches serve as an alternative in bio-fortification programmes. QTL analysis was conducted to determine the three QTL for grain Fe concentration (FeGC) and ten QTL Fe bioavailability (FeGB) from the Intermated B73×Mo17 (IBM) recombinant inbred (RI) population iron bio-fortification of maize grain (Lung'aho et al., 2011). Šimić et al. (2012) detected the QTLs for phosphorus (P), iron (Fe), zinc (Zn), and magnesium (Mg) concentrations in maize grain in a mapping population as well as QTLs for bioavailable Fe, Zn, and Mg. They identified three QTLs for Fe/P, Zn/P, and Mg/P on chromosome 3 linked with phytase genes ZM phys1 and phys2. The combined use of association mapping with linkage mapping is widely exploited for the dissection of complex traits in maize (Yan, Warburton, and Crouch, 2011). CIMMYT developed the carotenoid association-mapping (CAM) panel for the identification of correlation of carotenoid concentration in the maize endosperm. The CAM panel led to the identification of

SNPs linked with carotenoid content in maize endosperm (Suwarno et al., 2015). The significant association signals to the CRTRB1 gene linked with β-carotene concentration in a wide range of maize germplasm, revealed through association studies (Suwarno et al. 2015). The QTL co-localized with zeta-carotene desaturase (ZDS) and y1 genes, and led to the accumulation of carotenoids in maize endosperm (Wong et al., 2004). Harjes et al. (2008) mapped a QTL on chromosome 8 bin 5, and this QTL significantly affected the ratio of α to β branching in carotenoid biosynthesis pathway. The identification of QTLs for multiple traits offers breeding competitiveness among crop species in an efficient manner.

9.6 QPM maize

During 1960, CIMMYT initiated the use of opaque-2 gene for the development of lysine- and tryptophan-enriched maize known as "Quality Protein Maize", or QPM (Bjarnason & Vasal, 1992). QPM maize serves an alternative food to overcome the micronutrient deficiency in developing countries and can be essentially exchangeable with normal maize. The differences between maize and QPM maize lies in the fact that QPM has a 60–100% higher lysine and tryptophan content compared to normal maize, with a biological value of about 80% compared with 40–57% for normal maize (Bressani, 1992). It has been observed that the protein of normal maize has a biological value of about 40% while QPM is 90%, which leads to the development of superior nutritional value of QPM maize to meet the demand of dietary protein (Rahmanifar and Hamaker, 1999). QPM maize has been developed through three genetic systems; a) opaque-2 genetic system comprising of *Opaque-2* gene, recessive mutant alleles required for expression of a gene encoding the ribosomal inactivating protein and α-zein-coding (22 kDa) genes, b) En-modifier genetic system or endosperm hardness modifier genes, which results in the transformation of soft endosperm into hard and vitreous endosperm without loss of protein quality, and c) associated gene system or aa-modifier genetic system comprising of amino acid modifier genes that affect the lysine and tryptophan content of maize grain in segregating generations. Numerous QPM cultivars have been released in India, Indonesia, Mozambique, Perú, Uganda, Venezuela, Vietnam, and elsewhere throughout the world (Pixley et al., 2005). The QPM refers to maize homozygous for the o2 allele with enhanced lysine and tryptophan content, voiding the negative secondary effects of a soft endosperm. Moreover, the QPM appearance is similar to normal maize, and researchers have also been able to differentiate it in laboratory analysis (Krivanek et al., 2007). The clear distinction is based on the fact that QPM is the product of conventional breeding, avoiding a genetic engineering approach during QPM development. The QPM maize with opaque-2 mutation resulted in increased levels of the essential amino acids tryptophan and lysine. QPM with double the quantity of lysine and tryptophan enhanced the biological value of QPM protein as compared to normal maize (Hossain et al., 2019). It has been observed that QPM consumption led to an 8% increase in the rate of growth in height and a 9% increase in the rate of growth in weight in infants and young children (Gunaratna et al., 2008). Vivek et al. (2008) reported that the effect of QPM genotypes pollen contamination depends on the proximity of normal maize, flowering between normal and QPM genotypes, competitiveness of pollens for fertilization, wind speed, and wind direction. There is a need to conduct studies to estimate the extent of Zn in bio-fortified maize. Several varieties of QPM with high lysine and tryptophan (Machida et al., 2010), orange maize with high provitamin A, and Zn-enhanced maize (Chomba et al., 2015) have been commercialized globally. QPM breeding involves the manipulation of three distinct genetic systems. The first and central component is the presence of the recessive mutant allele of the Opaque-2 gene responsible for transcription factor of zein synthesis. The second distinct genetic system managed is the presence of the alleles of endosperm hardness modifier genes (referred as "enmodifiers") for the conversion of the soft/opaque mutant endosperm to a hard/vitreous endosperm with little loss of protein quality. Moreover, the third genetic system consists of distinct set of amino acid modifier genes (referred as "aa-modifiers") affecting the relative levels of lysine and tryptophan content in the maize grain endosperm (Krivanek et al., 2007). Based on the successfulness of QPM research that closed the yield gap between o2 genotypes and normal endosperm counterparts, simultaneously reducing ear rot and insect damage, researchers Surinder K. Vasal and Evangelina Villegas obtained a reward in the form of the World Food Prize in 2000 (Vasal, 2000). The technological advancements including high throughput, single seed-based DNA extraction, high density SNP genotyping strategies, and breeder-ready markers for adaptive traits in maize have the potential to enhance the efficiency and cost effectiveness of MAS in QPM breeding programmes.

9.7 RNA interference

RNA interference (RNAi) is a phenomenon for sequence-specific gene regulation through the introduction of dsRNA for the inhibition of transcription and translation mechanism in plants. The small RNA specifically interferes with the translation of target mRNA transcript leading to the suppression of gene expression. The researchers extensively employed this technique for nutritional enrichment in food crops. The RNAi reaction was based on the triggering of gene silencing reaction by long dsRNA, followed by activation of dicer enzyme that led to the cleavage of dsRNA into small interfering RNAs (siRNA) of ~21 nucleotides and that involved the binding with Argonaute proteins. The entire process results in the removal of one strand of the dsRNA, leaving the strand intact amenable to bind to messenger RNA target sequences, resulting in the silencing of target genes. RNAi is capable of inducing bio-fortification through the introduction of essential elements like Zn, Mg, Cu, Se, Ca, Fe, I, S, P, etc. in numerous plant species viz. tomato, brassica, potato, maize, wheat, barley, and cotton (Saurabh et al., 2014). Segal et al. (2003) utilized RNAi constructs derived from a 22-kDa α-zein and produced a dominant opaque phenotype. The silencing of carotenoid β-hydroxylases is efficient for conversion of β-carotene to zeaxanthin in maize endosperm. The silencing one of the zein storage protein genes allows the protein complement to be filled with lysine-rich storage proteins that enables the bio-fortification of lysine content in maize. Wu and Messing (2011) employed RNAi construct against both 22- and 19- kDa α-zeins for bio-fortification, and through silencing of both α–zeins found that dominance of Mo2s over an RNAi phenotype was vital for accelerating backcrossing breeding programmes. Basically, researchers Surinder K. Vasal and Evangelina Villegas showed that transgenic plants led to significant reduction in synthesis of zeins followed by the accumulation of high lysine concentration. Berman et al. (2017) used RNAi to silence the endogenous *ZmBCH1* and *ZmBCH2* genes, encoding two non-hemedi iron carotenoid β-hydroxylases and found that ZmBCH2 is responsible for the conversion of β-carotene to zeaxanthin in the maize endosperm. RNAi technology is one of the widely acceptable alternatives for the development of bio-fortified foods (Jagtap et al., 2011). Moreover, RNAi serve as one of the best seed-specific strategies for the development of dominant high-lysine corn through the suppression of the expression of zein storage proteins in an efficient manner.

9.8 Mutagenesis

Mutations are vital for the exploration of novel allelic variability for target traits or genes to revolutionize the plant breeding and biotechnological prospects for crop improvement. Since long ago, a mutagenesis approach has been employed for genetic improvement of different quantitative and qualitative traits in crop plants. The mutational approaches, including seed irradiation with X-rays, gamma-rays, and chemical mutagens like ethyl methane sulphate (EMS), result in the generation of random DNA changes and induce point mutations that can be analyzed through the analysis of loss of function mutants. Moreover, the subsequent generations and mapping studies result in the identification of mutational phenotypic effects depending upon the dominant and recessive nature of alleles. The natural mutations in the promoter region of the yellow1 gene encoding phytoene syntheses is linked with the up regulation of carotenoid expression in the endosperm of maize (Palaisa et al., 2003). The mutagenesis resulted in the reduction of phytic acid in maize (Raboy et al., 2000). Unger et al. (1993) created two mutant forms of the opaque-2 gene, that resulted in the inhibition of the expression of 22-kDa α-zein by ~ten-fold in maize endosperm suspension cells, followed by an increase in non-zein proteins that are rich in these essential amino acids. Vietmeyer (2000) reported that natural spontaneous mutation of maize develops soft and opaque grains in a Connecticut, US maize field, and they named it "opaque-2" (Singleton, 1939). The researchers considered the utility of seeds of opaque-2 maize for the protein quality improvement in maize accessions (Paes & Bicudo, 1994). Later on, the mutations in maize kernels, viz. opaque2 (o2), opaque6 (o6), opaque7 (o7), opaque11 (o11), floury2 (fl2), floury3 (fl3), mucronate (Mc), and defective endosperm (De-B30), resulted in the production of higher concentrations of lysine and tryptophan in the endosperm as compared to traditional maize (Boyer & Hannah, 2001). It has been observed that o2 mutant resulted in a two-fold increase of lysine and tryptophan as compared to normal maize (Mertz et al., 1964). It holds the title of one of the first major maize bio-fortification breakthroughs through breeding of high-lysine maize using genetic mutagenesis approach (Mertz et al., 1964). The mutational approach resulted in the reduction of lysine deficient zein proteins followed by enhanced synthesis of lysine-rich non-zein proteins (Habben et al., 1993). The bioavailability of micronutrients with the phytate:mineral molar ratios in the low-phytic acid mutants is efficient for micronutrient bio-fortification programmes (Aggarwal et al., 2018). EMS mutagenesis results in the identification of alleles of LCYE genes responsible for conversion of endosperm colour from yellow to orange leading to high vitamin A content (Harjes et al., 2008). The HPLC analysis also eases the quantification of the carotenoid content in mutant endosperms (Maqbool et al., 2018). The presence of genetic variation *Zea mays* crtRB1 enables the enhancement of β-carotene content in maize grain (Yan et al., 2010). The mutagenesis is an efficient approach resulting in two times more lysine and tryptophan in mutants than wild type (0.125% lysine and 0.035% tryptophan) in maize crop (Sarika et al., 2017). A micronutrient viz. iron was employed for bio-fortification through mutation-based endosperm-specific over-expression of soybean ferritin (Aluru et al., 2011; Curie et al., 2001). The reduction of phytic acid is an efficient approach to enhance the bio-fortification without compromising kernel viability (Raboy et al., 2000; Shi et al., 2005, 2007; Raboy, 2007). The researchers employed the inhibition of expression of 22-kDa α-zein followed by enhanced synthesis of lysine-rich non-zein proteins (Unger et al., 1993; Habben et al., 1993). With the utilization of linked marker, the mutant allele has the potential to transfer it to an elite maize inbred enabling the validation of potentiality of the marker in MAS for low-phytic acid (Sureshkumar et al. 2014). Marker-assisted introgression of lpa1 and lpa2 mutants in early maturing in breeds, viz. CM145 and V334 results in the development of proteinaceous maize crop. Mutagenic low-phytate maize enhances the improvement of zinc bio-availability in an efficient manner (Hambidge et al., 2004; Mazariegos et al., 2006; Miller et al., 2007).

9.9 Genome-wide association studies

Genome-wide association studies (GWAS) serve as a platform for the identification of allelic variation for genes and genomic regions to decipher the agronomic and nutritional traits in the plants. The process of decoding genotype-phenotype associations in plant species through GWAS is possible with the advancement in next-generation sequencing (NGS) technologies. The rare favourable alleles of lycopene epsilon cyclase (LcyE) and β-carotene hydroxylase 1 (crtRB1) for high provitamin A content was identified through GWAS-based candidate gene association analysis for maize grain carotenoid composition (Harjes et al., 2008; Yan et al., 2010). It serves as a programme based on high density and extensive marker coverage to increase the understanding of inheritance of target traits. The GWAS led to the identification of novel genes responsible for quantitative variation of grain carotenoid levels in a maize inbred panel suitable for maize bio-fortification programme. GWAS is influenced by linkage disequilibrium in association panel due to low mapping resolution in larger LD blocks due to slower rate of LD decay. GWAS studies have paved the way for the exploitation of natural variation in an efficient and controlled manner.

Harjes et al. (2008) employed association mapping, linkage mapping, and mutagenesis for the identification of allelic variability in a near-isogenic background of maize. The researchers identified the LCYE gene most linked with carotenoid activity and β-carotene and β-cryptoxanthin concentration. GWAS study led to the identification of a total of 58 candidate genes linked with biosynthesis and retention of carotenoids in maize and deciphered four major-effect loci, lcyE, crtRB1, zep1, and lut1, with maize grain carotenoids (Owens et al., 2014). The CIMMYT's maize breeding program employed candidate gene-based GWAS and identified pro- provitamin A enhancing alleles including LCYE and crtRB1 in maize. CIMMYT's maize carotenoid association-mapping panel composed of 380 diverse tropical, subtropical, and temperate inbred lines identified genes encoding hydroxylases and carotenoid cleavage dioxygenase 1 (CCD1) associated with carotenoid pathway, and carotenoids concentration has the potential for carotenoid bio-fortification in maize (Suwarno et al., 2015). The whole genome prediction models are vital for Fe and Zn bio-fortification in maize grain (Welch & Graham, 2002).

Maize kernels have low α-tocopherol content increasing the risk of vitamin E deficiency as human populations rely on maize crop. The ZmVTE1 (tocopherolcyclase) gene has an association between tocotrienol composition. GWAS also led to the identification of one 2.4-Mb genomic region significant for αt content in maize kernels (Li et al., 2012). Lipka et al. (2013) targeted tocochromanol levels through GWAS and identified that ZmHGGT1 (homogentisate geranylgeranyltransferase) and one prephenate dehydratase parolog (of four in the genome) that contribute to tocotrienol variation in the maize association panel. The genes DXS1, GGPS1, and GGPS2 are vital for the accumulation of precursor isoprenoids, HYD5, CCD1, ZEP1 involved in hydroxylation and carotenoid degradation (Suwarno et al., 2015).

The GWAS-based maize association panel suggested that the ZmVTE4 gene serves as a major gene causing natural phenotypic variation of α-tocopherol. GWAS deciphered that lcyE and zep1 genes affect carotenoid composition, dxs3 and dmes1 are involved in isoprenoid biosynthesis, ps1 and vp5 genes are specifically involved in carotenoid pathways and vp14 genes undergo cleavage of carotenoids in maize grain (Owens et al., 2019). Baseggio et al. (2019) analyzed the tocopherol pathways and identified that vte4 controls α-tocopherol which is linked with highest vitamin E activity, vte1 and hggt1 genes are associated with tocotrienols, and sh2 contributes towards ς- and γ-tocotrienols. Wu et al. (2021) identified that nas5 (nicotianamine synthase5) gene is involved in the synthesis of nicotianamine, a metal chelator, and has an association with both zinc and iron, suggesting a common genetic basis controlling the accumulation of these two metals in the maize grain. The exploitation of genes derived from GWAS studies has the potential to provide nutrition and to function as antioxidants in both plants and animals. Overall, the GWAS studies led to the identification of several genes showing significant associations with carotenoid levels efficient for bio-fortification.

9.10 Harvest-Plus program

Harvest-Plus program employed conventional breeding to improve the micronutrient content in the staple crops viz. wheat, rice, maize, cassava, pearl millet, beans and sweet potato in Asia and Africa (Zhao & Shewry, 2011). Breeding strategies rely on the limited genetic variation present in the germplasm that can be sourced from sexually compatible plants (Hirschi, 2009). The Harvest-Plus maize strategy serves as promising program that include the selection of increased micronutrient concentrations to increase and decrease the enzymatic content in the carotenoid biosynthetic pathway for the development of bio-fortified maize. The ultimate goal of Harvest-Plus is the reduction of micronutrient malnutrition among poor populations in Africa, Asia, and Latin America through improvement of food security with the enhancement of quality of life. The Harvest-Plus and Bio-Fort projects employed to increase the agronomic and provitamin A traits in maize led to the development of improved maize synthetic cultivars with 8.2 µg/g provitamin A content and 31.5 µg/g total carotenoid content (Schaffert et al. 2011). Since 2003, Harvest-Plus has been contributing towards bio-fortification to reduce micronutrient deficiency (Bouis and Saltzman, 2017) The Waite Analytical Services, University of Adelaide improved the Fe and Zn concentrations of more than 1000 CIMMYT-improved maize genotypes and 400 "core accessions" from different environments. CIMMYT is an improvement programme for Zn concentration. Harvest-Plus institute IITA are working on the enhancement of Fe and Zn enhancement in maize (Monasterio et al., 2007). The bio-fortification of crops through the combinatorial plan of action of mineral fertilizers with the plant breeding approach also has the potential to enhance mineral elements and improvement of crop yields (Pfeiffer & McClafferty, 2007). The Harvest-Plus has successfully installed 25 state-of-the-art micronutrient analytical laboratories at nine CGIAR centres, 12 at National Agricultural Research Stations and four at universities. Moreover, more than 100 laboratory staff have been trained in 13 countries in field sampling, sample preparation, equipment calibration, and operations. The initiative of Harvest-Plus is to provide services to the partners from public and private sectors for high-throughput quantification of minerals in different crops (Andersson et al., 2017).

9.11 Additional strategies

The plant breeding and molecular approaches including several additional strategies that have been employed for bio-fortification in maize have been summarized here briefly. Hoekenga et al. (2011) employed human cell culture (Caco-2)-based bioassay to improve the iron nutritional quality in maize, which is a phenotyping tool to guide genetic analysis of the trait using intermitted B73 x Mo17 recombinant inbred population. The next step is the consumer acceptance of yellow, provitamin A, bio-fortified maize to deal with the serious problem of vitamin A deficiency (Pillay et al., 2011). The maize seed bio-fortification with plant growth promoting bacteria *Azotobacter chroococcum, Azospirillum lipoferum +Azospirillum brasilense,* and *Pseudomonas fluorescence* improves the sustainability of agriculture (Hamidi et al., 2011). The determination of Fe nutrition in maize kernels was through integrated genetic, physiological, and biochemical strategy in the Intermitted B73×Mo17 recombinant inbred population of maize (Lee et al., 2002). Organic manure viz. farm yard manure, press mud, fisheries manure, and slaughter house waste in combination with Zn soil application and Zn foliar spray with recommended doses of N:P:K (140:100:60 kg ha^{-1}) improved the maize bio-fortification in a sustainable manner (Naveed et al., 2018). Maize bio-fortification enhanced with spirulina (Tuhy et al., 2015) enriched biomass of blackcurrant seeds (Samoraj et al., 2015), enriched with zinc (ZnII), manganese (MnII) and copper (CuII). The bio augmentation of Zn-solubilizing rhizobacteria (Mumtaz et al., 2017), zinc and iron fertilization as soil and foliar applications (Saleem et al., 2016), and ZnSO$_4$ (Imran & Rahim, 2016) improve the impact of bio-fortification. The zinc oxide ZnO-nano-particulates also improved the accumulation of zinc in maize grains (Umar et al., 2020). Quantitative techniques like atomic absorption spectrometry, inductively coupled plasma-optical emission spectrometry, ICP mass spectrometry, XRF spectrometry, synchrotron X-ray, elemental distribution maps, secondary ion mass spectrometry, fluorescence spectroscopy, micro-XRF spectroscopy and laser-induced breakdown spectroscopy ease the quantification of micronutrient contents for bio-fortification of crops. The genetic transformation-based biolistic method led to the development of multi-transgenic plants to increase the carotenoids and vitamin content in maize. The improvement in zinc and iron accumulation in maize grains through transgenic approach has been exploited and identified to show that ZmZIP5 play a vital role in Zn and Fe uptake and root-to-shoot translocation (Li et al., 2019).

9.12 Advantages

Bio-fortified maize maintained the vitamin A status in Mongolian gerbils, which proved that it is possible to enhance vitamin A content (Howe & Tanumihardjo, 2006). The bio-fortification-based supplementation programmes enable rural and urban poor in certain countries to have a healthy diet. The identification of genetic resources with high levels of the targeted micronutrients with high rate of heritability of the targeted trait micronutrient is the success. The exploration of the availability of nutrition-based genes through high-throughput screening tools allows a better understanding of genotype by environment interactions to be gained. The combinatorial approach of development of bio-fortified maize varieties with high yield potential, disease resistance, and consumer acceptability is a remarkable feature in the present scenario. The plant breeding strategies target micronutrient levels keeping in view the food intake, retention, and bioavailability related to food processing, anti-nutritional factors, and promoters. The strategies resulted in the bio-fortification of essential minerals and vitamins that are vital for the reduction of chronic micronutrient malnutrition (Lyddon, 2004). The advantages of bio-fortified foods include that they are inexpensive, locally adaptable, and a long-term solution to micronutrient deficiencies, but cultural preferences have limit the acceptance of bio-fortification-based foods. The advent of bio-fortification of these cereals with vitamins, minerals, and micronutrients has the power to combat malnutrition deficiency in crops. Bio-fortified maize genotypes have high yields compared to non-bio-fortified maize genotypes (Vivek et al., 2008). It has been reported that bio-fortified seeds have an indirect effect in providing better protection against different biotic and abiotic stresses leading to the enhancement of yield through the introduction of minerals in seeds (Welch & Graham, 2004).

9.13 Limitations

The crops face difficulty during uptake of micronutrients at the root-soil interface in the rhizosphere. The consideration of stimulation of root-cell processes, root-cell plasma membrane absorption mechanisms, through transporters and ion channels, and translocation to edible plant organs should be taken into account for micronutrient accumulation in plants. Reported that more than 50% loss of β-carotene occurs during preparation of bread using fortified maize flour. So, the main interest should be to reduce post-harvest losses of provitamin A in the bio-fortified maize. The limitation of transgenics includes regulatory and bio-safety concerns. The micronutrient deficiency content, maize production, importance as staple foods, household consumption scenario, and preparation methods are considerable factors for the development of bio-fortified crop varieties (Monasterio et al., 2007). End-use quality is a considerable factor, as maize grain hardness and factors affect the gelatinization and pasting properties of starch (Rooney and Suhendro, 1999). The carotenoid retention in bio-fortified maize should be must for post-harvest storage and packaging methods through effective control of moisture content and temperature of the kernels during storage (Taleon et al., 2017, Ortiz et al., 2016). The scrutiny of the influence of temperature and humidity enhances the stability of carotenoids in bio-fortified maize (*Zea mays* L.) genotypes during controlled post-harvest storage (Ortiz et al., 2016). International organizations developed iodized salt to eliminate iodine deficiency, but the excess amount of iodized salt in diet is not desirable (Hetzel, 2000). The breeding approach to cross each high provitamin A, iron, and zinc concentration genotype with elite cultivar has the potential to increase the provitamin A and micronutrient content, but these strategies are not feasible in terms of cost and complexity of laboratory analyses (Monasterio et al., 2007). Bio-fortification breeding programmes require expertise to take care of the different stages of breeding to avoid contamination either physical or genetic affecting the genetic basis of quality traits. The field trial management, line maintenance, and seed production and quantification of the nutritional contents of targeted should ensure the genetic gains in the bio-fortification programme. Xenia-effected pollen of normal non-bio-fortified maize resulted in the dilution of the nutritional concentrations.

9.14 Conclusion

The bio-fortification approach holds great promise in the present era for the improvement of the nutritive value of major crops. The advent of plant breeding and genetic engineering has the potential to increase the essential micronutrient and vitamin content in an efficient manner. It is a platform to inculcate minerals in food crops and make them available in the crops. The external fortification of these nutrients has limited value, as such bio-fortified food materials are generally available to the poor populations at an affordable cost. The policy regulation with "precautionary" principles will definitely benefit society at large scale to reach food security demand. The evaluation of clinical trials of bio-fortified foods led to enhanced bioavailability for nutritious health. The prevalence of natural or induced variation through conventional plant breeding and genetic engineering methods through molecular studies were capable of enhancing nutritional components of maize. The carotenoid biosynthesis pathway and key genes viz. PSY1, PDS, ZDS, LCYB, LCYE, and CRTRB1 contributing towards the provitamin A production viz. α-carotene, β-carotene, and β-cryptoxanthin and non-provitamin A carotenoids including lutein and zeaxanthin in maize endosperm modulate the carotenoid biosynthetic pathways. The genomics-based GWAS studies enable the understanding of underlying mechanism of natural variations in maize kernels that could be used for marker-assisted breeding in maize kernels for maize improvement programmes. The genome-wide association mapping coupled with high-resolution SNP density is effective in the identification of natural allelic variations in highly diverse maize germplasm for the identification of mineral content in maize kernels. Future research is based on the validation of effects and exploration of gene selection to increase the expression rates linked with carotenoid content (reported by Sekhon et al., 2011). The combinatorial approach of selection of darker orange maize grain with MAS for favourable QTL alleles is linked with carotenoid biosynthetic gene content viz. crtRB1 to increase the provitamin A carotenoid levels. The international breeding organizations of CIMMYT, IITA, and Harvest-Plus are partners efficient in developing the provitamin A dense maize varieties targeting nutritional properties in maize. The prevalent deficiency in two essential amino acids viz. lysine and tryptophan in maize resulted in poor net protein utilization and low biological value of traditional maize varieties. So, the exploitation of varieties through both plant breeding and biotechnological approaches is vital for the development of bio-fortified maize crop. The nutritional improvement in maize is based on the breakthrough theory during 1960s through the enhanced nutritional quality of the maize mutant opaque-2 (o2) (Mertz et al., 1964). The tocopherol content (alpha (α), beta (β), delta (δ) and gamma (γ)-tocopherol) are the group of four natural lipid-soluble antioxidant compounds essential for vitamin E. Although micronutrients like vitamins and minerals are vital for food safety, these are not available in sufficient amount in the human diet, so bio-fortification is the best approach to inculcate micronutrients in through plant breeding and biotechnological interventions. Plant breeding for the accumulation of micronutrients is a difficult approach, so the combinatorial approach of plant breeding and biotechnology serves as best tool to enhance the micronutrient content in crops. The exploitation of well-characterized carotenoid biosynthetic pathways results in the identification of number of crucial genes, multiple forms, variants, or alleles, capable of acting singly or in combination with other genes within or outside the pathway to enable the selection of final carotenoid profiles and concentrations in the maize kernels.

References

Abhijith, K. P., Muthusamy, V., Chhabra, R., Dosad, S., Bhatt, V., Chand, G., … Hossain, F. (2020). Development and validation of breeder-friendly gene-based markers for *lpa1-1* and *lpa2-1* genes conferring low phytic acid in maize kernel. *3 Biotechnology, 10*, 121. https://doi.org/10.1007/s13205-020-2113-x

Adams, C. L., Hambidge, M., Raboy, V., Dorsch, J. A., Sian, L., Westcott, J. L., & Krebs, N. F. (2002). Zinc absorption from a low–phytic acid maize. *American Journal of Clinical Nutrition, 76*(3), 556–559.

Aggarwal, S., Kumar, A., Bhati, K. K., Kaur, G., Shukla, V., Tiwari, S., & Pandey, A. K. (2018). RNAi-mediated downregulation of inositol pentakisphosphate kinase (ipk1) in wheat grains decreases phytic acid levels and increases Fe and Zn accumulation. *Frontiers in Plant Science, 9*, 259. https://doi.org/10.3389/fpls.2018.00259

Altpeter, F., Baisakh, N., Beachy, R., Bock, R., Capell, T., & Christou, P. (2005). Particle bombardment and the genetic enhancement of crops: Myths and realities. *Molecular Breeding, 15*(3), 305–327.

Aluru, M., Xu, Y., Guo, R., Wang, Z., Li, S., White, W., … Roderme, S. (2008). Generation of transgenic maize with enhanced provitamin A content. *Journal of Experimental Botany, 59*(13), 3551–3562.

Aluru, M. R., Rodermel, S. R., & Reddy, M. B. (2011). Genetic modification of low phytic acid 1–1 maize to enhance iron content and bioavailability. *Journal of Agricultural and Food Chemistry, 59*(24), 12954–12962.

Andersson, M. S., Saltzman, A., Virk, P. S., & Pfeiffer, W. H. (2017). Progress update: Crop development of bio-fortified staple food crops under HarvestPlus. *African Journal of Food, Agriculture, Nutrition and Development, 17*(2), 11905–11935.

Babu, R., Rojas, N. P., Gao, S., Yan, J., & Pixley, K. (2013). Validation of the effects of molecular marker polymorphisms in LcyE and CrtRB1 on provitamin A concentrations for 26 tropical maize populations. *Theoretical and Applied Genetics, 126*(2), 389–399.

Bai, C., Twyman, R. M., Farré, G., Sanahuja, G., Christou, P., Capell, T., & Zhu, C. (2011). A golden era—Provitamin A enhancement in diverse crops. *In Vitro Cellular and Developmental Biology, 47*(2), 205–221.

Bai, L., Kim, E., DellaPenna, D., & Brutnell, T. P. (2009). Novel lycopene epsilon cyclase activities in maize revealed through perturbation of carotenoid biosynthesis. *Plant Journal, 59*(4), 588–599.

Baseggio, M., Murray, M., Magallanes-Lundback, M., Kaczmar, N., Chamness, J., Buckler, E. S., … Gore, M. A. (2019). Genome-wide association and genomic prediction models of tocochromanols in fresh sweet corn kernels. *Plant Genome, 12*(1). http://doi.org/10.3835/plantgenome2018.06.0038

Baveja, A., Muthusamy, V., Panda, K. K., Zunjare, R. U., Das, A. K., Chhabra, R., … Hossain, F. (2021). Development of multinutrient-rich bio-fortified sweet corn hybrids through genomics-assisted selection of *shrunken2, opaque2, lcyE* and *crtRB1* genes. *Journal of Applied Genetics.* https://doi.org/10.1007/s13353-021-00633-4

Berman, J., Zorrilla-López, U., Sandmann, G., Capell, T., Christou, P., & Zhu, C. (2017). The silencing of carotenoid β-hydroxylases by rna interference in different maize genetic backgrounds increases the β-carotene content of the endosperm. *International Journal of Molecular Sciences, 18*(12), 2515. https://doi.org/10.3390/ijms18122515

Bjarnason, M., & Vasal, S. K. (1992). Breeding of quality protein maize (QPM). In J. Janick (Ed.), *Plant breeding reviews* (Vol. 9, pp. 181–216). Chichester: John Wiley & Sons.

Bouis, H. E., & Saltzman, A. (2017). Improving nutrition through biofortification: A review of evidence from HarvestPlus, 2003 through 2016. *Global Food Security, 12*, 49–58. https://doi.org/10.1016/j.gfs.2017.01.009.

Boyer, C. D., & Hannah, L. C. (2001). Kernel mutants of corn. In A. Hallauer (Ed.), *Specialty corn* (2nd ed., pp. 1–32). Boca Raton, FL: CRC.

Brenna, O. V., & Berardo, N. (2004). Application of near-infrared reflectance spectroscopy (NIRS) to the evaluation of carotenoids content in maize. *Journal of Agricultural and Food Chemistry, 52*(18), 5577–5582.

Bressani, R. (1992). Nutritional value of high-lysine maize in humans. In E. T. Mertz (Ed.), *Quality protein maize* (pp. 205–224). St. Paul: American Association of Cereal Chemists.

Carvalho, S. M. P., & Vasconcelos, M. W. (2013). Producing more with less: Strategies and novel technologies for plant-based food bio-fortification. *Food Research International, 54*(1), 961–971.

Caulfield, L., Richard, S., Rivera, J., Musgrove, P., & Black, R. (2005). Stunting, wasting, and micronutrient deficiency disorders. In D. T. Jamison, J. G. Breman, A. R. Measham, G. Alleyne, M. Claeson, D. B. Evans, Prabhat Jha, Anne Mills, and P. Musgrove (Eds.), *Disease control priorities in developing countries* (2nd ed., pp. 551–567). Washington, DC: World Bank Oxford University Press.

Chen, Z., Young, T. E., Ling, J., Chang, S.-C.,& Gallie, D. R. (2003). Increasing vitamin C content of plants through enhanced ascorbate recycling. *Proceedings of the National Academy of Sciences, United States of America, 100*(6), 3525–3530.

Chomba, E., Westcott, C. M., Westcott, J. E., Mpabalwani, E. M., Krebs, N. F., & Patinkin, Z. W. (2015). Zinc absorption from bio-fortified maize meets the requirements of young rural Zambian children. *Journal of Nutrition, 145*(3), 514–519.

Curie, C., Panaviene, Z., Loulergue, C., Dellaporta, S. L., Briat, J. F., & Walker, E. L. (2001). Maize yellow stripe1 encodes a membrane protein directly involved in Fe(III) uptake. *Nature, 409*(6818), 346–349.

Drakakaki, G., Marcel, S., Glahn, R. P., Lund, L., Periagh, S., Fischer, R., … Stoger, E. (2005). Endosperm specific co-expression of recombinant soybean ferritin and Aspergillus phytase in maize results in significant increases in the levels of bioavailable iron. *Plant Molecular Biology, 59*(6), 869–880.

Economic Survey of India. (2014). *Economic survey of India, food and agriculture* (pp. 137–149). New Delhi: Economic Survey of India.

Falco, S. C., Guida, T., Locke, M., Mauvais, J., Sanders, C., Ward, R. T., & Webber, P. (1995). Transgenic canola and soybean seeds with increased lysine. *Biotechnology, 13*(6), 577–582.

FAO. (2008). Food and Agriculture Organization. www.fao.org

Fiedler, J. L., Zulu, R., Kabaghe, G., Lividini, K., Tehinse, J., & Bermudez, O. (2013). Assessing Zambia's industrial fortification options: Getting beyond changes in prevalence and cost effectiveness. *Food Nutrition Bulletin, 33*, 501–519.

Flint-Garcia, S. A., Thuillet, A.-C., Yu, J., Pressoir, G., Romero, S. M., Mitchell, S. E., … Buckler, E. S. (2005). Maize association population: A high-resolution platform for quantitative trait locus dissection. *Plant Journal, 44*(6), 1054–1064.

Gama, J., Sylos, C., & Dufosse, L. (2005). Major carotenoid composition of Brazilian valencia orange juice: Identification and quantification by HPLC. *Food Research International, 38*(8–9), 899–903.

Garcia-Casal, M. N., Layrisse, M., Solano, L., Baron, M. A., Arguello, F., Llovera, D., Ramirez, J., Leets, I., & Tropper, E. (1998). Vitamin A and beta-carotene can improve nonheme iron absorption from rice, wheat and corn by humans. *Journal of Nutrition, 128*(3), 646–650.

Gardner, H. W., & Inglett, G. E. (1971). Food products from corn germ: Enzyme activity and oil stability. *Journal of Food Science, 36*(4), 645–648.

Gibson, R. S. (1994). Zinc nutrition in developing countries. *Nutrition Research Reviews, 7*(1), 151–173.

Giuliano,G. (2017). Provitamin A biofortification of crop plants: A gold rush with many miners. *Current Opinion in Biotechnology, 44*, 169–180.

Gómez-Galera, S., Rojas, E., Sudhakar, D., Zhu, C., Pelacho, A. M., Capell, T., & Christou, P. (2010). Critical evaluation of strategies for mineral fortification of staple food crops. *Transgenic Research, 19*(2), 165–180.

Gonzales, J. W., Rhodes, A. M., & Dickinson, D. B. (1976). Carbohydrate and enzymatic characterization of a high-sucrose sugary inbred line of sweet corn. *Plant Physiology, 58*(1), 28–32.

Goredema-Matongera, N., Ndhlela, T.,Magorokosho, C., Kamutando, C. N., van Biljon, A., & Labuschagne, M. (2021). Multinutrientbio-fortification of maize (*Zea mays* L.) in Africa: Current status, opportunities and limitations. *Nutrients, 13*(3), 1039. https://doi.org/10.3390/nu13031039.

Graham, R. D., Welch, R. M., Saunders, D. A., Ortiz-Monasterio, I., Bouis, H. E., Bonierbale, M., … Twomlow, S. (2007). Nutritious subsistence food systems. *Advances in Agronomy, 92*, 1–74.

Gunaratna, N. S., Groote, H. D., & McCabe, G. P. (2008). Evaluating the impact of bio-fortification: A meta-analysis of community-level studies on quality protein maize (QPM). 2008 International Congress, August 26-29, 2008, Ghent, Belgium 44166, European Association of Agricultural Economists, pp. 1–9.

Gupta, H. S., Hossain, F., & Muthusamy, V. (2015). Biofortification of maize: An Indian perspective. *Indian Journal of Genetics and Plant Breeding, 75*(1), 1–22. https://doi.org/10.5958/0975-6906.2015.00001.2.

Habben, I. E., Kirleis, A. W., & Larkins, B. A. (1993). The origin of lysine-containing proteins in opaque-2 maize endosperm. *Plant Molecular Biology, 23*(4), 825–838.

Hagh, E. D., Mirshekari, B., Ardakani, M. R., Farahvash, F., & Rejali, F. (2016). Maize bio-fortification and yield improvement through organic biochemical nutrient management. *Idesia, 34*(5), 37–46.

Halilu, A. D., Ado, S. G., AbaInuwa, D. A., & Usman, S. (2016). Genetics of carotenoids for provitamin A bio-fortification in tropical-adapted maize. *Crop Journal, 4*(4), 313–322.

Hambridge, K. M. (2000). Human zinc deficiency. *Journal of Nutrition, 130*, S1344–S1349.

Hambidge, K. M., Huffer, J. W., Raboy, V., Grunwald, G. K., Westcott, J. L., Sian, L., … Krebs, N. F. (2004). Zinc absorption from low-phytate hybrids of maize and their wild-type isohybrids. *American Journal of Clinical Nutrition, 79*(6), 1053–1059.

Hamidi, A., Asgharzadeh, A., Chaokan, R., & Khalvati, M. A. (2011). Maize (*Zea mays* L.) seed bio-fortification by plant growth promoting bacteria

(PGPB). *International Journal of Agronomy and Plant Production*, *2*(5), 194–205.

Harjes, C. E., Rocheford, T. R., Bai, L., Brutnell, T. P., Kandianis, C. B., Sowinski, S. G., ... Buckler, E. S. (2008). Natural genetic variation in lycopene epsilon cyclase tapped for maize bio-fortification. *Science*, *319*(5861), 330–333.

Harris, D., Rashid, A., Miraj, G., Arif, M., & Shah, H. (2007). On-farm seed priming with zinc sulphate solution – A cost-effective way to increase the maize yields of resource-poor farmers. *Field Crops Research*, *10*, 119–127.

Hetzel, B. S. (2000). Iodine and neuropsychological development. *Journal of Nutrition*, *130*(Suppl 2), 493S–495S. https://doi.org/10.1093/jn/130.2.493S.

Hirschi, K. D. (2009). Nutrient bio-fortification of food crops. *Annual Review of Nutrition*, *29*, 401–421.

Hoekenga, O. A., Lungaho, M. G., Tako, E., Kochian, L. V., & Glahn, R. P. (2011). Iron bio-fortification of maize grain. *Plant Genetic Resources: Characterization and Utilization*, *9*(2), 327–329.

Hossain, F., Muthusamy, V., Zunjare, R. U., & Gupta, H. S. (2019). Bio-fortification of maize for protein quality and provitamin-A content. In P. Jaiwal, A. Chhillar, D. Chaudhary, & R. Jaiwal (Eds.), *Nutritional quality improvement in plants: Concepts and strategies in plant sciences* (pp. 115–136). Cham: Springer.

Hotz, C. (2009). The potential to improve zinc status through bio-fortification of staple food crops with zinc. *Food and Nutrition Bulletin*, *30*(1_suppl1), S172–S178.

Howe, J. A., & Tanumihardjo, S. A. (2006). Carotenoid-bio-fortified maize maintains adequate vitamin A status in Mongolian gerbils. *Journal of Nutrition*, *136*(10), 2562–2567.

Hurrell, R. F., Juillerat, M. A., Reddy, M. B., Lynch, S. R., Dassenko, S. A., & Cook, J. D. (1992). Soy protein, phytate and iron absorption in man. *American Journal of Clinical Nutrition*, *56*(3), 573–578.

Imran, M., & Rehim, A. (2016). Zinc fertilization approaches for agronomic bio-fortification and estimated human bioavailability of zinc in maize grain. *Archives of Agronomy and Soil Science*, *63*(1), 106–116.

Islam, S. N. (2004). *Survey of carotenoid variation and quantitiative trait loci mapping for carotenoid and tocopherol variation in maize*. M.Sc. Thesis, University Illinois, Urbana-Champaign, USA.

Jagtap, U. B., Gurav, R. G., & Bapat, V. A. (2011). Role of RNA interference in plant improvement. *Naturwissenschaften*, *98*(6), 473–492.

Jaiwal, P. K., Chhillar, A. K., Chaudhary, D., & Jaiwal, R. (2019). [Concepts and strategies in plant sciences]. Nutritional quality improvement in plants || Bio-fortification of maize for protein quality and provitamin-A content. Chapter 5. 115–136.

Jia, Y., & Jannink, J. L. (2012). Multiple-trait genomic selection methods increase genetic value prediction accuracy. *Genetics, 192*(4), 1513–1522.

Johnson, J. (2018b). First zinc maize variety launched to reduce malnutrition in Colombia. Retrieved from https://www.cimmyt.org/first-zinc-maize-variety-launched-to-reduce-malnutrition-in colombia/#

Kemper, E. L., Neto, C. G., Papes, F., Moraes, M. K. C., Leite, A., & Arruda, P. (1999). The role of opaque2 in the control of lysine-degrading activities in developing maize endosperm. *Plant Cell, 11*(10), 1981–1993.

Kerkeb, L., Mukherjee, I., Chatterjee, I., Lahner, B., Salt, D. E., & Connolly, E. L. (2008). Iron-Induced turnover of the Arabidopsis iron-regulated transporter1 metal transporter requires lysine residues. *Plant Physiology, 146*(4), 1964–1973.

Khush, G. S., Lee, S., Cho, J. I., & Jeon, J. S. (2012). Bio-fortification of crops for reducing malnutrition. *Plant Biotechnology Reports, 6*(3), 195–202.

Krivanek, A., Groote, H., Gunaratna, N., Diallo, A., & Freisen, D. (2007). Breeding and disseminating quality protein maize for Africa. *African Journal of Biotechnology, 6*, 312–324.

Lee, M., Sharopova, N., Beavis, W. D., Grant, D., Katt, M., Blair, D., & Hallauer, A. (2002). Expanding the genetic map of maize with the intermated B73 × Mo17 (IBM) population. *Plant Molecular Biology, 48*(5–6), 453–461.

Li, Q., Yang, X., Xu, S., Cai, Y., Zhang, D., Han, Y., … Yan, J. (2012). Genome-wide association studies identified three independent polymorphisms associated with α-Tocopherol content in maize kernels. *PLOS ONE, 7*(5), e36807.

Li, S., Liu, X., Zhou, X., Li, Y., Yang, W., & Chen, R. (2019). Improving zinc and iron accumulation in maize grains using the zinc and iron transporter ZmZIP5. *Plant and Cell Physiology, 60*(9), 2077–2085.

Li, S., Nugroho, A., Rocheford, T., & White, W. S. (2010). Vitamin A equivalence of the β-carotene in β-carotene–bio-fortified maize porridge consumed by women. *American Journal of Clinical Nutrition, 92*(5), 1105–1112.

Lipka, A. E., Gore, M. A., Magallanes-Lundback, M., Mesberg, A., Lin, H., Tiede, T., … DellaPenna, D. (2013). Genome-wide association study and pathway-level analysis of tocochromanol levels in maize grain. *G3 Genes Genomes Genetics, 3*(8), 1287–1299.

Liu, L., Jeffers, D., Zhang, Y., Ding, M., Chen, W., Kang, M. S., & Fan, X. (2015). Introgression of the crtRB1 gene into quality protein maize inbred lines using molecular markers. *Molecular Breeding, 35*(8), 154. https://doi.org/10.1007/s11032-015-0349-7

Lung'aho, M. G., Mwaniki, A. M., Szalma, S. J., Hart, J. J., Rutzke, M. A., Kochian, L. V., … Hoekenga, O. A. (2011). Genetic and physiological analysis of iron bio-fortification in maize kernels. *PLOS ONE, 6*(6), e20429. https://doi.org/10.1371/journal.pone.0020429

Lyddon, C. (2004). Flour millers join to fight malnutrition. *World Grain, 23*(9), 42–45.

Machida, L., Derera, J., Tongoona, P., & MacRobert, J. (2010). Combining ability and reciprocal cross effects of elite quality protein maize inbred lines in subtropical environments. *Crop Science, 50*(5), 1708–1717.

Malik, K. A., & Maqbool, A. (2020). Transgenic crops for biofortification. *Frontiers in Sustainable Food Systems.* https://doi.org/10.3389/fsufs.2020.571402.

Maqbool, M. A. (2017). *Heterosis estimation of indigenous maize (Zea mays L.) hybrids and stability analysis of exotic accessions for provitamin A and yield components.* PhD thesis, Dept. Plant Breeding and Genetics, University of Agriculture Faisalabad, Pakistan.

Maqbool, M. A., Aslam, M., Beshir, A., & Khan, M. S. (2018). Breeding for provitamin A biofortification of maize (*Zea mays* L.). *Plant Breeding, 137*(4), 451–469.

Maqbool, M. A., & Beshir, A. (2018). Zinc biofortification of maize (*Zea mays* L.): Status and challenges. *Plant Breeding, 138,* 1–28.

Mazariegos, M., Hambidge, M. K., Krebs, N. F., Westcott, J. E., Lei, S., Grunwald, G. K., ... Solomons, N. W. (2006). Zinc absorption in Guatemalan school children fed normal or low-phytate maize. *American Journal of Clinical Nutrition, 83*(1), 59–64.

Mertz, E. T., Bates, L. S., & Nelson, O. E. (1964). Mutant gene that changes protein composition and increases lysine content of maize endosperm. *Science, 145*(3629), 279–280. https://doi.org/10.1126/science.145.3629.279.

Miller, L. V., Krebs, N. F., & Hambidge, K. M. (2007). A mathematical model of zinc absorption in humans as a function of dietary zinc and phytate. *Journal of Nutrition, 137*(1), 135–141.

Mumtaz, M. Z., Ahmad, M., Jamil, M., & Hussain, T. (2017). Zinc solubilizing *Bacillus* spp. potential candidates for bio-fortification in maize. *Microbiological Research, 202,* 51–60.

Muthusamy, V., Hossain, F., Thirunavukkarasu, N., Choudhary, M., Saha, S., Bhat, J. S., ... Gupta, H. S. (2014). Development of β-carotene rich maize hybrids through marker-assisted introgression of β-carotene hydroxylase allele. *PLOS ONE, 9*(12), 1–22.

Naqvi, S., Zhu, C., Farre, G., Ramessar, K., Bassie, L., Breitenbach, J., ... Christou, P. (2009). Transgenic multivitamin corn through bio-fortification of endosperm with three vitamins representing three distinct metabolic pathways. *Proceedings of the National Academy of Sciences of the United States of America, 106*(19), 7762–7767.

Natesan, S., Duraisamy, T., Pukalenthy, B., Chandran, S., Nallathambi, J., Adhimoolam, K., ... Rajasekaran, R. (2020). Enhancing β -carotene concentration in parental lines of CO6 Maize hybrid through marker-assisted backcross breeding (MABB). *Frontiers in Nutrition, 7,* 1–12.

Naveed, S., Rehim, A., Imran, M., Bashir, M. A., Anwar, M. F., & Ahmad, F. (2018). Organic manures: An efficient move towards maize grain bio-fortification. *International Journal of Recycling of Organic Waste in Agriculture, 7*(3), 189–197.

Nuss, E. T., & Tanumihardjo, S. A. (2010). Maize: A paramount staple crop in the context of global nutrition. *Comprehensive Reviews in Food Science and Food Safety, 9*(4), 417–436.

Oikeh, S. O., Menkir, A., Maziya-Dixon, B., Welch, R., & Glahn, R. P. (2003). Assessment of concentrations of iron and zinc and bioavailable iron in grains of early-maturing tropical maize varieties. *Journal of Agricultural and Food Chemistry, 51*(12), 3688–3694.

Olson, R. A., & Frey, K. J. (Eds.). (1987). Nutritional quality of cereals grains: Genetic and agronomic improvement. *Agronomy, 28*, 1–10.

Ortiz, D., Rocheford, T., & Ferruzz, M. G. (2016). Influence of temperature and humidity on the stability of carotenoids in bio-fortified maize (*Zea mays* L.) genotypes during controlled postharvest storage. *Journal of Agricultural and Food Chemistry, 64*(13), 2727–2736.

Ortiz-Monasterio, J. I., Palacios-Rojas, N., Meng, E., Pixley, K., Trethowan, R., & Pena, R. J. (2007). Enhancing the mineral and vitamin content of wheat and maize through plant breeding. *Journal of Cereal Science, 46*(3), 293–307.

Owens, B. F., Lipka, A. E., Magallanes-Lundback, M., Tiede, T., Diepenbrock, C. H., Kandianis, C. B., … Rocheford, T. (2014). A foundation for provitamin a bio-fortification of maize: Genome-wide association and genomic prediction models of carotenoid levels. *Genetics, 198*(4), 1699–1716.

Owens, B. F., Mathew, D., Diepenbrock, C. H., Tiede, T., Wu, D., Mateos-Hernandez, M., … Rocheford, T. (2019). Genome-wide association study and pathway-level analysis of kernel color in maize. *G3 Genes Genomes Genetics, 9*(6), 1945–1955.

Paes, M. C. D., & Bicudo, M. H. (1994). Nutritional perspectives of quality protein maize. In *Quality protein maize: 1964–1994. Proceedings of the international symposium on quality protein maize*, SeteLagoas, MG, Brazil, December 1–3, 1994 (pp. 65–78).

Palaisa, K., Morgante, M., Tingey, S., & Rafalski, A. (2004). Long-range patterns of diversity and linkage disequilibrium surrounding the maize Y1 gene are indicative of an asymmetric selective sweep. *Proceedings of the National Academy of Sciences, 101*(26), 9885–9890.

Palaisa, K. A., Morgante, M., Williams, M., & Rafalski, A. (2003). Contrasting effects of selection on sequence diversity and linkage disequilibrium at two phytoene synthase loci. *Plant Cell, 15*(8), 1795–1806.

Pfeiffer, W. H., & McClafferty, B. (2007). HarvestPlus: Breeding crops for better nutrition. *Crop Science, 47*, S88–S105.

Pillary, K., Derera, J., Siwela, M., & Veldman, F. J. (2011). Consumer acceptance of yellow, provitamin A-Bio-fortified maize in KwaZulu-Natal. *South African Journal of Clinical Nutrition, 24*(4), 186–191.

Pillay, K., Derera, J., Siwela, M., & Veldman, F. J. (2011). Consumer acceptance of yellow, provitamin A-bio-fortified maize in KwaZulu-Natal, *24*(4). https://doi.org/10.1080/16070658.2011.11734386

Pixley, K., Palacios, N. R., Babu, R., Mutale, R., Surles, R., & Simpungwe, E. (2013a). Bio-fortification of maize with provitamin A carotenoids. In S. A. Tanumihardjo (Ed.), *Carotenoids in human health* (pp. 271–292). New York: Springer Science and Business Media.

Pixley, K., Rojas, N. P., Babu, R., Mutale, R., Surles, R., & Simpungwe, E. (2013b). Bio-fortification of maize with provitamin A carotenoids. In S. Tanumihardjo (Ed.), *Carotenoids and human health. Nutrition and health* (pp. 115–136). Totowa, NJ: Humana Press.

Prasanna, B. M. (2014). Maize research-for-development scenario: Challenges and opportunities for Asia. In Prasanna, ... K. Pixley (Eds.). *Proceedings of the 12th Asian maize conference and expert consultation on maize for food, feed and nutritional security, book of extended summaries* (pp. 2–11). Bangkok, Thailand.

Prasanna, B. M., Palacios-Rojas, N., Hossain, F., Muthusamy, V., Menkir, A., Dhliwayo, T., ... Fan, X. (2020). Molecular breeding for nutritionally enriched maize: Status and prospects. *Frontiers in Genetics, 10,* 1392. https://doi.org/10.3389/fgene.2019.01392

Raboy, V. (2007). The ABCs of low-phytate crops. *Nature Biotechnology, 25*(8), 874–875.

Raboy, V., Gerbasi, P., Young, K., Stoneberg, S., Pickett, S., Bauman, A., ... Ertl, D. (2000). Origin and seed phenotype of maize low phytic acid 1-1 and low phytic acid 2-1. *Plant Physiology, 124*(1), 355–368.

Rahmanifar, A., & Hamaker, B. R. (1999). Potential nutritional contribution of quality protein maize: A close-up on children in poor communities. *Ecology of Food and Nutrition, 38*(2), 165–182.

Ribaut, J. M., & Hoisington, D. A. (1998). Marker-assisted selection: New tools and strategies. *Trends in Plant Science, 3*(6), 236–239.

Rooney, L. W., & Suhendro, E. L. (1999). Perspectives on nixtamalization (alkaline cooking) of maize for tortillas and snacks. *Cereal Foods World, 44,* 466–470.

Saleem, I., Javid, S., Bibi, F., Ehsan, S., Niaz, A., & Ahmad, Z. A. (2016). Bio-fortification of maize grain with zinc and iron by using fertilizing approach. *Journal of Agriculture and Ecology Research International,* 1–6. https://doi.org/10.9734/jaeri/2016/24532

Sarika, K., Hossain, F., Muthusamy, V., Baveja, A., Zunjare, R., Goswami, R., ... Gupta, H. (2017). Exploration of novel opaque16 mutation as a source for high lysine and tryptophan in maize endosperm. *Indian Journal of Genetics and Plant Breeding, 77*(1), 59–64.

Saurabh, S., Vidyarthi, A. S., & Prasad, D. (2014). RNA interference: Concept to reality in crop improvement. *Planta, 239*(3), 543–564.

Schaffert, R. E., Paes, M. C. D., & Guimarães, P. E. O. (2011). Results of the maize bio-fortification research actions in the HarvestPlus and BioFort projects. In *Embrapa Milho e Sorgo-Resumo em anais de congresso (ALICE).* In: REUNIÃO DE BIO-FORTIFICAÇÃO NO BRASIL, 4. 2011. Teresina. Palestras e resumos... Rio de Janeiro: Embrapa Agroindústria de Alimentos; Teresina: Embrapa Meio-Norte.

Sekhon, R. S., Lin, H., Childs, K. L., Hansey, C. N., Buell, C. R., de Leon, N., & Kaeppler, S. M. (2011). Genome-wide atlas of transcription during maize development. *Plant Journal, 66*(4), 553–563.

Segal, G., Song, R., & Messing, J. (2003). A new opaque variant of maize by a single dominant RNA interference-inducing transgene. *Genetics, 165*(1), 387–397.

Shi, J., Wang, H., Hazebroek, J., Ertl, D. S., & Harp, T. (2005). The maize *low-phytic acid 3* encodes a *myo*-inositol kinase that plays a role in phytic acid biosynthesis in developing seeds. *Plant Journal, 42*(5), 708–719.

Shi, J., Wang, H., Schellin, K., Li, B., Faller, M., Stoop, J. M., … Glassman, K. (2007). Embryo-specific silencing of a transporter reduces phytic acid content of maize and soybean seeds. *Nature Biotechnology, 25*(8), 930–937.

Shiferaw, B., Prasanna, B. M., Hellin, J., & Banziger, M. (2011). Crops that feed the world 6. Past successes and future challenges to the role played by maize in global food security. *Food Security, 3*(3), 307–327.

Shrestha, J., & Karki, T. B. (2014). Provitamin A maize development: A strategy for fighting against Malnutrition in Nepal. *Our Nature, 12*(1), 44–48. http://doi.org/10.3126/on.v12i1.12256.

Šimić, D., Drinić, S. M., Zdunić, Z., Jambrović, A., Ledenčan, T., Brkić, J., … Brkić, I. (2012). Quantitative trait loci for bio-fortification traits in maize grain. *Journal of Heredity, 103*(1), 47–54.

Singleton, W. R. (1939). Recent linkage studies in maize: V. Opaque endosperm-2 (o2). Connecticut Experiment Station, New Haven. Genetics 24: 61.

Stein, A. J. (2010). Global impact of human mineral malnutrition. *Plant and Soil, 335*(1–2), 133–154.

Stein, A. J., Meenakshi, J. V., Qaim, M., Nestel, P., Sachdev, H. P. S., & Bhutta, Z. A. (2005). Analysing health benefits of bio-fortified staple crops by means of the disability-adjusted life years approach—A handbook focusing on iron, zinc and vitamin A. HarvestPlus Technical Monograph No. 4. *Harvestplus, Washington, D.C.* pp. 115–136.

Stoltzfus, R. J., & Dreyfuss, M. L. (1998). *Guidelines for the use of iron supplements to prevent and treat iron deficiency anemia.* Washington, DC: ILSI Press.

Stoltzfus, R. J., Mullany, L., & Black, R. E. (2004). Iron deficiency anemia. In M. Ezzati, A. D. Lopez, A. Rodgers, & C. J. L. Murray (Eds.), *Comparative quantification of health risks: Global and regional burden of disease attribution to selected major risk factors, vol. I.* Geneva: World Health Organization, pp. 163–210.

Sureshkumar, S., Tamilkumar, P., Senthil, N., Nagarajan, P., Thangavelu, A. U., Raveendran, M., Vellaikumar, S., Ganesan, K. N., Balagopal, R., Vijayalakshmi, G., & Shobana, V. (2014). Marker assisted selection of low phytic acid trait in maize (*Zea mays* L.). *Hereditas, 151*(1), 20–27. https://doi.org/10.1111/j.1601-5223.2013.00030.x.

Suwarno, W. B., Pixley, K. V., Palacios-Rojas, N., Kaeppler, S. M., & Babu, R. (2014). Formation of heterotic groups and understanding genetic effects in a provitamin a bio-fortified maize breeding program. *Crop Science*, *54*(1), 14–24.

Suwarno, W. B., Pixley, K. V., Palacios-Rojas, N., Kaeppler, S. M., & Babu, R. (2015). Genome-wide association analysis reveals new targets for carotenoid bio-fortification in maize. *Theoretical and Applied Genetics*, *128*(5), 851–864.

Taleon, V., Mugode, L., Cabrera-Soto, L., & Palacios-Rojas, N. (2017). Carotenoid retention in bio-fortified maize using different post-harvest storage and packaging methods. *Food Chemistry*, *232*, 60–66.

Tuhy, Ł., Samoraj, M., Witkowska, Z., & Chojnacka, K. (2015). Bio-fortification of maize with micronutrients by Spirulina. *Open Chemistry*, *13*, 1119–1126.

Umar, W., Hameed, M. K., Aziz, T., Maqsood, M. A., Bilal, H. M., & Rasheed, N. (2020). Synthesis, characterization and application of ZnO nanoparticles for improved growth and Zn bio-fortification in maize. *Archives of Agronomy and Soil Science*, 1–13. https://doi.org/10.1080/03650340.2020.1782893

Unger, E., Parsons, R. L., Schmidt, R. J., Bowen, B., & Roth, B. A. (1993). Dominant negative mutants of Opaque2 suppress transactivation of a 22-kD zein promoter by Opaque2 in maize endosperm cells. *Plant Cell*, *5*(8), 831–841.

Vallabhaneni, R., Gallagher, C. E., Licciardello, N., Cuttriss, A. J., Quinlan, R. F., & Wurtzel, E. T. (2009). Metabolite sorting of a germplasm collection reveals the hydroxylase3 locus as a new target for maize provitamin A bio-fortification. *Plant Physiology*, *151*(3), 1635–1645.

Vallabhaneni, R., & Wurtzel, E. T. (2009). Timing and biosynthetic potential for carotenoid accumulation in genetically diverse germplasm of maize. *Plant Physiology*, *150*(2), 562–572.

Vasal, S. K. (2000). The quality protein maize story. *Food and Nutrition Bulletin*, *21*, 445–450.

Vietmeyer, N. D. (2000). A drama in three long acts: The story behind the story of the development of quality-protein maize. *Diversity*, *16*, 29–32.

Vignesh, M., Hossain, F., Nepolean, T., Saha, S., Agrawal, P. K., Guleria, S. K.; Prasanna, B. M., & Gupta, H. S. (2012). Genetic variability for kernel β-carotene and utilization of crtRB1 3'TE gene for biofortification in maize (*Zea mays* L.). *Indian Journal of Genetics*, *72*, 189–194.

Vivek, B. S., Krivanek, A. F., Palacios-Rojas, N., Twumasi-Afiriye, S., & Diallo, A. O. (2008). *Breeding quality protein maize (QPM) cultivars: Protocols for developing QPM cultivars*. Mexico: CIMMYT.

Welch, R. M., & Graham, R. D. (2002). Breeding crops for enhanced micronutrient content. In J. J. Adu-Gyamfi (Ed.), *Food security in nutrient-stressed environments: Exploiting plants' genetic capabilities* (pp. 267–276). Netherlands, Dordrecht: Springer.

Welch, R. M., & Graham, R. D. (2004). Breeding for micronutrients in staple food crops from a human nutrition perspective. *Journal of Experimental Botany, 55*(396), 353–364.

White, P. J., & Broadley, M. R. (2005). Bio-fortifying crops with essential mineral elements. *Trends in Plant Science, 10*(12), 586–593.

White, P. J., & Broadley, M. R. (2009). Biofortification of crops with seven mineral elements often lacking in human diets – Iron, zinc, copper, calcium, magnesium, selenium and iodine. *New Phytologist, 182*(1), 49–84.

WHO (2002). World Health Organization. www.who.int

Wong, J. C., Lambert, R. J., Wurtzel, E. T., & Rocheford, T. R. (2004). QTL and candidate genes phytoene synthase and ζ-carotene desaturase associated with the accumulation of carotenoids in maize. *Theoretical and Applied Genetics, 108*(2), 349–359.

Wu, D., Tanaka, R., Li, X., Ramstein, G. P., Cu, S., Hamilton, J. P., Buell, C. R., Stangoulis, J., Rocheford, T., & Gore, M. A. (2021). High-resolution genome-wide association study pinpoints metal transporter and chelator genes involved in the genetic control of element levels in maize grain. *G3 (Bethesda, MD), 11*(4), jkab059. https://doi.org/10.1093/g3journal/jkab059.

Wu, Y., & Messing, J. (2011). Novel genetic selection system for quantitative trait loci of quality protein maize. *Genetics, 188*(4), 1019–1022.

Yadav, O. P., Hossain, F., Karjagi, C. G., Kumar, B., Zaidi, P. H., Jat, S. L., Chawla, J. S., Kaul, J., Hooda, K. S., Kumar, P., Yadava, P., & Dhillon, B. S. (2015). Genetic improvement of maize in india: Retrospect and prospects. *Agricultural Research, 4*, 325–338. https://doi.org/10.1007/s40003-015-0180-8.

Yadava, D. K., Choudhury, P. R., Hossain, F., & Kumar, D. (2017). *Bio-fortified varieties*. New Delhi: Sustainable Way to Alleviate Malnutrition. Indian Council of Agricultural Research.

Yan, J., Kandianis, B. C., Harjes, E. C., Bai, L., Kim, H. E., Yang, X., … Rocheford, T. (2010). Rare genetic variation at *Zea mays* crtRB1 increases β-carotene in maize grain. *Nature Genetics, 42*(4), 322–327.

Yan, J., Warburton, M., & Crouch, J. (2011). Association mapping for enhancing maize (*Zea mays* L.) genetic improvement. *Crop Science, 51*(2), 433–449.

Zanga, D., Capell, T., Slafer, G. A., Christou, P., & Savin, R. (2016). A carotenogenic mini pathway introduced into white corn does not affect development or agronomic performance. *Scientific Reports, 6*, 1–12.

Zhao, F. J., & Shewry, P. R. (2011). Recent developments in modifying crops and agronomic practice to improve human health. *Food Policy, 36*, S94–S101.

Zhu, C. F., Naqvi, S., Breitenbach, J., Sandmann, G., Christou, P., & Capell, T. (2008). Combinatorial genetictransformation generates a library ofmetabolic phenotypes for the carotenoid pathway in maize. *Proceedings of the National Academy of Sciences, United States of America, 105*(47), 18232–18237.

Zunjare, R. U., Chhabra, R., Hossain, F., Baveja, A., Muthusamy, V., & Gupta, H. S. (2018b). Molecular characterization of 5' UTR of the *lycopene epsilon cyclase* (*lcyE*) gene among exotic and indigenous inbreds for its utilization in maize bio-fortification. *3 Biotech, 8*(1), 75. https://doi.org/10.1007/s13205-018-1100-y

Zunjare, R. U., Hossain, F., Muthusamy, V., Baveja, A., Chauhan, H. S., Bhat, J. S., … Gupta, H. S. (2018a). Development of bio-fortified maize hybrids through marker-assisted stacking of β-carotene hydroxylase, lycopene-ε cyclase and opaque2 genes. *Frontiers in Plant Science, 9,* 178. https://doi.org/10.3389/fpls.2018.00178

Chapter 10

Storage of maize and its products

Maninder Kaur, Gurveer Kaur, Preeti Birwal, Ramandeep Kaur, and Sandhya

DOI: 10.1201/9781003245230-10

10.1 Introduction

Maize (*Zea mays* L.) is one of the world's most important cereal crops. It is one of the most versatile emerging crops having extensive adaptability. Maize crop can be grown in diverse conditions. Maize is also known as the queen of cereals because of its potential of highest genetic yield. The different types of maize are normal yellow/white grain, baby corn, popcorn, sweet corn, waxy corn, high amylase corn, etc. Maize is also one of the most significant industrial raw materials having great opportunity for increasing value of the crop. It is widely consumed by both humans and animals. Starch is the major component of the corn kernel which can be utilized either in its natural form or after alteration from its chemical form. Starch can also be transformed into glucose or fructose for use as a diet sweetener. Glucose can also be fermented into ethanol for its use in beverages, fuel, or other chemicals.

Maize is cultivated worldwide on nearly 197 m ha with production of 1148 m tonnes and productivity of 5823.8 kg/ha, having a broad diversity of climate, soil, biodiversity, and management practices. It is the third most important cereal crop in India after rice and wheat and is cultivated in a wide range of environments. India produced 30 MT in an area of 9.9 million hectares in 2020–2021. Maize accounts for about 10% of total food grain production in addition to being a principal food for human beings and quality fodder for animals.

10.2 Types of corn

The different types of corn include flour, flint, dent, pop, sweet, and pod corn. Excluding pod corn, this classification is based on the quality, quantity, and shape of endosperm composition in the kernel (Brown & Darrah, 1985). Dent corn is categorized by the presence of corneous, horny endosperm at the sides and back of the kernels, though the central core is a soft, floury endosperm spreading to the crown of the endosperm. In addition to its use as animal feed, it can also be used as a staple food and raw material for industries. The flint corns typically have a thick, hard, vitreous or corneous endosperm layer enclosed by small, soft granular centre. Usually the kernels are smooth and rounded, and the ears are long and slender. The floury maize types consist of soft starch all over, with no firm and vitreous endosperm, resulting in opaque kernel phenotype. In sweet corn, the sugary gene retards the conversion of sugar into starch during endosperm growth, and the kernel collects a water-soluble polysaccharide called "phytoglycogen". The water-soluble polysaccharide enhances quality in terms of texture along with sweetness.

Popcorn is described as having a very rigid, hornlike endosperm containing only a small portion of soft starch. It has basically flint-type mall kernels. Pod corn is more of an ornamental type but may be sweet, dent, pop, waxy, or floury in endosperm features. The endosperm in waxy corn gives a waxy appearance. The commonly used corn starch consists of about 73% amylopectin and 27% amylose, whereas waxy starch has only amylopectin. Products manufactured from waxy corn can be used by the food industry as retorted foods, and stabilizers and thickeners for puddings, sauces, gravies, salad dressings, etc. Waxy grain corn is also grown as a feed for cattle. The starch from high-amylose corn can be utilized in the textile industry, in gum candies, and as an adhesive in the production of corrugated cardboard.

10.3 Post-harvest management

10.3.1 Losses

Post-harvest loss comprises the food loss throughout the whole food supply chain i.e. from harvesting to its consumption (Aulakh et al., 2013). The losses can be weight loss, quality loss, seed viability loss, and nutritional loss. The causes of post-harvest loss include limited accessibility of suitable varieties for processing, lack of suitable processing technologies, and insufficient commercialization of new technologies. Post-harvest loss accounts for direct physical and quality losses that decrease the economic value of crop or may make it inappropriate for human consumption. These losses can be up to 80% of the total production in severe cases (Fox, 2013).

To increase the storage period of maize, post-harvest losses need to be reduced. However, many challenges make it hard to store grain for a longer period without spoilage. These challenges include mould growth, pest attacks, and insect infestation which not only result in economic losses but also cause a health risk for consumers.

Two ways to reduce post-harvest losses in grains include:

1. Storing maize at the right moisture content. The storage of maize at the right moisture content decreases the activity of mould and insect infestation in the grain
2. Using hermetic bags for storage of maize. Use of hermetic bags ensures that maize remains at a constant moisture content throughout storage. As compared to regular jute bags or polypropylene bags used to store maize, hermetic bags generally contain an inner rubber lining in which grains are placed and sealed so that they are airtight. This breaks oxygen and moisture movement between

the outside air and the stored maize. This leads to the trapping of insects within the bag which will die after some time due to lack of oxygen.

10.3.2 Processing status

Maize is generally processed by two different methods i.e. wet milling and dry milling. Dry-milling units having capacity of 10 MT/day produce grits, corn flour, and a small quantity of corn meal (www.entrepreneurindia.co/ project-and-profile-details/Maize%20Processing%20Unit). This technology has been standardized by the Central Food Technological Research Institute (CFTRI) in Mysore. The setting up of these units depends upon the availability of land, raw material, and power requirements. The wet-milling industries in India are constrained to certain areas such as Gujarat, Punjab, Maharashtra, Madhya Pradesh, Chattisgarh, and Karnataka. There are about 17 wet-milling units with a crushing capacity of 3400 MT of maize/day. The average processing capacity of the units available in India is about 200 MT of maize/day.

10.4 Harvesting methods according to storage

The methods commonly used for maize harvesting are given as follows.

10.4.1 Cob plucking

The ears of maize crop are detached from the standing plants. The stalks may be used as green fodder also.

10.4.2 Stalk cutting

The stalks are cut and heaped up in the shade. The cobs are generally removed after two or three days of harvesting. The dried plants can also be used for haymaking. The sun drying of ears can be done for four to five days. Then the grains are shelled at 12–15% moisture content, again dried and stored at about 8–10% moisture. The grains should be stored in sealed containers to reduce damage caused by insects and rodents.

Therefore, it is necessary to harvest the crop at full maturity to obtain good quality maize grains. Maize harvesting is done by manual workers in small maize growing areas. However, appropriate machinery can be used for harvesting and threshing on a large scale.

10.4.3 Cleaning and grading with standards

Impurities like stones, dust, metal parts, foreign matter, etc. present in grain are removed in cleaning segments through perforated metals sheets and air blowers. Grading is the process of separating produce on the basis of different grades or groups. For maize, the quality features such as moisture content, other food grains, damaged grains, foreign matter, admixture of other varieties, immature grains, and weevilled and shrivelled grains are measured while grading (https://agmarknet.gov.in/CommodityProfiles/preface-maize.aspx).

The term "standard" mentions the criteria that serve as a basis for comparing the accuracy of different samples. The three different types of standards are:

- Standard specification which describes a subject
- Standard test method for testing of specification
- Grading standard which classifies a subject into different categories

There are about 330 specifications for cereals and their products at national and international levels of which 12 are relevant worldwide (www.fao.org/3/t1838e/T1838E0i.htm). Standard specifications describe the nature of a product on the basis of pass or fail. Many countries have institutions for national standards which issue specifications for products in addition to different testing methods. Several countries adopt or amend international standards, like the International Organization for Standardization (ISO) standards, into their national system. Another basis is the Codex Alimentarius Commission (Codex) which functions as a committee to frame standards on cereals, pulses, and legumes. Also, among 420 standard test methods for cereals and their products at national and international level, about 75 are applicable internationally. The different organizations issuing standard testing methods include the International Organization for Standardization (ISO) standards, Association of Official Analytical Chemists (AOAC), and the American Association of Cereal Chemists (AACC).

10.5 Storage

10.5.1 Objective

Food grain preservation is essential to the economies of both developed and developing countries. Customers must have access to high-quality food grains for use in various goods and marketing, and farmers must produce and develop healthy cereals and pulse grains. This demands continuous agricultural output, which would aid in the economic survival of any country. Therefore, grains are stored for whole year and progressively delivered for retail or direct marketing throughout off-season times to fulfil the need for an ample supply, which helps to keep a price steady. Furthermore, store cleanliness and grain wetness must be maintained throughout this time to keep grain pests, insects, rodents, and birds at bay. Nukenine (2010) stated that "storage is a technique or practice for preserving agricultural goods or items for future use".

10.5.2 Need and importance

In developing countries, a quarter of farmers kept their produce in the village level. Conventionally, grain storage methods are viewed as effective or satisfactory, and they continue to develop to safeguard grains from harm. However, the total proportion of crop production kept on the farm and the time of storage is influenced by yield per acre, size of farm, marketing patterns, consumption patterns, labour payment, future crop expectations, and loan availability. There are various structures for grain storage, ranging from mud huts to sophisticated containers inside, outside, or underground. The containers used for storage are made of local available materials and in various forms, shapes, sizes, and functions. Some fundamental precautions to make when storing maize grains are as follows:

- Hygiene and sanitation are essential from harvest through storage in eliminating infection sources and decreasing contamination levels. The storage facility should be constructed on an elevated platform that is well-drained. It should be easy to locate. There should be no excessive heat, dampness, water logging, termites, insects, and rodents, or other pests.
- The measurements for storage structure must be determined according to volume of stored maize or products of maize and storage duration. In godowns, for better aeration there should be adequate space between walls and stacks. Maize should be maintained in a sealed, airtight container or structure to reduce oxygen concentration and limit the presence of aerobic organisms. Maize should be maintained in an airtight sealed container or in a structure where the oxygen concentration is reduced to remove aerobic organisms.
- The grains' temperature, environment of surroundings, and relative humidity must be carefully monitored during storage, particularly in the early stages, to yield the finest quality grain. Because

of the reduced moisture content and temperature, grains may be stored for longer periods without becoming infected by mould or insects.

- Before storage, grains should be dried properly and cleaned for prevention of quality degradation. For reduction of deterioration shortly after harvest, maize should be dried at 14% or less moisture content in tropical and subtropical countries.
- The storage structure or godowns must be free of cracks, fractures, and holes, be cleaned properly and fumigated before maize storage. Clean, fumigate, or separate the maize grain as soon as insects and fungi are found. Grain must be sorted or cleaned after harvesting to remove foreign, broken corn kernels, which stimulate the growth of bugs and fungi on the grains.
- Every time, use a new gunny bag. For disinfection bags must be boiled for three to four minutes in malathion solution (1%) and then dried fully. Fresh grains must be separated from old grains to reduce contamination. This in turn avoids transfer of pests from one lot to another.

10.6 Factors affecting storage

Biological and abiological variables influence the shelf life of stored maize grains. Grain moisture level and temperature play the biggest part in impacting grain quality, biochemical reactions, dry matter losses, permissible storage durations, and overall grain storage management.

10.6.1 Moisture content

Moisture is required for biological and metabolic processes. As a result, the moisture level of both grains and adjacent environment air should be decreased and observed for safe grain storage. Like other stored goods, maize grains are hygroscopic grains with consistent dry matter content, but their water content varies. According to Brewbaker (2003), moisture content is important in grain storage because it causes the grain to heat up, which leads to mould deterioration. A rise in grain moisture content causes mould and insect degradation and also makes it more sensitive for deterioration.

10.6.2 Temperature and relative humidity

The relative humidity is the percentage amount of water vapour in the air as a percentage of the amount of water vapour necessary to saturate the air at the same temperature. Several research studies have showed the link between relative humidity and temperature of grains in tropical regions. The results indicated a clear link between them, indicating that the grain loses moisture to the air of the surroundings when the temperature rises, in turn relative humidity also rises (Devereau et al., 2002). Studies discovered that every 10 °C increase in temperature induces a 3% increase in relative humidity in most cereal grains (ACDI/VOCA, 2003). Changes in temperature and relative humidity increase mould formation and result in significant grain nutrient losses. When maize grains are open in air, moisture content starts exchanging between the grains and their surroundings. It occurs only until equilibrium level is attained (Samuel et al., 2011). Furthermore, in tropical regions, the temperature and relative humidity fluctuations accelerate the swift multiplication of insects and moulds, facilitating faster spoilage of grains.

10.6.3 Dry matter loss with respiration

In storage structures, living things such as viable grains, insects, mites, moulds, and other creatures are present, which can breathe with oxygen, carbon dioxide, and water. During the process of respiration, generated heat induced a rise in the temperature of the stored grains owing to dry matter loss (Bern et al., 2013). Several researchers have utilized carbon dioxide to measure the degradation of maize grain over time. The stored grains' respiration creates moisture, carbon dioxide, and heat, which encourage a rise in dry matter loss with temperature (Lee, 1999). The quality or soundness of the stored grain has a significant impact on its respiration activity.

10.6.4 Moulds and fungi

In fields, most of the fungus and mould species formed on maize that cause contamination are considered one of the most severe safety issues in tropical regions and worldwide. Field fungi and storage fungi are two types of toxin-producing fungi that infest maize grains.

 a. *Field fungi*: Before harvest and threshing, grains got infested by field fungi and released toxin on grains. They can thrive in conditions of moisture (22–33%), relative humidity (over 80%), and a wide temperature range (10–35 °C) (Montross et al., 1999).
 b. *Storage fungus:* This infests grain predominantly throughout storage and requires equilibrium moisture content with relative humidity of 70% to 90%. Field moulds invading/contaminating maize before harvest are replaced by storage moulds (Reed et al., 2007).

The major fungal species are *Tilletia* spp., *Fusarium* spp., *Rhizopus* spp., *Penicillium* spp., and *Aspergillus* spp., which are found usually in stored grains. When storage fungus infects maize grain, it causes discolouration, loss of dry matter, nutritional and chemical alterations, and a general decrease in maize grain quality. Fractured maize and foreign objects enhance the growth of storage moulds because fungus can more easily enter broken kernels than undamaged kernels. In addition, mould and fungus development is influenced by two major environmental factors: moisture content and temperature. Maize grain is usually collected when it has a moisture level of about 18% to 20% and then dried. If the grain is not properly dried, moulds and fungus thrive, resulting in a considerable reduction in grain quality and quantity.

10.6.5 Mycotoxins

Moulds that develop on maize grains pose a serious hazard, particularly since they produce secondary metabolites (mycotoxins). Maize cultivated in warm, humid, tropical, and subtropical climates is plagued by mycotoxins. Moulds and fungal diseases can contaminate mycotoxin production at any stage of the process, including growing, harvesting, storing, and processing. Cereal grains have been infested with major common mycotoxins such as aflatoxins, zearalenone, ochratoxins, trichothecenes, and fumonisins in tropical as well as subtropical areas. The most dangerous and frequent mycotoxins chemicals found on maize are aflatoxin and fumonisin. Aflatoxin is primarily an issue in cereal grains, notably maize.

10.6.6 Insects and pests

During storage, insects are the leading cause of infestation for loss of grains across the world. Several insects thrive in grain storage, consuming grain nutrients and contaminating it with bug pieces and faeces. Storage structures for grains are ecologically unstable because they include organisms which are very reproductive and can harm grains quickly. Insect damage of stored grains is estimated to account for 1–5% in developed countries, and 20–50% in developing countries; more than 500 insect species have been linked to grain, and out of these, 250 of them are linked directly to maize grains, in the field as well as in storage (Jian & Jayas, 2012). Internal feeds are insects that grow inside grain kernels, whereas external feeds are insects whose eggs hatch on the grain kernels' surface. The maize weevil, *Sitophilus zeamais*, is one of the most damaging insects in maize storage. Because of its capacity to damage a full-grain kernel, S. zeamais is categorized as a primary or major pest. When maize grains are stored, they are subjected to various complicated ecological variables, the most important of which are moisture and temperature, which impact on pest development and quality of grains. Other than these, environmental factors like relative humidity and temperature influence quick growth of the insects and pests, and endanger the stored grain quality and quantity.

It has been discovered that most grain storage insects thrive in 25–30 °C and 70–80% relative humidity. Studies showed hermetic storage conditions could manage insect infestation problems at temperatures of 10–27 °C and moisture content 6–16% (Yakubu et al., 2011).

10.7 Pest control during storage of maize

10.7.1 Characteristics of insect pests

The most common pests in the tropics are beetles and moths, which cause significant crop loss and degradation. As illustrated in Figure 10.1, pests

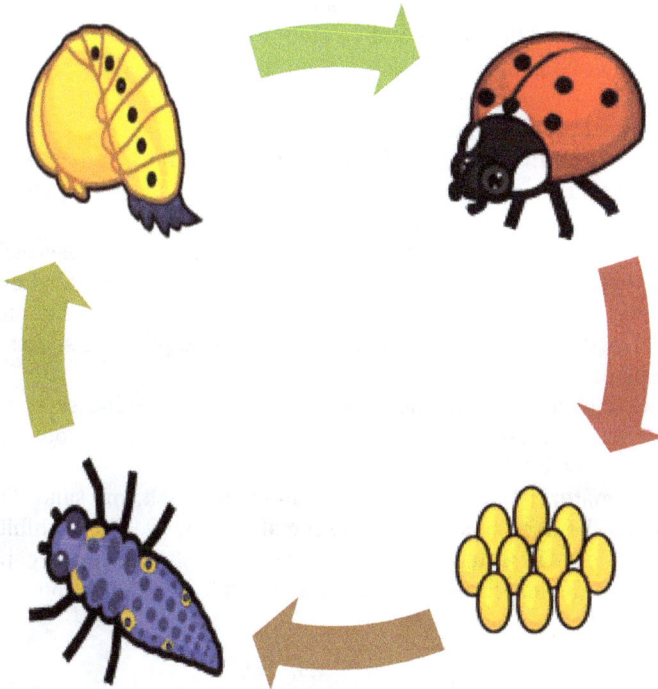

Figure 10.1 The cycle of life of a typical storage insect pest

generally have a life cycle of four stages: egg, larva, pupa, and ultimately adult.

The eggs are deposited on the grain kernel's surface (usually on the physically damaged section of the grain) or within a small hole partially drilled on the kernel by the parent. The eggs develop into larvae that eat their way through the grain, causing the most agricultural damage. Pupae, a dormant, non-feeding stage, develops as the larvae mature. In domestic storage, the Prostephanu-struncatus or larger grain borer (LGB) caused losses ranging from 10 to 35% in five to six months, and up to 60% in nine months. The adult stage of the moth is short and it does not feed, but their larvae are voracious eaters that contaminate stored goods with frass and webbing. The damage caused by insects and pests may be divided into two categories: ones that bore holes in the grain and ones that cause the loss of a substantial section of the endosperm.

10.7.2 Insect and pest control strategy

Insect management for maize must begin before harvest, not after they have been discovered in storage. As a result, consider the following points for better control. The most effective method entails timely harvesting and appropriate drying and cleaning of the maize before storage.

1. Conduct practices that are unique.
2. Use material like ashes (for their abrasive and metal impact on the insect's cuticle), oils, and minerals, as these create a physical hurdle for insects and control their growth.
3. Use entire or portions of plants with natural fungicidal, insecticidal, or repelling properties (mainly terpenes, alcohols, and alkaloids).
4. Always store a dried crop. If an infestation is discovered in stored maize, then dried it again with maize grains kept under sheaths for husk protection.
5. Use unpleasant native herbs and plants (Nim ground seed, acanthaceae, acardiaceae, annonaceae, myrtaceae, various plant extracts, etc.).
6. Inert materials like wood, crushed limestone, ash, and sand. These should be used in a granular space at 1–5% w/w. They inhibit the activities of insects, diatomaceous earth, etc. It causes insect pests to become dehydrated, resulting in death from desiccation.

10.7.3 Chemical control methods

Sometimes there is a need for chemical methods to control infestation despite some of their shortcomings.

10.7.3.1 Fumigation with fumigants

Fumigants release vapours or gases which enter the insect's body throughout its respiratory system, causing it to die. Fumigants have a non-residual impact, yet they can permeate stacks or bulk goods, destroying all insect life cycle stages. They are very toxic and, if not handled correctly, can cause death. In addition, they also do not protect against re-infestation of the grain. Fumigation should only be carried out by well-trained individuals who have a licence to do so.

10.7.3.2 Contact insecticide

Poisonous substances can pierce the insect's cuticle and enter the bodily tissue, causing death. Smallholder farmers will use contact insecticides as admixtures or as a surface treatment. These pesticides have a long-term influence on the environment. They do, however, have a particular impact on some insect species.

10.7.3.3 Sandwich method: layer by layer dusting of maize cobs

When the cleaned and dried maize cobs are placed in storage, they should be treated layer by layer (granary). This treatment is effective against most traditional maize storage insect pests, however, it may not be effective against LGB. An overdose can be dangerous to one's health. Different insecticides such as malathion 2%, gardona 3.25%, methacrifos 2%, actellic 1%, and etrimfos 1%, can be used with 50 grams/90 kg maize cobs application rate.

10.7.3.4 Pest control treatment of shelled maize

Shelled maize responds better than maize cobs to chemical methods, and it is also easier to manage insects and rats. In this condition losses decline and it makes storing more manageable. When maize is shelled and treated with synthetic pyrethroids like permethrin and deltamethrin, the LGB is simpler to manage. Permethrin and deltamethrin can be applied at rates of 100 g/90 kg grain and 50 g/90 kg grain, respectively, with actellic at 0.5% and bromophos at 2%. It's worth noting that the pesticide may be applied to a concrete surface by dusting it evenly with a scoop. Another option is to combine all the ingredients in the drum with maize and shake it properly.

10.8 Storage structures

10.8.1 Traditional/rural storage structures

Granaries: Thatched granaries were the most common structures in primitive times and are a traditional practice to store maize after harvesting.

Granaries were usually built in sunny areas with rooves made of straw in a conical form. The average recommended size of small granary is 2–5 m in width and 1–2 m in height. Sometimes neem leaves with thin layers of ash along with proper fumigation serve to create a hostile environment for insect pests. It is required to construct granaries at about 1.2–1.6 m raised from the ground to keep away rodents. Schematic diagrams of different types of granaries have been represented in Figures 10.2–10.5.

Mud Bin: Mud bin, as the name indicates, is constructed out of bricks laid down with mud or straw along with cow dung. Various capacities of mud bins are available and are usually cylindrical in shape.

Bamboo reed bin: These structures are made out of bamboo splits which are plastered with cow dung and a mud mixture.

Gunny bags: These are the most common storage materials, they are made of jute and are popular nowadays for storing maize and other cereal grains.

Thekka: These structures are rectangular in shape and are made by winding cotton cloth or gunny material around a wooden support.

10.9 Improved storage structures

The production of maize is increasing gradually, which has increased a demand to improve traditional storage structures to store, preserve, and handle large amounts of grain. Storage of grains is a process through which they are kept safely for future use (Nukenine, 2010), protected from environmental factors like relative humidity, temperature, and presence and growth of insects, pests etc.

Figure 10.2 Low granary

Figure 10.3 High granary

Figure 10.4 Woven wall granary

Improved storage structures are designed to strengthen traditional storage structures by the means of capacity and duration of storage with some kinds of modern inputs. These structures are also called advanced storage structures.

The improved storage structures have been generally used to reduce the qualitative and quantitative losses by technical means, and they are

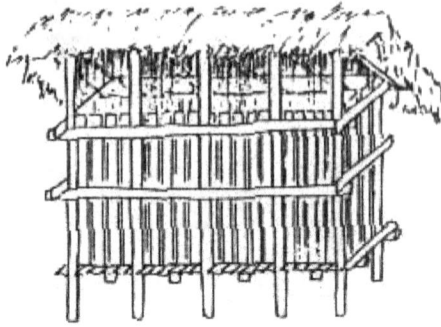

Figure 10.5 Rectangular crib

more effective and efficient for storing maize grains. These types of structures are used by maize processing industries, maize mills, whole sellers etc. These storage structures store the maize grains for longer duration very safely and economically by preventing excessive losses during handling and storage.

A number of modern storage structures have been designed and developed like the Pusa bin, PAU bins, cover and plinth structures (CAP), etc. for storing the grains. Selection of these structures depends upon some parameters as follows:

   i. Availability and quantity of grains
   ii. Suitability and acceptability
   iii. Place of storage
   iv. Environmental parameters like relative humidity, temperature, etc.
   v. Duration of storage
   vi. Demand throughout the storage period

Storage structures like the Pusa bin, PAU bin, brick bin, and Hapur tekka have been used for small quantity storage. Storage systems for maize are mainly divided into three types: bags, cribs, and bulk storage (Montross et al., 1999). These structures have generally 1.5–150 tons of storage capacity in the form of bulk as well as bag storage in different rural and urban regions of India. Bins are preferred for storage inside the house.

10.9.1 Pusa bin

The Pusa bin is a small storage structure constructed by mud bricks as a traditional storage structure. The design and fundamental concept of this bin was proposed by the Indian Agricultural Research Institute. A Pusa

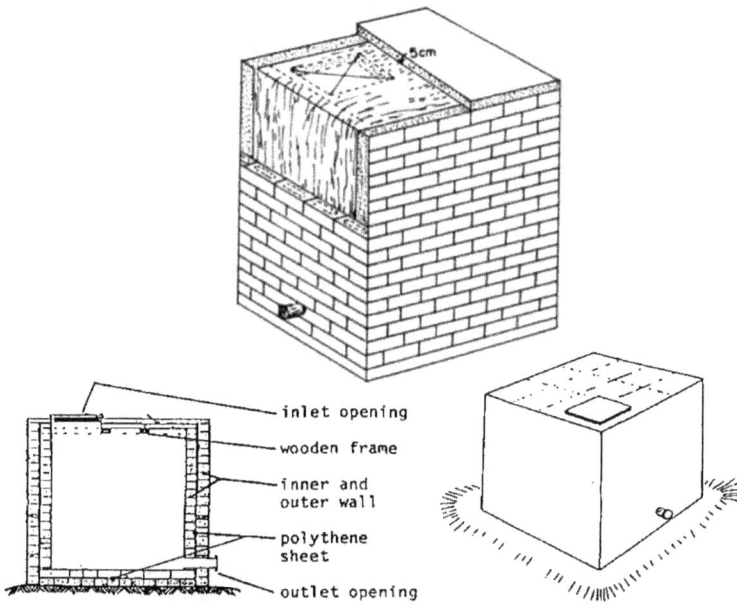

Figure 10.6 Pusa bin

bin is shown in Figure 10.6. The Pusa bin is a moisture proof storage structure, as a plastic sheet/film is used on the inner wall of the bin to provide a
sealing action. Polythene sheets of 700 gauges, i.e. low density polythene,
are used on platforms of mud bricks in such a way that they overlaps the
platform by at least 6 cm on all sides, and on the polythene sheet a layer
of mud bricks with 7 cm thickness is laid. Burnt bricks are used to make
the upper roof, and it is covered with polythene sheet. For loading of maize
grains, an open space of 0.5 m × 0.5 m is left at the top and the remainder
of the roof is sealed using mud plaster. Once the grains are filled in the bin,
the open space is also covered by a polythene sheet to avoid entry of the air.
An inclined steel or wooden pipe is fixed to unload the grains by means of
gravity, and the pipe mouth is covered well to avoid the entry of insects or
moisture/air into the bin.

10.9.2 Brick and cement bin

Brick and cement bins for maize storage are constructed using brick and
cement as plaster materials, so these structures are strong, and the different seasons have a negligible effect. The storage capacity of the bin is 1.5
to 60 tonnes. The bin is constructed on a raised level bed 0.60 m above the

ground level. The thickness of walls is kept nearly about 0.23 m and walls are plastered with cement on both sides. The roof of the bin is made of reinforced cement concrete and a hole of 0.60 m diameter and a ladder on one side of wall is kept for loading purposes. For the purpose of unloading, an inclined base is made with an outlet. For uniform loading, complete unloading, and cleaning of the bin, iron ring steps are provided inside the bin for entry and exit of a person.

10.9.3 PAU bin

The PAU bin is made up of galvanized iron sheets and designed by the Punjab Agricultural University with a storage capacity of 0.15 to 1.5 tons. The bin protects against moisture and insect growth.

10.9.4 Bunker storage

A large volume of grains is stored in this type of improved structure for longer periods of time. The bunker storage structure is used to store the grains economically, safely, and securely by means of controlling moisture and insects. A plastic sheet is used at the base as well as on top to store the grains, and to drain the rain water a drainage pipe is provided. Losses can be reduced by 0.5% in bunker storage.

10.9.5 Hapur Thekka

Hapur tekkas are constructed locally with galvanized iron/aluminium sheet sides and metal tubes at the base in cylindrical shape and designed by the Indian Grain Storage Institute, Hapur. Hapur tekka is a more strong and durable improved structure than traditional structures. The storage capacity of Hapur tekka is 2–10 tons. A small rectangular or circular inlet and outlet is provided for loading at the top and for unloading at the bottom of the structure.

10.9.6 Cover and plinth storage (CAP) structure

The CAP structure is designed for storage of gains at intermediate or large scale. Cover means top cover of grains and plinth means plinth from the bottom. CAP is a type of open storage structure in which grains are stored in bags for a short period of time. CAP is constructed in rectangular shape with the help of 1000 gauge polythene sheet having only five sides as the bottom side of the structure is left open. Wooden crates are fixed or placed on the bottom having nearly 1.17 feet height from the ground, or brick pillars are constructed from the ground to 1.17 feet height into stacks (Said & Pradhan, 2014). Bags of maize grains are arranged on stacks, and stacks of the bags are covered with the polythene cover. The stacks are normally made with 9.11 m × 6.1 m space along with 18 bags height. The cover of the structure has 9.4 m × 6.4 m × 5.5 m dimensions for covering the bags which weighs around 52 kg. The CAP storage structure with above dimensions provides the storage capacity of nearly about 150 tons (Naik & Kaushik, 2011). The cover with smaller dimensions is used when stacks are covered in the covered veranda of the godown. These covers are also called as "veranda covers". Maize grains under veranda covers are stored where height of the stack is limited to 7 bags, and average storage capacity of this type is 24 tons (Karthikeyen et al., 2009). The CAP storage structure is the most economical and frequently used type by the Food Corporation of India (FCI), as it is most suitable for bag storage of maize grains up to 6–12 months, and the structure can be built within three weeks.

The CAP storage structure needs regular inspection and management to avoid severe losses, as this type of storage structure gets easily damaged by high winds. Quality control should be carefully maintained with regular sampling.

10.9.7 Modified domestic bricks bins

Clay bins are improved to become modified domestic brick or cement bins, having a cylindrical shape. In this type of bin generally grains are stored for domestic use. These bins are generally constructed using cement, burnt bricks etc. to strengthen the structure and to make it more protected against moisture and insects due to its airproof structure, and grains can be stored for a good period of time without any quality deterioration (Mehrotra et al., 1987).

10.9.8 Granary room

A granary room is a modern permanent type of storage structure, which is constructed inside the house with sand, concrete, cement, and bricks. It is also known as roof almirah, Du-chhatti, or Kotha in India. It is designed by providing RCC slab below the roof in the corner of the wall or just above the door, and one mini door is provided for easy handling of maize grains. This type of storage structure was observed to have minimum damage from rodents because grains are stored at nearly 6.5 to 7.5 feet from the floor. Grains can be stored in bags or bulk in this type of storage structure.

10.10 Commercial/modern storage structures

A large volume of maize grain is stored commercially in terms of bulk in silo and in terms of bag storage in sheds/warehouses/godowns. Modern storage structures are generally known as permanent structures as their life of storing grains is very long. The cost of designing and installation is comparatively higher than the improved storage structures but automation in inspection, environmental control, and fumigation is easily possible in these structures. That is the reason the modern storage structures should be preferred for longer and safe storage of grains based on quality first and then cost considerations. Silos are made of either reinforced concrete or steel. Walls of godowns or warehouses are constructed with stone masonry or bricks, and rooves are made of corrugated galvanized iron (CGI) or asbestos sheets with adequate slope. Silos are designed to join to the large capacity processing plants and also for easy and low cost handling of grains. The following are the types of modern permanent storage structures.

10.10.1 Warehouse or Shed Storage

Warehouses are a safe and scientific storage for bulk grain storage in India. The warehouses are generally constructed for protection of quality as well as quantity of stored grains. The warehouses are operated and owned by FCI, the Central Warehousing Corporation (CWC), or the State Warehousing Corporation (SWC). The CWC, India's largest public warehouse operator, provides reliable and safe storage facilities for nearly 120 industrial and agricultural commodities. The CWC provides services in terms of procurement and distribution, cleaning, handling and transportation, insurance, fumigation and disinfection, security and safety, and documentation.

The SWC has constructed state-wide warehouses for safe storage of grains of particular states. State warehousing procures the grains from different districts of the particular state, and loading and unloading of the grains is performed according to the requirement and storage area.

Apart from SWC and CWC, the FCI has also constructed storage facilities (Figure 10.7). The FCI is the largest corporation of 26.62 million tons capacity.

Warehouses are used to store large quantities of maize grains in bulk or bags for longer periods of time. Warehouses are generally constructed with the use of cement concrete, bricks, and metals, and 0.45–1 m of the foundation is kept to prevent rodent entry from the ground. On the foundation within the warehouse structure, wooden or metal pallets are placed for arranging the bags in the stacks to avoid migration of the moisture from floor to grains. There is a standard distance between two consecutive stacks for easy transportation, inspection, and handling operations. The

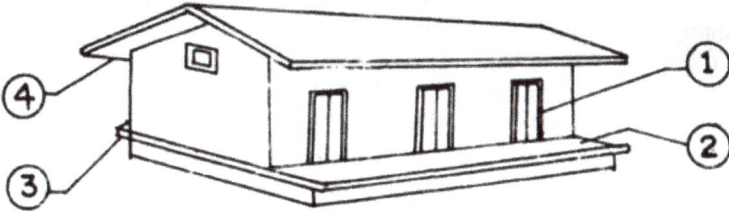

Figure 10.7 Warehouse shed: 1) sealed doors, 2) floor 3), rat-proof slab, 4) airproof roof

doors of the warehouses need proper sealing and controlled ventilation for air and light movement.

The most important things for the scientific and safe storage of grains is selecting the location and site of storage structure, cleaning and moisture maintenance operations, proper and regular inspection, and fumigation and aeration practices. Environmental conditions like temperature and relative humidity, moisture of grains, storage period, processing, and hygienic conditions decide the frequency of fumigation for pest control and infestation. Curative as well as preventive measures are taken for safe storage of grains in the warehouses.

10.10.2 Silo

Grain storage in bulk is stored in structures with circular or conical bottoms known as silos. Silos are modern and large capacity storage structures. Bulk storage structures, bins, or silos are made of corrugated galvanized sheets, reinforced concrete, aluminium sheets, mild steel, asbestos sheets, etc. But RCC bins and mild steel bins are commonly preferred over concrete silos in India to store the grains in bulk. These modern bulk storage bins have the following advantages:

i. Easy handling due to mechanization and quality control of grains
ii. Less land space requirement

345

    iii. Less expensive and removes cost of bags

    iv. Protection from rodents, birds, and insects losses

    v. Easy provision of automation for quality control

The storage capacity of grains in silos is approximately 25 thousand tons. Bins or silos are conical at the top and bottom and cylindrical in the middle, in which mechanization is done to aerate, fumigate, and inspect (Flinn et al., 1997). The chances of moisture migration, insect growth, and hot spot creation are prevented by proper aeration and temperature control in the silos (Minjinyawa, 2010). Temperature and moisture in the storage structure is controlled to avoid spoilage of grains (Navarro and Noyes, 2001).

Bins/silos are generally classified into two types based on the relative dimensions of the structure. The capacity of the silo depends on diameter/width and height of the silo. Based on the dimensions of width or diameter and depth, the storage structure is termed to be a shallow or deep bin, and that is decided by the concept of plane of rupture. The plane of rupture is the surface down a wedge of material bounded by free surface, one wall face. If the bound wall was to move, the plane of rupture would start moving. Depending on the specific dimensions of the container, silos are classified into two types i.e., deep bins and shallow bins.

10.10.3 Hermetic storage structures

Hermetic storage (HS) is a novel technology of storing the grains in airtight storage, which plays an important role in preventing the growth of yeasts and moulds. Hermetic storage is also called sealed storage, airtight storage, bunkers, cocoons, or bio-generated modified atmospheric storage (Navarro et al., 1984). The construction of the HS structure majorly depends on the quantity of grains to be stored, and when storage capacity of this structure is 0.59–1 tons then the structure is called a super grain bag. When the storage capacity varies from 5–30 thousand tons, it is called a cocoons or bunker. In the HS, the oxygen level is lowered up to 1–2% by means of a vacuum or by increasing the level of carbon dioxide. The hermetic storage of maize grains is environment friendly technique that is suitable, sustainable, and safe as it eliminates the fumigation, chemical treatments, or any other kind of climate control. Bunkers have the storage capacity of 10–30 thousand tons of grains hermetically, and bunkers have been used safely for grain storage of 12.5% moisture content without any qualitative degradation for at least four years (Donahaye et al., 1997). HS structures provide protection from insect pests and maintain the quality of stored grains along with easy portability during period of storage (Figure 10.8). The following three goals needs to be achieved for successful hermetic storage of grains:

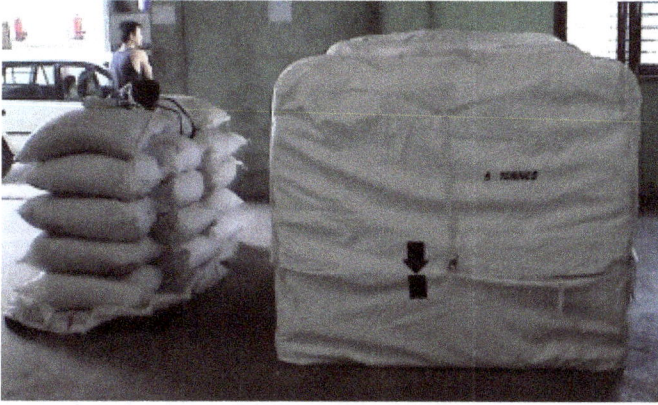

Figure 10.8 Hermetic storage of grains

   i. Prohibition of moisture entry from the surrounding environment
  ii. High carbon dioxide and low oxygen atmosphere to prevent oxidation and the growth of moulds or insects
 iii. Protection of rodents

10.11 Maize products

A range of products are derived from maize. There are several products processed and developed domestically and industrially. A large number of industries are commercialized and benefit from single production of corn-related products only. Maize is more commonly known as corn. Therefore, products available are also known as baby corn, corn syrup, corn starch, corn oil, cornflakes, sweet corn, etc. (Figure 10.9a and b). Generally, at the industrial level, two types of processing, i.e. dry milling and wet milling, are employed for maize to transform it into various products. In wet milling, the maize is completely purified to produce starch, protein, oil, and fibre. These products are further subjected to processing before consumption. In drying milling after the screening, the maize is subjected to either whole de-germination or partial de-germination to remove the germ and fibre from the maize. Flour mills like hammer mill, roller mill, and stone mill with sifters, screeners, gravity separators, and aspirators are commonly used for dry-milling processing. In the case of de-germination, separate dehullers, peelers, and degerminators are required (Gwirtz and Garcia, 2014). In this section, there will be a discussion on the different types of products and their production methods. Different steps can be seen in Figure 10.9a.

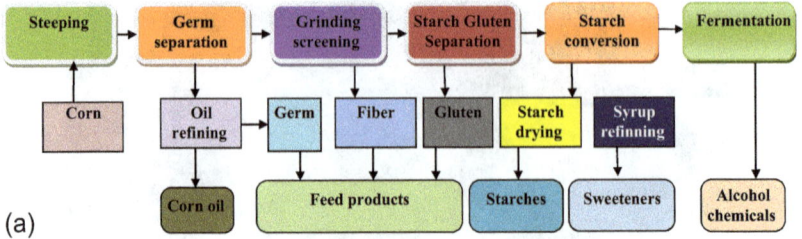

Figure 10.9 (a) Production of different corn products (Mangia, M., 2010). (b) Various maize products (www.vectorstock.com/royalty-free-vector/corn-food-set-maize-meal-oil-vector-37265208)

10.11.1 Corn starch

Corn starch derived from maize has multiple uses. Corn starch powder is widely used in the soup which covers major restaurant preparations. It is used in the preparation of gravies, sauces, and custard. Also, high maltose corn syrup, dextrose monohydrate, and maltodextrin are available in the market, which are used as sweeteners and in confectionery industries. Corn starch also gives strength to ice cream cones and helps in moisture-holding in bread. Corn starch is a very good source as a dispersing agent for instant drinks. Corn starch is used as an adhesive and coating in the food industry and has incredible absorbent properties. Thickening, binding, and gelling are major properties of starch which help in the manufacturing of food products. Sometimes it is also used as an electrical conductor for electrical appliances. There are three types of starches obtained from the corn with the different unit operations. Once the starch slurry is obtained with steeping and milling, the slurry is subjected to modification and pregelatinizing to obtain modified starch and pregelatinized starch respectively. The process is shown in Figure 10.10.

10.11.2 Cornflakes

Cornflakes are one of the major breakfast cereals. Different types of cornflakes are available in the market, such as like choco flakes, honey flakes, roasted flakes, and sugar-coated flakes. Also, plain flakes are used in the preparation of many namkeens. There is a high export demand too. Factors like temperature, cooking time, shear applied, and moisture content are very important for the perfect production of flakes. There could be fortification or mixing of other oilseeds and cereals for manufacturing flakes. One of the examples with peanut flour is shown in Figure 10.11. Peanuts flour has a high protein content (Miller 1994).

10.11.3 Baby corn

Baby corn is harvested at a premature stage. Baby corn is just like an organic food as it is free from many pesticides' effects due to preharvest. Baby corn is also a very good option for crop diversification and great value addition to the food processing and production industries due to rich nutritive values (Rani et al., 2017). Baby corn is a great source of fibrous protein and it is very easily digestible compared to other corn products. Many products have been developed from baby corn e.g. corn burfi, corn pakoda, corn murraba, corn chutney, frozen corn, canned corn, pickle, corn halwa, corn

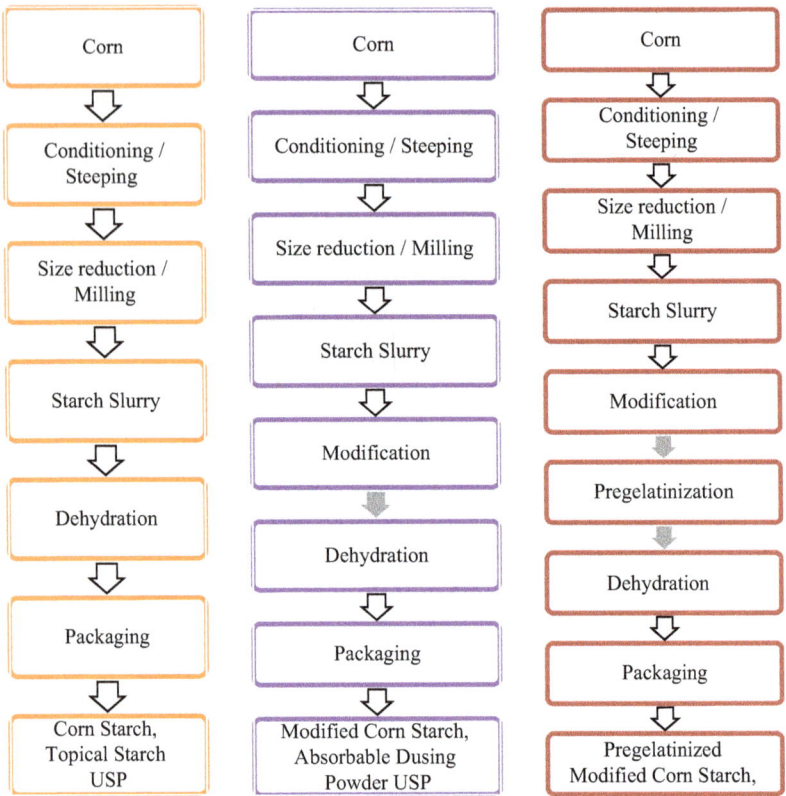

Figure 10.10 Flow diagram for different corn starch production (www.grain-processing.com/nutraceutical/starch-production-flow-chart.html)

cutlets, and dehydrated corn (Asaduzzaman et al., 2014). Apart from these many food products, baby corn is a rich source of nutrition for animals also. Dry leaves and cob can be converted to fuel also.

10.11.4 Corn syrup

Corn syrup is a mixture of short, medium, and long-chain saccharides which are obtained from high heat and pressure. The high heat and pressure break down the glycosidic bonds, and with additional acid and enzymes, it is converted to the slurry. The initial step for the manufacture of corn syrup is to gelatinize the obtained starch under heat and pressure.

```
┌─────────────────────────┐
│   Dry ingredients-Corn  │
│   meal, sugar, salt etc.│
└─────────────────────────┘
             ↓
┌─────────────────────────┐
│     Mixing (20 min)     │
└─────────────────────────┘
             ↓
┌─────────────────────────┐
│       Extruding         │
└─────────────────────────┘
             ↓
┌─────────────────────────┐
│        Collets          │
└─────────────────────────┘
             ↓
┌─────────────────────────┐
│        Cooling          │
└─────────────────────────┘
             ↓
┌─────────────────────────┐
│        Flaking          │
└─────────────────────────┘
             ↓
┌─────────────────────────┐
│   Toasting (170°C for   │
│        3 mn)            │
└─────────────────────────┘
             ↓
┌─────────────────────────┐
│   Cooling and Packing   │
└─────────────────────────┘
```

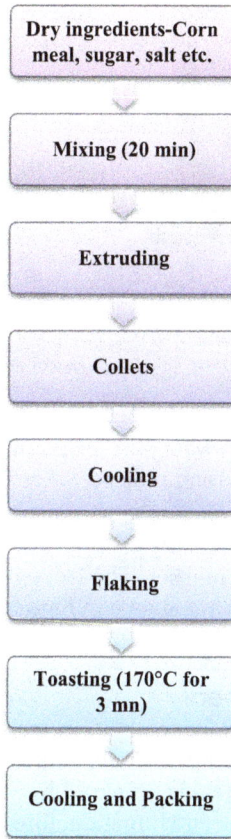

Figure 10.11 Processing of corn cereal flakes (Heewapramong, 2002)

10.11.5 Maize protein

Maize protein is also in commercial products as protein and has high biological value and good digestibility compared to other complex proteins. Protein gives a great diversification to food formulation and dietary plans. This protein can be taken by all age groups. Also, a great export potential for India is present.

10.11.6 Sweet corn

Sweet corn is widely acceptable by all age groups. Highly nutritious and easily digestible. Sweet corn is used in soups, salads, snacks, condiments, etc.

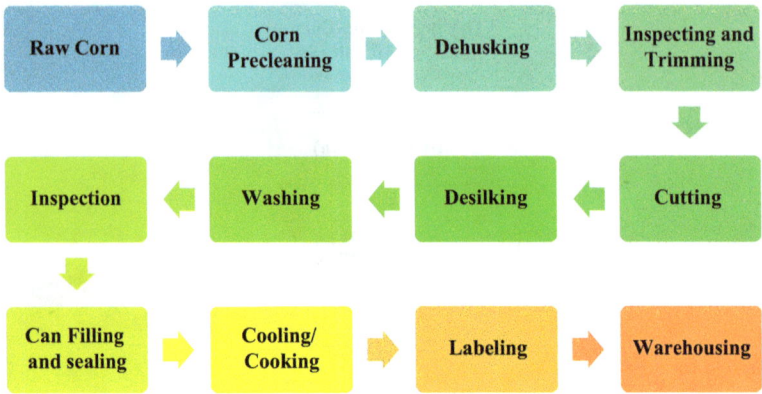

Figure 10.12 Processing of sweet corn (http://niftem.ac.in/site/pmfme/processingnew/sweetcornprocessing.pdf)

Frozen sweet corn is commercially present with a better shelf life. Canned corn is also available in the market. The frozen one is subjected to various processing steps. The steps are shown in Figure 10.12.

10.11.7 Corn gluten meal

The corn gluten meal has a high protein content, nutrient density, and is a good energy source. The meal is prepared by subjecting the gluten slurry obtained from the mill to re-centrifugation, filtering, and drying. It is one of the good sources of metabolizable energy with a good content of vitamins and minerals (Yigit et al., 2012).

10.11.8 Corn oil

Corn oil is obtained by crushing the maize, and in proper refinement processes, the oil is converted to edible oil for human consumption. The steps are degumming, neutralization, bleaching, winterization, and deodorization. The obtained oil could be further subjected to hydrogenation, interesterification, and fractionation depending on the use (Barrera-Arellano et al., 2019). The oil is used as a fortification in the bakery and confectionery industry, also as a cooking medium at home. The corn germ oil has a high amount of the dietary antioxidants, tocotrienols, and tocopherols, which enhance the oxidative stability of the oil. Corn oils also have great sensorial and nutritional characteristics for diverse food applications (O'Brien, 2008).

10.11.9 High maltose corn syrup

High maltose corn syrup is obtained by subjecting the starch slurry to dextrinization and liquefaction. The syrup obtained is used for the preparation of jam, jellies, canned fruits, and chewing gums. As maltose has low viscosity, its handling is easy. The candies produced are smooth, bright, and transparent. The product is homogenous and palatable due to its non-crystallizing properties. High maltose corn syrup is used to extend the shelf life of products as it has many advantages over cane sugar.

References

ACDI/VOCA (Agricultural Cooperative Development International and Volunteers in Overseas Cooperative Assistance). (2003). Staple crops storage handbook. USAID- East Africa. Retrieved November, 2012, from http://www. acdivoca.org/site/Lookup/StorageHandbook/$fil e/StorageHandbook.pdf

Agmark Grading Statistics, 2004–05, directorate marketing and inspection, Faridabad. Retrieved from https://agmarknet.gov.in/ CommodityProfiles/preface-maize.aspx

Asaduzzaman, M., Biswas, M., Islam, M. N., Rahman, M. M., Begum, R., Sarkar, M. A. R., & Asaduzzaman, M. (2014). Variety and N-fertilizer rate influence the growth, yield and yield parameters of baby corn (Zea mays L.). *Journal of Agricultural Science, 6*(3), 118–131.

Aulakh, J., Regmi, A., Fulton, J. R., & Alexander, C. (2013). Estimating post-harvest food losses: Developing a consistent global estimation framework. Proceedings of the Agricultural & Applied Economics Association's 2013 AAEA & CAES Joint Annual Meeting; Washington, DC, USA. 4–6 August 2013.

Barrera-Arellano, D., Badan-Ribeiro, A. P., & Serna-Saldivar, S. O. (2019). Corn oil: Composition, processing, and utilization. In O. Serna-Saldivar (Ed.), *Corn* (pp. 593–613). Woodhead Publishing and AACC International Press, Elsevier.

Bern, C., Hurburgh, C. R., & Brumm, T. J. (2013). Managing grain after harvest. Course works, agricultural and biosystems engineering department, Iowa state university bookstore.

Brewbaker, J. L. (2003). Corn production in the tropics. The Hawaii experience. College of tropical agriculture and human resources university of Hawaii at Manoa. Retrieved from http://www.ctahr.hawaii.edu/oc/ freepubs/pdf/corn2003.pdf

Brown, W. L., & Darrah, L. L. (1985). Origin, adaptation, and types of corn. *National Corn Handbook- Cooperative Extension Service, Iowa State University of Science and Technology and the United States Department* of Agriculture cooperating. Robert L. Crom, director, Ames, Iowa. Distributed in furtherance of the Acts of Congress of May 8 and June 30, 1914., pp. 1–6.

Devereau, A. D., Myhara, R., & Anderson, C. (2002). Chapter 3: Physical factors in post-harvest quality. In P. Golob, G. Farrell, J. E. Orchard (Eds.), *Crop post-harvest: Science and technology: Principles and practice* (Vol. 1, pp. 62–92). Blackwell Science Ltd. ISBN:9780470751015. DOI:10.1002/9780470751015

Donahaye, J., Navarro, S., & Varnava, A. (1997). *Proceedings of the international conference controlled atmosphere and fumigation in stored products* (pp. 183–190). Nicosia, Cyprus: Printco Ltd.

Flinn, P. W., Hagstrum, D. W., & Muir, W. E. (1997). Effects of time of aeration, bin size, and latitude on insect populations in stored wheat: A simulation study. *Journal of Economic Entomology, 90*(2), 626–651.

Fox, T. (2013). Global food: Waste not, want not. In *Institution of Mechanical Engineers*. London, UK: Westminster.

Gwirtz, J. A., & Garcia-Casal, M. N. (2014). Processing maize Flour and corn meal food products. *Annals of the New York Academy of Sciences, 1312*(1), 66.

Jian, F., & Jayas, D. S. (2012). The ecosystem approach to grain storage. *Agricultural Research, 1*(2), 148–156.

Karthikeyen, C., Veeraragavathatham, D., Karpagam, D., & Ayisha Firdouse, S. (2009). Indiginous storage structures. *Indian Journal of Traditional Knowledge, 8*(2), 225–229.

Lee, S. (1999). Low-temperature damp corn storage with and without chemical preservatives, Doctoral (PhD) dissertation. The University of Guelph.

Maize Outlook Report-Jan to May 2021, Agricultural Market Intelligence Centre, ANGRAU, Lam. Retrieved from https://angrau.ac.in/downloads/AMIC/MAIZE%20OUTLOOK%20REPORT%20-%20January%20to%20May%202021.pdf

Maize Processing Unit. Retrieved from https://www.entrepreneurindia.co /project-and-profile-details/Maize%20Processing%20Unit

Mangia, M. (2010). Free and hidden fumonisins in maize and gluten-free products. *Molecular Nutrition & Food Research, 53*, 492–499.

Mehrotra, S. N., Verma, N., Datta, A., & Batra, Y. K. (1987). Building research note, central building research institute, India, small capacity grain storage bins for rural areas. Central Building Research Institute. Roorkee, India, p. 120.

Miller, R. C. (1994). Breakfast cereal extrusion technology. In N. D. Frame (Ed.), *The technology of extrusion cooking* (pp. 73–109). New York: Blackie Academic & Professional.

Minjinyawa, Y. (2010). *Food and crop storage technology, farm structures* (2nd ed.). Ibadan, Nigeria: Ibadan University Press, p. 110.

Montross, J. E., Montross, M. D., & Bakker-Arkema, F. W. (1999). Part 1.4 Grain storage 46–59. In F. W. Bakker-Arkema, D. P. Amirante, M. Ruiz-Altisent, & C. J. Studman (Eds.), *CIGR hand book of agricultural engineering* (Vol. IV). Agro-Processing Engineering. St. Joseph, MI.

Naik, S. N., & Kaushik, G. (2011). Grain storage in India: An overview. *Centre for Rural Development & Technology*, IIT Delhi, India. p. 119.

Navarro, S., Donahaye, E., Kashanchi, Y., Pisarev, V., & Bulbul, O. (1984). Airtight storage of wheat in a PVC covered bunker. In B. E. Ripp et al. (Eds.), *Controlled atmosphere and fumigation in grain storages* (pp. 591–604). Amsterdam: Elsevier.

Navarro, S., & Noyes, R. (2001). *The mechanics and physics of modern grain aeration management* (S. Navarro & R. Noyes Eds.). Boca Raton, FL: CRC Press, p. 627.

Nukenine, E. N. (2010). Stored product protection in Africa: Past, present and future. *Julius-Kühn-Archiv, 425*, S-26.

O'brien, R. D. (2008). *Fats and oils: Formulating and processing for applications*. Boca Raton, FL: CRC Press.

Rani, R., Sheoran, R. K., Soni, P. G., Kaith, S., Sharma, A., & Corn, B. (2017). A wonderful vegetable. *International Journal of Science, Environmental and Technology, 6*(2), 1407–1412.

Reed, C., Doyungan, S., Ioerger, B., & Getchel, A. (2007). Response of storage molds to different initial moisture contents of maize (corn) stored at 25°C, and effect on respiration rate and nutrient composition. *Journal of Stored Products Research, 43*(4), 443–458.

Said, P. P., & Pradhan, R. C. (2014). Food grain storage practices - A review. *Journal of Grain Processing and Storage, 1*(1), 1–5.

Samuel, A., Saburi, A., Usanga, O. E., Ikotun, I., & Isong, I. U. (2011). Post-harvest food losses reduction in maize production in Nigeria. *African Journal of Agricultural Research, 6*(21), 4833–4839.

Yakubu, A., Bern, C. J., Coats, J. R., & Bailey, T. B. (2011). Hermetic on-farm storage for maize weevil control in East Africa. *African Journal of Agricultural Research, 6*(14), 3311–3319.

Yigit, M., Bulut, M., Ergün, S., Güroy, D., Karga, M., Kesbiç, O. S., & Güroy, B. (2012). Utilization of corn gluten meal as a protein source in diets for gilthead sea bream (Sparus aurata L.) juveniles. *Journal of FisheriesSciences, 6*(1), 63.

Index

For Product Safety Concerns and Information please contact our EU
representative GPSR@taylorandfrancis.com
Taylor & Francis Verlag GmbH, Kaufingerstraße 24, 80331 München, Germany

9 7 8 1 0 3 2 1 5 6 6 7 5